Responses of Soil Carbon and Nitrogen Dynamics and GHG Fluxes in Forest Ecosystems to Climate Change and Human Activity

Responses of Soil Carbon and Nitrogen Dynamics and GHG Fluxes in Forest Ecosystems to Climate Change and Human Activity

Guest Editor

Xingkai Xu

MDPI

Basel • Beijing • Wuhan • Barcelona • Belgrade • Novi Sad • Cluj • Manchester

Guest Editor
Xingkai Xu
State Key Laboratory of
Atmospheric Boundary Layer
Physics and Atmospheric Chemistry
Chinese Academy of Sciences
Beijing
China

Editorial Office
MDPI AG
Grossspeteranlage 5
4052 Basel, Switzerland

This is a reprint of the Special Issue, published open access by the journal *Forests* (ISSN 1999-4907), freely accessible at: www.mdpi.com/journal/forests/special_issues/carbon_nitrogen_GHG.

For citation purposes, cite each article independently as indicated on the article page online and using the guide below:

Lastname, A.A.; Lastname, B.B. Article Title. *Journal Name* **Year**, *Volume Number*, Page Range.

ISBN 978-3-7258-3058-9 (Hbk)
ISBN 978-3-7258-3057-2 (PDF)
https://doi.org/10.3390/books978-3-7258-3057-2

Cover image courtesy of Xingkai Xu

Contents

About the Editor . . . . **vii**

Xingkai Xu
Responses of Soil Carbon and Nitrogen Dynamics and GHG Fluxes in Forest Ecosystems to Climate Change and Human Activity
Reprinted from: *Forests* **2024**, *15*, 1235, https://doi.org/10.3390/f15071235 . . . . **1**

Yang Song, Shuai Huang, Haiying Zhang, Qin Wang, Lin Ding and Yanjie Liu
The Impact of Permafrost Change on Soil Organic Carbon Stocks in Northeast China
Reprinted from: *Forests* **2024**, *15*, 14, https://doi.org/10.3390/f15010014 . . . . **7**

Yue Wu, Quanxin Zeng, Xianchu Su, Wei Zheng, Qiufang Zhang and Yuehmin Chen
The Different Factors Driving SOC Stability under Different N Addition Durations in a *Phyllostachys edulis* Forest
Reprinted from: *Forests* **2023**, *14*, 1890, https://doi.org/10.3390/f14091890 . . . . **29**

Xingkai Xu, Tingting Xu and Jin Yue
Effect of In Situ Large Soil Column Translocation on CO_2 and CH_4 Fluxes under Two Temperate Forests of Northeastern China
Reprinted from: *Forests* **2023**, *14*, 1531, https://doi.org/10.3390/f14081531 . . . . **43**

Xingkai Xu, Haohao Wu, Jin Yue, Shuirong Tang and Weiguo Cheng
Effects of Snow Cover on Carbon Dioxide Emissions and Their $\delta^{13}C$ Values of Temperate Forest Soils with and without Litter
Reprinted from: *Forests* **2023**, *14*, 1384, https://doi.org/10.3390/f14071384 . . . . **62**

Baishu Kong, Jilei Zhou, Liguo Qi, Shuying Jiao, Lujie Ma and Wenwen Geng et al.
Effects of Nitrogen Deposition on Leaf Litter Decomposition and Soil Organic Carbon Density in Arid and Barren Rocky Mountainous Regions: A Case Study of Yimeng Mountain
Reprinted from: *Forests* **2023**, *14*, 1351, https://doi.org/10.3390/f14071351 . . . . **83**

Xingkai Xu
Effect of Changes in Throughfall on Soil Respiration in Global Forest Ecosystems: A Meta-Analysis
Reprinted from: *Forests* **2023**, *14*, 1037, https://doi.org/10.3390/f14051037 . . . . **98**

Yuhua Kong, Anran Qu, Erpeng Feng, Rui Chen, Xitian Yang and Yong Lai
Seasonal Dynamics of Soil Enzymatic Activity under Different Land-Use Types in Rocky Mountainous Region of North China
Reprinted from: *Forests* **2023**, *14*, 536, https://doi.org/10.3390/f14030536 . . . . **118**

Erpeng Feng, Liwei Zhang, Yuhua Kong, Xingkai Xu, Ting Wang and Caifeng Wang
Distribution Characteristics of Active Soil Substances along Elevation Gradients in the Southern of Taihang Mountain, China
Reprinted from: *Forests* **2023**, *14*, 370, https://doi.org/10.3390/f14020370 . . . . **128**

Enhui Wang, Ning Yuan, Shaofeng Lv, Xiaoping Tang, Gang Wang and Linlin Wu et al.
Biochar-Based Fertilizer Decreased Soil N_2O Emission and Increased Soil CH_4 Uptake in a Subtropical Typical Bamboo Plantation
Reprinted from: *Forests* **2022**, *13*, 2181, https://doi.org/10.3390/f13122181 . . . . **143**

Binjie Li, Guangsheng Chen, Xiaochen Lu and Hongbo Jiao
Effects of Nitrogen and Phosphorus Additions on Soil N_2O Emissions and CH_4 Uptake in a Phosphorus-Limited Subtropical Chinese Fir Plantation
Reprinted from: *Forests* **2022**, *13*, 772, https://doi.org/10.3390/f13050772 . . . . **162**

About the Editor

Xingkai Xu

Xingkai Xu has been a full professor at the Institute of Atmospheric Physics, Chinese Academy of Sciences (CAS), since 2005. He received his B.S. degree in Agricultural Science from Anhui Agricultural University and Ph.D. degree in Soil Science from the Institute of Applied Ecology, CAS, in 1993 and 1998, respectively. From November 1996 to April 1997, Dr. Xu worked as a research fellow at the Faculty of Agricultural and Applied Biological Science, University of Ghent, Belgium. From July 1998 to June 2000, Dr. Xu worked as a postdoctoral fellow at the Research Center for Eco-environmental Sciences, CAS. From January to April 2000, he worked as a research fellow at the Delft Technical University, The Netherlands. From November 2002 to November 2004, he worked as a JSPS (Japan Society for the Promotion Science) research fellow at Chiba University, Japan. In March 2005, he was an awardee of the CAS Hundred Talents Program. From August to September 2012, he worked as a JSPS-bridge research fellow at Chiba University, Japan. From January to March 2020, he worked as a JSPS invitational fellow at Yamagata University, Japan. He is mainly interested in terrestrial carbon and nitrogen biogeochemistry, particularly carbon and nitrogen cycling in soil and GHG fluxes. He is an editorial board member of *International Agrophysics* and *Forests*, respectively. He won the Pollyanna Chu Outstanding Teacher Award at the University of Chinese Academy of Sciences in 2021.

forests

Editorial

Responses of Soil Carbon and Nitrogen Dynamics and GHG Fluxes in Forest Ecosystems to Climate Change and Human Activity

Xingkai Xu [1,2]

[1] State Key Laboratory of Atmospheric Boundary Layer Physics and Atmospheric Chemistry, Institute of Atmospheric Physics, Chinese Academy of Sciences, Beijing 100029, China; xingkai_xu@mail.iap.ac.cn

[2] Department of Atmospheric Chemistry and Environmental Science, College of Earth and Planetary Sciences, University of Chinese Academy of Sciences, Beijing 100049, China

Abstract: Forest soils are considered the largest carbon and nitrogen pools in soil organic matter among terrestrial ecosystems, and soil carbon and nitrogen dynamics and greenhouse gas (GHG) emissions are normally affected by climate change and human activity. The collection of recent research on this scientific theme would provide a basis for understanding the responses of soil carbon and nitrogen dynamics and GHG fluxes in forest ecosystems to climate change and human activity. A Special Issue was, thus, organized to discuss recent research achievements, including a total of nine research articles and one review. This Special Issue includes the effects of climate changes such as changes in throughfall, snow cover, and permafrost degradation; human activities such as nitrogen and/or phosphorus addition and the use of biochar; and soil–plant interactions on soil carbon and nitrogen dynamics and GHG fluxes in forest ecosystems. Although this collection of papers reflects only a small part of this scientific theme, it can, to some extent, provide a basis for understanding some important research aspects related to the future of forest soil carbon and nitrogen dynamics and GHG fluxes in a changing world, thereby enabling sustainable development and the mitigation of climate change.

Keywords: climate change; forest ecosystem; GHG flux; human activity; soil carbon and nitrogen dynamics

Citation: Xu, X. Responses of Soil Carbon and Nitrogen Dynamics and GHG Fluxes in Forest Ecosystems to Climate Change and Human Activity. *Forests* **2024**, *15*, 1235. https://doi.org/10.3390/f15071235

Received: 6 July 2024
Accepted: 9 July 2024
Published: 16 July 2024

As the largest carbon and nitrogen sinks among terrestrial ecosystems, forest ecosystems account for more than 60% of carbon and nitrogen pools in terrestrial soil organic matter at 0–100 cm depths [1]. A small change in soil carbon and nitrogen pools due to climate change and human activity can even lead to an important change in the carbon and nitrogen balance of forest ecosystems, probably resulting in impacts on global carbon and nitrogen cycling, atmospheric greenhouse gas (GHG) concentration, and the climate system [2–4]. Both climate change (e.g., changes in precipitation and snowfall, global warming, and changes in dry–wet and freeze–thaw processes) and human activity (e.g., increased atmospheric nitrogen deposition, forest management, deforestation and afforestation, land use, and increased carbon dioxide and ozone concentrations) can affect the construction and functions of vegetation in forest ecosystems as well as the soil environment via its physical, chemical, and biological processes [5–8] (Figure 1). These variations following climate change and human activity and the associated soil–plant interactions can have a substantial impact on soil carbon and nitrogen dynamics and GHG fluxes in forest ecosystems [9–16] (Figure 1). In this Special Issue, a collection of articles has been published covering a wide range of topics related to forest soil carbon and nitrogen dynamics and GHG fluxes under climate change and human activity, which can, to some extent, improve our understanding these dynamics and fluxes in forest ecosystems in the changing world.

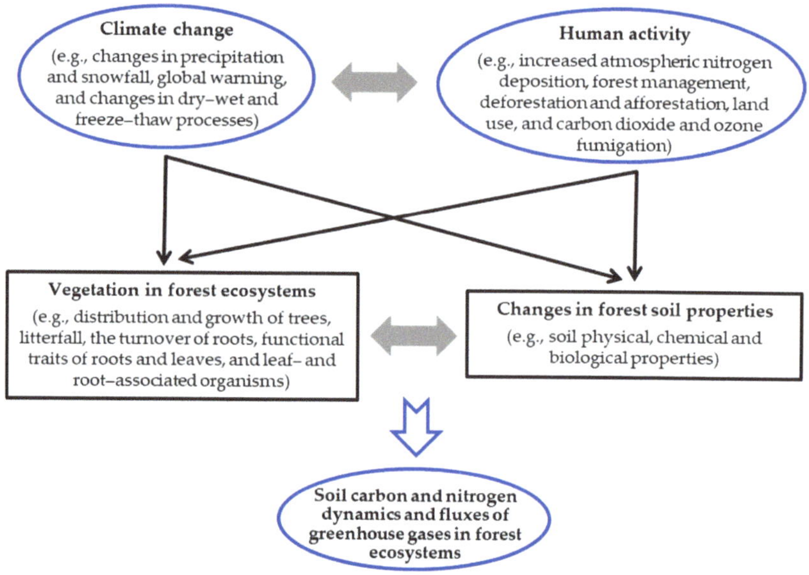

Figure 1. Conceptual model explaining effects of climate change and human activity on soil carbon and nitrogen dynamics and GHG fluxes in forest ecosystems. Double-headed gray arrows represent interactions between components. The unidirectional black and blue arrows represent unilateral impact.

Changes in precipitation and global climate warming can affect the structure and functions of forest ecosystems by impacting forest vegetation and soil environments (Figure 1). In the last three decades, precipitation manipulation experiments via throughfall transferring have been conducted to study the dynamics of soil respiration and soil nutrient status in forest ecosystems in different climate zones around the world [11,17–22]. In a study by Xu [23], a meta-analysis was performed to explore the effects of experimental categories such as climate types, forest types, soil texture, and the area size of changes in throughfall manipulation on the changes in altered throughfall-induced annual soil carbon dioxide emissions. Based on the results of the meta-analysis, it can be reasonably concluded that the responses of annual soil carbon dioxide emissions to the altered throughfall become more sensitive in temperate forests than those in tropical and subtropical forests. Moreover, a relatively large positive response in soil carbon dioxide emissions to increased throughfall occurs in Mediterranean forests [23]. It was pointed out that fine root biomass and soil microbial biomass, together with the changes in annual precipitation, would substantially affect the altered throughfall-induced annual soil carbon dioxide emissions in global forest ecosystems [23]. Besides changes in precipitation manipulation in the context of climate change, some experimental simulated apparatus such as soil heating cables, infrared heaters, and open-top chambers have been widely used to perform in situ soil warming experiments to study the responses of soil carbon and nitrogen dynamics and GHG fluxes to globe climate warming in forest ecosystems [13,24]. Compared with these artificial, simulated apparatus manipulation experiments, translocation experiments with a small gradient in micrometeorological factors (e.g., temperature and rainfall) can be used to simulate climate change in nature as much as possible. In a study by Xu et al. [25], four-year field measurement of in situ soil column translocation experiments indicated that a small increase in climate warming would increase soil carbon dioxide emission and methane uptake in the temperate forests of northeastern China, especially during the non-growing season. A robust estimation illustrated that a moderate warming of 1 °C would increase annual soil heterotrophic respiration and annual methane uptake in the temperate forests in northeastern China by a magnitude of 44.536 and 0.0168 Tg carbon, respectively [25], which should be incorporated into soil-carbon-cycling-related models to reasonably predict the responses of soil carbon fluxes at the soil–atmosphere interface in forest ecosystems to global climate warming.

Global climate warming can affect the growth of vegetation and the structure of vegetation communities and also results in an increase in the active layer's thickness of permafrost zones and changes in surface soil hydrothermal conditions [26,27]. These variations would, in turn, impact soil carbon and nitrogen dynamics and GHG fluxes in terrestrial ecosystems [28–30]. Using a regression model, TTOP model, and related datasets, Song et al. [30] analyzed the spatial distribution of permafrost and soil organic carbon (SOC) density at 0–30 cm depths in northeast China in the recent four decades, as well as annual changes in SOC stocks. The results reported by Song et al. [30] indicate that permafrost degradation due to global climate warming significantly reduces SOC stocks in northeast China in the short term but increases SOC stocks in the long term, mainly resulting from changes in land use and land cover. This study would provide a valuable scientific reference for understanding the impacts of alterations in permafrost degradation and land use on SOC changes in northeast China and other regions of the world.

Terrestrial ecosystems at high-latitude and/or high-altitude regions are currently experiencing variations in soil freezing and thawing events caused by reduced winter snowfall under global climate warming [31]. Soil freezing and thaw alterations, and their combination with litter decomposition and soil moisture can, to some extent, affect soil carbon and nitrogen dynamics and GHG fluxes (Figure 1) [15,32–34]. In a study by Xu et al. [34], simulations of autumn freeze–thaw, winter freeze, spring freeze–thaw, and the growing season were sequentially performed under laboratory conditions using undisturbed large soil columns collected from two adjacent temperate forest stands in northeastern China, to explore how the presence of litter and winter snow cover can influence carbon dioxide emissions from forest soils and its $\delta^{13}C$ values. The results reported by Xu et al. [34] indicated that the impacts of snow cover on soil heterotrophic respiration and its $\delta^{13}C$ values in cool temperate forests would vary with the different seasons of the year and the presence of aboveground litter.

As one of the main anthropogenic activities, chemical fertilizers and organic fertilizers (and their combination) can be selected as one of main effective management pathways to improve forest productivity and mitigate global warming potential in plantation forest ecosystems. Wang et al. [35] conducted a 12-month field study to investigate the impacts of biochar-based fertilizer, chemical fertilizer, and their combination on GHG fluxes from a subtropical typical bamboo plantation. The results reported by Wang et al. [35] indicate that the application of biochar-based fertilizers can be considered an environmentally friendly soil management method to mitigate GHG fluxes from subtropical bamboo plantations. Following the long-term application of chemical fertilizers such as nitrogen-based fertilizers and/or under enhanced atmospheric nitrogen deposition, both nutrient imbalance and phosphorus limitation normally occur in many tropical and subtropical forest ecosystems around the world, which can result in changes in nitrous oxide and methane fluxes from forest soils. In a study by Li et al. [36], manipulation experiments lasting one and half years were conducted to explore the effects of nitrogen and phosphorus addition at different levels on soil nitrous oxide and methane fluxes in a nitrogen-rich and phosphorus-limited subtropical Chinese fir plantation. The results reported by Li et al. [36] suggest that the impacts of nitrogen and phosphorus addition on soil nitrous oxide and methane fluxes may depend on the soil nutrient status and plant–microbial competition for soil nutrients in subtropical forests. Besides changes in soil GHG fluxes, enhanced atmospheric nitrogen deposition and/or extra nitrogen input may have impacts on SOC stability and fractions in forest ecosystems. Based on the measurement of a 7-year nitrogen addition experiment, Wu et al. [37] pointed out whether and how nitrogen addition of different durations affects soil particulate organic carbon and mineral-associated organic carbon in a subtropical *Phyllostachys edulis* forest. The results reported by Wu et al. [37] indicate the key role of microbial characteristics (e.g., microbial biomass, microbial carbon utilization efficiency, and soil enzyme activity) in regulating SOC stability and fractions following different durations of nitrogen addition in subtropical forests; this deserves further investigation at more specific sites around the world in the future. Following the addition of extra nitrogen, changes in leaf litter decomposition and the release of nutrients into the soil may complicate the changes in SOC density and stability in forest ecosystems with different nitrogen limitations (Figure 1). Based on the measurement of leaf litter decomposition and soil and litter characteristics in the Yimeng mountainous forests of China, Kong et al. [38] demonstrated that nitrogen addition can alter leaf litter decomposition, nutrient cycling, and SOC density in arid and barren mountainous regions with the limitation of available nitrogen, which can be closely associated with the tree species and leaf litter characteristics. This study highlights the importance of litter decomposition and nitrogen input in regulating soil nutrient dynamics and SOC sequestration in forest ecosystems with poor nutrient status.

Soil microbial properties such as soil enzyme activity and microbial biomass are closely associated with soil carbon and nitrogen dynamics and GHG fluxes in the changing world (Figure 1). A study by Kong et al. [39] showed that different land use types in the rocky, mountainous region of north China would have an important impact on the activities of soil enzymes such as urease, catalase, β-glucosidase, and protease, along with relatively large values in farmland cultivation and *Platycladus orientalis* (L.) Franco plantation. To improve soil fertility and enhance the adaptability of trees, the results reported by Kong et al. [39] indicate that *Platycladus orientalis* would be selected as the dominant tree species for plantation to achieve sustainable development in the barren, rocky, mountainous region of north China. Along elevation gradients, changes in vegetation properties and soil hydrothermal conditions are normally proposed to have an important impact on the availability of soil substances and the complicate plant–soil interactions in alpine ecosystems. Based on field investigations and laboratory analysis, Feng et al. [40] reported changes in vegetation diversity and soil biophysical and biochemical properties at three soil layers as well as their relationships along the five elevation gradients in the southern Taihang mountains, China. A study by Feng et al. [40] strongly indicates that the contents of active soil substances in alpine ecosystems would be greatly influenced by elevation gradients, soil layers, and plant–soil relationships.

In summary, this Special Issue contains some recent research results in the field of soil carbon and nitrogen dynamics and GHG fluxes in forest ecosystems following climate change and human activity. Although this collection only reflects a small part of this important scientific theme, it is our aim to promote international scientific researchers to further study these dynamics and fluxes in the changing world for sustainable development and the mitigation of climate change. Long-term perspectives are essential because changes in soil carbon and nitrogen pools, especially mineral-associated pools, are normally slow and the full influence of climate change and human activities (e.g., enhanced atmospheric nitrogen deposition and changes in land management practices) on soil carbon and nitrogen dynamics and GHG fluxes in terrestrial ecosystems may take decades to become obvious [1,11,41–43]. Furthermore, additional studies should be conducted to explore soil carbon responses to the varying nitrogen availability in forest ecosystems around the world, including carbon responses at different soil layers [10,41,43,44]. Regarding the related conceptual model (Figure 1), to date, there have been few studies focusing on the differences in these dynamics and fluxes under short-term and long-term manipulation of climate change and human activities [42,44] or under natural gradient changes and anthropogenic abrupt change conditions [45]. The upscaling of methods from sites to ecosystem and regional scales is also lacking [46–48]. These study aspects should be fully taken into consideration in the future.

As guest editor, I would like to thank all the authors for their valuable contributions to this Special Issue. I also wish to express my sincere appreciation to all reviewers for their insightful comments, which improved the quality of earlier versions of each paper. I would also like to sincerely thank the editorial office of *Forests* for their valuable assistance throughout the publication process.

Funding: This research was funded by the National Natural Science Foundation of China (grant number 41175133, 41775163, 41975121, and 42275130).

Conflicts of Interest: The authors declare no conflicts of interest.

References

1. Nieder, R.; Benbi, D.K. *Carbon and Nitrogen in the Terrestrial Environment*; Spring Science + Business Media B.V.: Berlin/Heidelberg, Germany, 2008.
2. Cox, P.M.; Betts, R.A.; Jones, C.D.; Spall, S.A.; Totterdell, I.J. Acceleration of global warming due to carbon-cycle feedbacks in a coupled climate model. *Nature* **2000**, *408*, 184–187. [CrossRef] [PubMed]
3. Zaehle, S.; Bondeau, A.; Carter, T.R.; Cramer, W.; Erhard, M.; Prentice, I.C.; Reginster, I.; Rounsevell, M.D.A.; Sitch, S.; Smith, B.; et al. Projected changes in terrestrial carbon storage in Europe under climate and land-use change, 1990–2100. *Ecosystems* **2007**, *10*, 380–401. [CrossRef]
4. Bond-Lamberty, B.; Bailey, V.L.; Chen, M.; Gough, C.M.; Vargas, R. Globally rising soil heterotrophic respiration over recent decades. *Nature* **2018**, *560*, 80–83. [CrossRef] [PubMed]
5. Ceulemans, R.; Janssens, I.A.; Jach, M.E. Effect of CO_2 enrichment on trees and forests: Lessons to be learned in view of future ecosystem studies. *Ann. Bot.* **1999**, *84*, 577–590. [CrossRef]
6. Xu, X.K.; Luo, X.B. Effect of wetting intensity on soil GHG fluxes and microbial biomass under a temperate forest floor during dry season. *Geoderma* **2012**, *170*, 118–126. [CrossRef]
7. Assefa, D.; Rewald, B.; Sandén, H.; Rosinger, C.; Abiyu, A.; Yitaferu, B.; Godbold, D.L. Deforestation and land use strongly effect soil organic carbon and nitrogen stock in Northwest Ethiopia. *Catena* **2017**, *153*, 89–99. [CrossRef]
8. Mayer, M.; Prescott, C.E.; Abaker, W.E.A.; Augusto, L.; Cécillon, L.; Ferreira, G.W.D.; James, J.; Jandl, R.; Katzensteiner, K.; Laclau, J.P.; et al. Tamm Review: Influence of forest management activities on soil organic carbon stocks: A knowledge syhthesis. *Forest Ecol. Manag.* **2020**, *466*, 118127. [CrossRef]
9. Rasmussen, C.; Southard, R.J.; Horwath, W.R. Litter type and soil minerals control temperate forest soil carbon response to climate change. *Glob. Chang. Biol.* **2008**, *14*, 2064–2080. [CrossRef]
10. Xu, X.K.; Han, L.; Luo, X.B.; Han, S.J. Effects of nitrogen addition on dissolved N_2O and CO_2, dissolved organic matter, and inorganic nitrogen in soil solution under a temperate old-growth forest. *Geoderma* **2009**, *151*, 370–377. [CrossRef]
11. Xu, X.K.; Duan, C.T.; Wu, H.H.; Luo, X.B.; Han, L. Effects of changes in throughfall on soil GHG fluxes under a mature temperate forest, northeastern China. *J. Environ. Manag.* **2021**, *294*, 112950. [CrossRef]
12. Janssens, I.A.; Dieleman, W.; Luyssaert, S.; Subke, J.A.; Reichstein, M.; Ceulemans, R.; Ciais, P.; Dolman, A.J.; Grace, J.; Matteucci, G.; et al. Reduction of forest soil respiration in response to nitrogen deposition. *Nat. Geosci.* **2010**, *3*, 315–322. [CrossRef]
13. Lu, M.; Zhou, X.H.; Yang, Q.; Li, H.; Luo, Y.Q.; Fang, C.M.; Chen, J.K.; Yang, X.; Li, B. Responses of ecosystem carbon cycle to experimental warming: A meta-analysis. *Ecology* **2013**, *94*, 726–738. [CrossRef] [PubMed]
14. Wang, B.; Shugart, H.H.; Lerdau, M.T. Sensitivity of global greenhouse gas budgets to tropospheric ozone pollution mediated by the biosphere. *Environ. Res. Lett.* **2017**, *12*, 084001. [CrossRef]
15. Xu, X.K. Effect of freeze-thaw disturbance on soil C and N dynamics and GHG fluxes of East Asia forests: Review and future perspectives. *Soil. Sci. Plant Nutr.* **2022**, *68*, 15–26. [CrossRef]
16. Andresen, L.C.; Ambus, P.; Beier, C.; Michelsen, A. Moderate nitrogen retention in temperate heath ecosystem after elevated CO_2, drought and warming through 7 years. *Eur. J. Soil. Sci.* **2023**, *74*, e13397. [CrossRef]
17. Cattânio, J.; Davidson, E.; Nepstad, D.; Verchot, L.; Ackerman, I. Unexpected results of a pilot throughfall exclusion experiment on soil emissions of CO_2, CH_4, N_2O, and NO in eastern Amazonia. *Biol. Fertil. Soils* **2002**, *36*, 102–108.
18. Johnson, D.W.; Todd, D.E.; Hanson, P.J. Effects of throughfall manipulation on soil nutrient status: Results of 12 years of sustained wet and dry treatments. *Glob. Chang. Biol.* **2008**, *14*, 1661–1675. [CrossRef]
19. Cleveland, C.C.; Wieder, W.R.; Reed, S.C.; Townsend, A.R. Experimental drought in a tropical rain forest increases soil carbon dioxide losses to the atmosphere. *Ecology* **2010**, *91*, 2313–2323. [CrossRef]
20. Deng, Q.; Hui, D.; Zhang, D.; Zhou, G.; Liu, J.; Liu, S.; Chu, G.; Li, J. Effects of precipitation increase on soil respiration: A three-year field experiment in subtropical forests in China. *PLoS ONE* **2012**, *7*, e41493. [CrossRef]
21. Hinko-Najera, N.; Fest, B.; Livesley, S.J.; Arndt, S.K. Reduced throughfall decreases autotrophic respiration, but not heterotrophic respiration in a dry temperate broadleaved evergreen forest. *Agric. For. Meteorol.* **2015**, *200*, 66–77. [CrossRef]
22. Liu, Y.C.; Liu, S.R.; Wan, S.Q.; Wang, J.X.; Luan, J.W.; Wang, H. Differential responses of soil respiration to soil warming and experimental throughfall reduction in a transitional oak forest in central China. *Agric. For. Meteorol.* **2016**, *226*, 186–198. [CrossRef]
23. Xu, X. Effect of changes in throughfall on soil respiration in global forest ecosystems: A meta-analysis. *Forests* **2023**, *14*, 1037. [CrossRef]
24. Yang, J.Y.; Jia, X.Y.; Ma, H.Z.; Chen, X.; Liu, J.; Shuangguan, Z.P.; Yan, W.M. Effects of warming and precipitation changes on soil GHG fluxes: A meta-analysis. *Sci. Total Environ.* **2022**, *827*, 154351. [CrossRef] [PubMed]
25. Xu, X.; Xu, T.; Yue, J. Effect of in situ large soil column translocation on CO_2 and CH_4 fluxes under two temperate forests of northeastern China. *Forests* **2023**, *14*, 1531. [CrossRef]

26. Schuur, E.A.G.; Crummer, K.G.; Vogel, J.G.; Mack, M.C. Plant species composition and productivity following permafrost thaw and thermokarst in Alaskan tundra. *Ecosystems* **2007**, *10*, 280–292. [CrossRef]
27. Lin, D.L.; Xia, J.Y.; Wan, S.Q. Climate warming and biomass accumulation of terrestrial plants: A meta-analysis. *New Phytol.* **2010**, *188*, 187–198. [CrossRef]
28. Schuur, E.A.G.; Vogel, J.G.; Crummer, K.G.; Lee, H.; Sickman, J.O.; Osterkamp, T. The effect of permafrost thaw on old carbon release and net carbon exchange from tundra. *Nature* **2009**, *459*, 556–559. [CrossRef]
29. Adiya, S.; Dalantai, S.; Wu, T.H.; Wu, X.D.; Yamkhin, J.; Bao, Y.H.; Sumiya, E.; Yadamsuren, G.; Avirmed, D.; Dorjgotov, B. Spatial and temporal change patterns of near-surface CO_2 and CH_4 concentrations in different permafrost regions on the Mongolian Plateau from 2010 to 2017. *Sci. Total Environ.* **2021**, *800*, 149433. [CrossRef]
30. Song, Y.; Huang, S.; Zhang, H.; Wang, Q.; Ding, L.; Liu, Y. The impact of permafrost change on soil organic carbon stocks in northeast China. *Forests* **2024**, *15*, 14. [CrossRef]
31. Henry, H.A.L. Climate change and soil freezing dynamics: Historical trends and projected changes. *Clim. Chang.* **2008**, *87*, 421–434. [CrossRef]
32. Wu, H.H.; Xu, X.K.; Cheng, W.G.; Han, L. Dissolved organic matter and inorganic N jointly regulate greenhouse gases fluxes from forest soils with different moistures during a freeze-thaw period. *Soil. Sci. Plant Nutr.* **2020**, *66*, 163–176. [CrossRef]
33. Wu, H.H.; Xu, X.K.; Fu, P.Q.; Cheng, W.G.; Fu, C.S. Responses of soil WEOM quantity and quality to freeze-thaw and litter manipulation with contrasting soil water content: A laboratory experiment. *Catena* **2021**, *198*, 105058. [CrossRef]
34. Xu, X.; Wu, H.; Yue, J.; Tang, S.; Cheng, W. Effects of snow cover on carbon dioxide emissions and their $\delta^{13}C$ values of temperate forest soils with and without litter. *Forests* **2023**, *14*, 1384. [CrossRef]
35. Wang, E.; Yuan, N.; Lv, S.; Tang, X.; Wang, G.; Wu, L.; Zhou, Y.; Zhou, G.; Shi, Y.; Xu, L. Biochar-based fertilizer decreased soil N_2O emission and increased soil CH_4 uptake in a subtropical typical bamboo plantation. *Forests* **2022**, *13*, 2181. [CrossRef]
36. Li, B.; Chen, G.; Lu, X.; Jiao, H. Effects of nitrogen and phosphorus additions on soil N_2O emissions and CH_4 uptake in a phosphorus-limited subtropical Chinese fir plantation. *Forests* **2022**, *13*, 772. [CrossRef]
37. Wu, Y.; Zeng, Q.; Su, X.; Zheng, W.; Zhang, Q.; Chen, Y. The different factors driving SOC stability under different N addition durations in a Phyllostachys edulis Forest. *Forests* **2023**, *14*, 1890. [CrossRef]
38. Kong, B.; Zhou, J.; Qi, L.; Jiao, S.; Ma, L.; Geng, W.; Zhao, Y.; Gao, T.; Gong, J.; Li, K.; et al. Effects of nitrogen deposition on leaf litter decomposition and soil organic carbon density in arid and barren rocky mountainous regions: A case study of Yimeng mountain. *Forests* **2023**, *14*, 1351. [CrossRef]
39. Kong, Y.; Qu, A.; Feng, E.; Chen, R.; Yang, X.; Lai, Y. Seasonal dynamics of soil enzymatic activity under different land-use types in rocky mountainous region of north China. *Forests* **2023**, *14*, 536. [CrossRef]
40. Feng, E.; Zhang, L.; Kong, Y.; Xu, X.; Wang, T.; Wang, C. Distribution characteristics of active soil substances along elevation gradients in the southern of Taihang mountain, China. *Forests* **2023**, *14*, 370. [CrossRef]
41. Neff, J.C.; Townsend, A.R.; Gleixner, G.; Lehman, S.J.; Turnbull, J.; Bowman, W.D. Variable effects of nitrogen additions on the stability and turnover of soil carbon. *Nature* **2002**, *419*, 915–917. [CrossRef]
42. Melillo, J.M.; Frey, S.D.; DeAngelis, K.M.; Werner, W.J.; Bernard, M.J.; Bowles, F.P.; Pold, G.; Knorr, M.A.; Grandy, A.S. Long-term pattern and magnitude of soil carbon feedback to the climate system in a warming world. *Science* **2017**, *358*, 101–104. [CrossRef]
43. Yu, M.X.; Wang, Y.P.; Baldock, J.A.; Jiang, J.; Mo, J.M.; Zhou, G.Y.; Yan, J.H. Divergent responses of soil organic carbon accumulation to 14 years of nitrogen addition in two typical subtropical forests. *Sci. Total Environ.* **2020**, *707*, 136104. [CrossRef]
44. Rocci, K.S.; Lavallee, J.M.; Stewart, C.E.; Cotrufo, M.F. Soil organic carbon response to global environmental change depends on its distribution between mineral-associated and particulate organic matter: A meta-analysis. *Sci. Total Environ.* **2021**, *793*, 148569. [CrossRef]
45. Fukami, T.; Wardle, D.A. Long-tern ecological dynamics: Reciprocal insights from natural and anthropogenic gradients. *Proc. R. Soc. B* **2005**, *272*, 2105–2115. [CrossRef] [PubMed]
46. Shrestha, R.K.; Strahm, B.D.; Sucre, E.B. Greenhouse gas emissions in response to nitrogen fertilization in managed forest ecosystems. *New Forests* **2015**, *46*, 167–193. [CrossRef]
47. Phillips, C.L.; Bond-Lamberty, B.; Desai, A.R.; Lavoie, M.; Risk, D.; Tang, J.W.; Todd-Brown, K.; Vargas, R. The value of soil respiration measurements for interpreting and modeling terrestrial carbon cycling. *Plant Soil.* **2017**, *413*, 1–25. [CrossRef]
48. Clément, R.; Pärn, J.; Maddison, M.; Henine, H.; Chaumont, C.; Tournebize, J.; Uri, V.; Espenberg, M.; Günther, T.; Mander, U. Frequency-domain electromagnetic induction for upscaling greenhouse gas fluxes in two hemiboreal drained peatland forests. *J. Appl. Geophys.* **2020**, *173*, 103944. [CrossRef]

forests

Article

The Impact of Permafrost Change on Soil Organic Carbon Stocks in Northeast China

Yang Song [1], Shuai Huang [2], Haiying Zhang [1], Qin Wang [1], Lin Ding [3] and Yanjie Liu [1,*]

1 School of Civil Engineering, Heilongjiang University, Harbin 150080, China; 2212647@s.hlju.edu.cn (Y.S.); 2222724@s.hlju.edu.cn (H.Z.); 2222735@s.hlju.edu.cn (Q.W.)
2 School of Forestry, Northeast Forestry University, Harbin 150080, China; s_hwang@nefu.edu.cn
3 Heilongjiang Province Hydraulic Research Institute, Harbin 150080, China; dinglin@hlju.edu.cn
* Correspondence: 1996058@hlju.edu.cn

Abstract: Climate warming has resulted in significant changes in permafrost in Northeast China, leading to notable alterations in soil organic carbon (SOC) stocks. These changes are crucial for both the global carbon cycle and climate change, as well as directly impacting the sustainable development of ecosystems. In order to examine the SOC dynamics and the impact of permafrost changes on SOC, we investigate the changes of permafrost extent based on a regression model and TTOP (top temperature of permafrost) model and the relationship between land use and land cover (LULC), SOC stocks, and permafrost changes in Northeast China. The results showing a shrinking permafrost area from 37.43×10^4 km^2 to 16.48×10^4 km^2 during the period from the 1980s to the 2010s in Northeast China, and the SOC stock decreased by 24.18 Tg C from the 1980s to the 1990s and then rapidly increased by 102.84 Tg C in the 2000s. Permafrost degradation speeds up the succession of LULC, impacting about 90% of the SOC in permafrost regions. The relationship between permafrost changes and SOC in Northeast China shows that permafrost degradation significantly reduces SOC stocks in the short term but increases SOC stocks in the long term, and that LULC play a crucial role in regulating this relationship. The goals of this study are to acquire an understanding of permafrost status and deepening insights into the dynamics of SOC. Simultaneously, the study aims to furnish valuable scientific references for shaping policies on sustainable land use and management in the future, all the while advancing the cause of ecological equilibrium and sustainable development in Northeast China and other areas.

Keywords: land use and land cover; permafrost changes; soil organic carbon stocks; temperature at the top of permafrost

check for updates

Citation: Song, Y.; Huang, S.; Zhang, H.; Wang, Q.; Ding, L.; Liu, Y. The Impact of Permafrost Change on Soil Organic Carbon Stocks in Northeast China. *Forests* **2024**, *15*, 14. https://doi.org/10.3390/f15010014

Academic Editor: Timo Domisch

Received: 10 November 2023
Revised: 11 December 2023
Accepted: 15 December 2023
Published: 20 December 2023

1. Introduction

Soil organic carbon (SOC) is a significant contributor to greenhouse gas emissions and has a considerable impact on the global carbon cycle [1] and represents the largest carbon (C) pool within terrestrial ecosystems [2]. Globally, there is an estimated SOC pool ranging from approximately 700 Pg to 2946 Pg [3]. Against the backdrop of global warming, the relationship between SOC, climate, and the environment has become a focal point of research [4]. Climate and land cover drive SOC [5]. Temperature, precipitation, and humidity, among other factors, impact SOC by affecting plant litter, soil respiration, and microbial decomposition processes [6]. It is worth noting that the cryosphere is a critical component of the Earth's ecosystem, covering approximately 14% of the land surface [7]. Permafrost, as one of the components of the cryosphere, is also one of the most vulnerable carbon (C) pools [8]. It is estimated that the global permafrost contains approximately 1320 ± 200 Pg of SOC [9,10]. In the Intergovernmental Panel on Climate Change (IPCC) Sixth Assessment Report [11], it was stated with high confidence that permafrost temperatures have risen to record levels, leading to increased release of methane (CH_4) and carbon dioxide (CO_2) from permafrost regions. The rate of permafrost warming, as exemplified by the Qinghai-Tibet Plateau at 0.3 °C per decade, significantly exceeds the global average warming rate of 0.12 °C per decade [12]. During a warming trend, permafrost undergoes significant degradation [13–15], which affects the process of SOC decomposition [16,17]. Alterations in the SOC pool within permafrost can have a significant impact on atmospheric CO_2 concentrations, further accelerating the effects of climate change caused by global warming [9].

The relationship between permafrost changes and SOC is intricately linked to the spatial distribution of permafrost. Permafrost simulation models primarily include the Stefan model [18], the frost number model [19], the altitude model [20], and the top temperature of permafrost (TTOP) model [21]. The TTOP model, specifically, has garnered significant attention for its exceptional performance and applicability. It links surface climate with subsurface thermal features and establishes a correlation between compensation effect and climatic factors [22]. With the development of remote sensing technologies such as optical remote sensing, thermal infrared remote sensing, and microwave remote sensing, observation and mapping of perennial permafrost are carried out by developing statistical learning [23], constructing models [24,25], and logical discriminations [26]. The application of TTOP modeling has thus led to many research results [27–30]. Permafrost degradation is typically marked by a reduction in the permafrost's extent, an elevation in soil temperature, and an increase in the active layer's thickness [11,31,32]. This alters surface drainage patterns and the structure of vegetation communities, which impacts the physical properties of soil and ecosystems [33]. The degradation characteristics of permafrost are obvious, but the resulting environmental effects and impacts on soil organic carbon (SOC) unfold over a long period of time. The response mechanisms are multifaceted, creating complex and interconnected relationships. Permafrost degradation can hasten microbial decomposition of organic matter, rendering SOC more vulnerable to exposure and release, which could result in a decrease in SOC [34,35]. In contrast, increased soil temperatures and soil moisture resulting from permafrost degradation promote vegetation growth in the short run [36]. The carbon sequestration from the increased vegetation can offset or even exceed carbon losses [10,37]. Such uncertainty provides a cushion against SOC release, making the study of the impact of permafrost change on SOC dynamics particularly crucial.

In Northeast China, permafrost has higher temperatures compared to the permafrost in high-altitude and Arctic regions, with a thinner permafrost thickness, making it more sensitive to climate warming [38,39]. According to a report [40], the mean annual air temperature in Northeast area has risen by 0.31 °C per decade from 1961 to 2017. Climate warming has resulted in a substantial degradation of permafrost, with continuous permafrost being transformed into discontinuous permafrost and the disappearance of many isolated permafrost patches and permafrost islands [14]. Many researchers have conducted extensive research on the extent of permafrost degradation in Northeast China, utilizing multiple sources of data such as meteorological station data, measurement data from boreholes, and remote sensing data (Landsat and MODIS inverted surface temperature data, interferometric synthetic aperture radar data, etc.) [41–43]. However, comparing the latest results of permafrost distribution in Northeast China during the same period, it is found that the area of permafrost varies greatly, reaching 12.81×10^4 km^2, which indicates that further research on the distribution of permafrost in Northeast China is needed. Furthermore, since 2003, Chinese meteorological stations have switched from manual to automated measurements, resulting in significant variations in recorded ground surface temperature (GST) over different periods. This has created a lack of uniformity in GST data after this transition. Undoubtedly, permafrost is mainly concentrated in the northern part of Northeast China, where there is abundant vegetation cover, and an amount of SOC is stored in permafrost [44]. Changes in permafrost could potentially affect SOC density, stock, and succession of land type, thus impacting ecosystem evolution [45]. Land use and land cover (LULC) have a direct influence on the density and stock of SOC. Their capacity to release or sequester SOC differs significantly [46]. However, the precise distribution of permafrost in Northeast China and its impact on SOC remains unexplored, as well as the impact of LULC types on SOC in permafrost degradation zones.

2. Materials and Methods

2.1. Study Area

Northeast China (Figure 1) is located in the southeastern region of the Eurasian. The permafrost in this region represents a transitional type between high-mountain permafrost and high-latitude permafrost, making it a unique form of permafrost known as the 'Xing'an-Baikal' permafrost. The permafrost in Northeast China is influenced by both environmental and climatic factors. The Northeast China region covers an area of approximately 14.2 × 10^5 km^2. The area is situated in both the temperate and cold temperate zones, and features a continental monsoon climate.

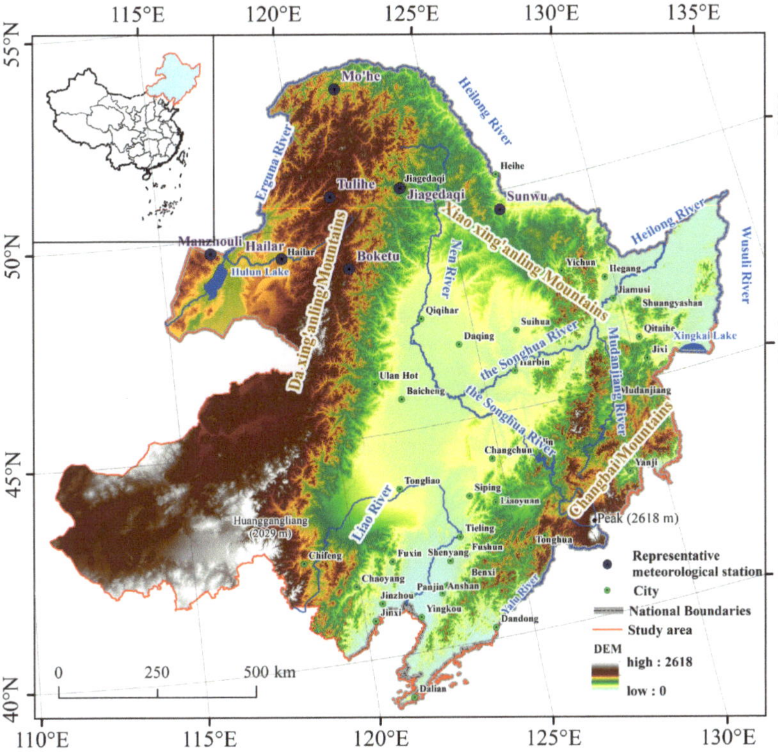

Figure 1. Digital Elevation Model (DEM) Map of Northeast China.

2.2. Data Sources

2.2.1. Meteorological Data

In this study, daily meteorological data including air temperature, ground surface temperature (GST), precipitation, relative humidity, air pressure, and sunshine duration were collected from 263 ground meteorological stations for the period from 1980 to 2020. The data were sourced from the National Meteorological Information Center (https://data.cma.cn, accessed on 20 May 2023). Additionally, daily snow cover data with a spatial resolution of 25 km from 1979 to 2020 were obtained from the China Snow Depth Long-term Time Series dataset available at the National Tibetan Plateau Data Center (https://data.tpdc.ac.cn, accessed on 20 May 2023).

During the period from 2003 to 2005, there was a transition in the monitoring methods of Chinese meteorological stations from manual to automatic measurements. This transition caused several systematic errors between snow-covered and non-snow-covered ground, due to the difference in GST between manual measurements (taken at the surface level of snow cover) and automatic measurements (which measure temperature at a depth of 5–10 cm in the ground) [47].

2.2.2. DEM

The Digital Elevation Model (DEM) data with a spatial resolution of 1 km were extracted from United States Geological Survey (https://earthexplorer.usgs.gov, accessed on 20 May 2023). These data were introduced as covariates when performing spatial interpolation using the Australian National University Spline (ANUSPLIN) software (ANUSPLIN 4.2, Canberra, Australia).

2.2.3. Land Use and Land Cover

Land use and land cover data with a spatial resolution of 1 km from 1980, 1990, 2000, 2010, and 2020 were obtained from the Resource and Environment Science and Data Center (https://www.resdc.cn, accessed on 20 May 2023).

2.2.4. Soil Organic Carbon (SOC)

The global 1 km surface (0–30 cm) SOC pool product for the years 1981 to 2019 was obtained from the National Earth System Science Data Center (http://www.geodata.cn, accessed on 20 May 2023). These data simulate the spatial and temporal distribution of global SOC pools from 1981 to 2019 based on global soil sample point SOC density data, combining static and dynamic covariates, and utilizing integrated machine learning and RothC process models. The production of SOC products at 1 km resolution was realized through the Random Forest Spatial and Temporal Substitution model, and the validation results showed an accuracy of R^2 = 0.422, RMSE = 27.00, and MRE = 30.9%.

2.3. Data Processing Methods

2.3.1. TTOP Model

The TTOP (top temperature of permafrost) model as described below, developed by Smith and Riseborough in 1996, is an equilibrium equation used to describe the relationship between permafrost and climate:

$$\mathrm{TTOP} = \frac{r_k \cdot \mathrm{DDT} - \mathrm{DDF}}{\mathrm{P}} \tag{1}$$

$$\mathrm{DDT} = \sum\nolimits_{i=1}^{N_T} Ti,\ T_i < 0\ ^{\circ}\mathrm{C} \tag{2}$$

$$\mathrm{DDF} = \sum\nolimits_{i=1}^{N_F} |T_i|_{i,}\ T_i < 0\ ^{\circ}\mathrm{C} \tag{3}$$

$$r_k = \frac{\lambda_t}{\lambda_f} \tag{4}$$

where DDT is ground surface thawing index (°C·d); DDF is ground surface freezing index (°C·d); T_i is temperature on the ith day (°C); N_F and N_T are the number of days in the year with surface temperature below and above 0 °C, respectively (d); λ_t is the thawing soil thermal conductivity coefficient (W/m·°C); λ_f is the freezing soil thermal conductivity coefficient (W/m·°C); P is total number of days in a year (d); and r_k represents assigned values based on LULC types, as shown in Table 1.

Table 1. r_k (the ratio of thermal conductivity of soil under thawed to frozen states) value in different LULC types.

LULC Types	r_k Value	LCCS Classification System Number
Bare areas	0.95	140, 150, 152, 153, 200, 201, 202
Grasslands and croplands	0.75	10, 11, 12, 20, 130
Shrubs	0.80	120, 121, 122
Deciduous forest	0.95	50, 60, 61, 62, 70, 71, 72, 80, 81, 82
Wetlands	0.55	180
Urban land	0.70	190

The permafrost distribution type is determined by the following equation:

$$\begin{cases} \text{TTOP} \leq 0, \text{ Permafrost} \\ \text{TTOP} > 0, \text{ Seasonally frozen ground} \end{cases} \tag{5}$$

2.3.2. Regression Model [48,49]

The ground freeze-thaw index (DDT/DDF) of the TTOP model is calculated based on GST. Currently, there are two methods for obtaining GST data: remote-sensing-based retrieval and meteorological station measurements. Among them, thermal infrared remote sensing collects and records the thermal infrared information of ground objects by means of on-board or airborne sensors and uses such thermal infrared information to identify the ground objects and invert the ground surface temperature, while the ground meteorological stations obtain the ground surface temperature by means of a temperature meter or temperature sensor observation. However, in Northeast China, which has dense vegetation cover and thick winter snow cover, the retrieval results often represent the temperature of the vegetation canopy or snow surface [50]. Therefore, in this study, GST data from meteorological stations were used as input parameters of the TTOP model. Due to the systematic error in GST measurements before 2003, the GST data before 2003 were used as the basis for training and validating historical data to obtain the feature coefficients in the model. Subsequently, the model was used to output daily GST data from 2003 onwards.

Regression model:

$$y = b_0 + b_1 x_1 + b_2 x_2 + \cdots + b_n x_n \tag{6}$$

Here, y represents the target variable (GST), b_0 is the intercept, and b_1 to b_n are the coefficients of the features, while x_1 to x_n represent the feature values.

The regression model is a simple yet effective method, and it is well-suited for establishing relationships between meteorological factors. Utilizing a large amount of data enhances the model's fitting capability, improves its robustness, and reduces errors. This model was built and calculated using the Sklearn (Scikit-learn) library in Python 3.10. The features used as input include longitude, latitude, elevation, air temperature, precipitation, relative humidity, air pressure, sunshine duration, and snow depth, with GST as the target variable.

2.3.3. Local Thin-Plate Smoothing Spline Interpolation Method

The results of the TTOP were interpolated using the ANUSPLIN software. The model calculations were interpolated utilizing longitude and latitude as independent variables, altitude as a covariate, and TTOP value as the dependent variable. The model expression is as follows:

$$Z_i = f(x_i) + b^T y_i + e_i (i = 1, 2, 3 \cdots, N) \tag{7}$$

where Z_i represents the dependent variable for spatial point i; x_i represents the d-dimensional spline independent variables; f is the unknown smoothing function related to x_i; y_i is the p-dimensional independent covariate; b is the p-dimensional coefficients for y_i; and e_i is the random error associated with the independent variables; T is a symbol for transpose.

The function f and coefficients b are estimated using the least squares method:

$$min : \sum_{i=1}^{n} \left[\frac{Z_i - f(x_i) - b^T y_i}{w_i} \right] + \rho J_m(f) \tag{8}$$

where $J_m(f)$ represents the mth-order partial derivative of the function f and ρ is a positive smoothing parameter; w_i is the known local relative coefficient of variation used as a weight.

3. Results

3.1. Regression Model Adjusted GST

We selected seven representative national meteorological stations in the northern part of Northeast China (Mo'he, Tulihe, Jiagedaqi, Manzhouli, Hailar, Sunwu, and Boketu) to represent the permafrost regions (Table 2), and compared the mean annual ground surface temperatures and mean annual air temperatures in China, Northeast China, and permafrost regions. As illustrated in Figure 2a, the mean annual air temperature warming rates from 1980 to 2020 for China, Northeast China, and permafrost regions stand at 0.36 °C/10a, 0.33 °C/10a and 0.37 °C/10a, respectively. During the period from 1980 to 2002, Northeast China and permafrost regions exhibited mean annual ground surface temperature warming rates of 0.61 °C/10a and 0.77 °C/10a, respectively. Notably, the mean annual ground surface temperature warming rates for Northeast China and permafrost regions between 2003 and 2020 are recorded at 1.1 °C/10a and 1.48 °C/10a, respectively.

Table 2. Basic information on representative national meteorological stations of the permafrost region.

Station	Longitude (°E)	Latitude (°N)	Elevation (m)
Mo'he	122.52	52.97	439.7
Tulihe	121.68	50.48	733.7
Jiagedaqi	124.12	50.40	371.7
Manzhouli	117.32	49.58	661.8
Hailar	119.70	49.25	649.6
Sunwu	127.35	49.43	234.5
Boketu	121.92	48.77	739.7

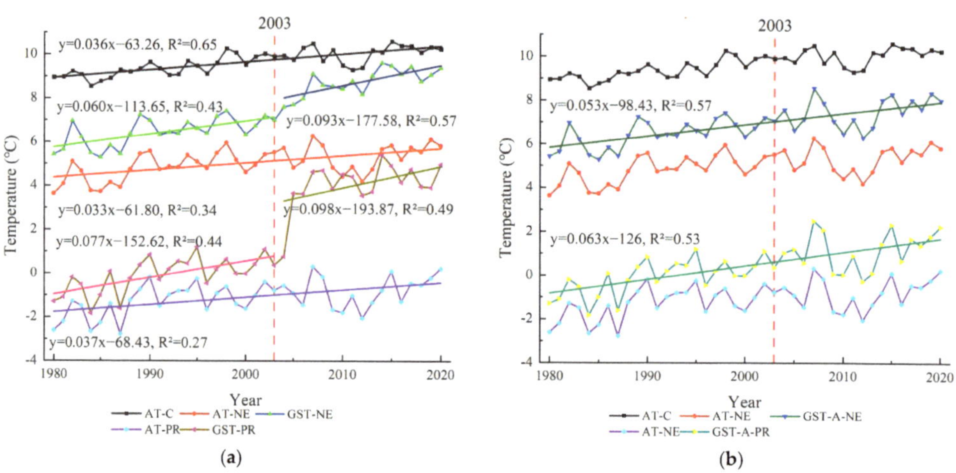

Figure 2. The mean annual air temperatures and mean annual ground surface temperatures in representative areas of Northeast China and permafrost regions from 1980 to 2020, and the trend of

mean annual air temperature change in China. (**a**) Actual observed data. (**b**) Adjusted data. AT-C: mean annual air temperature of China; AT-NE: mean annual air temperature of Northeast China; AT-PR: mean annual air temperature of permafrost regions; GST-NE: mean annual ground surface temperature of Northeast China; GST-PR: mean annual ground surface temperature of permafrost regions; GST-A-NE: adjusted mean annual ground surface temperature of Northeast China; GST-A-PR: adjusted mean annual ground surface temperature of permafrost regions (Table 3).

Table 3. Maximum, minimum, and average of GST at different time intervals in Northeast China and permafrost regions.

		Year	Maximum (°C)	Minimum (°C)	Average (°C)
Northeast China	Actual observed data	1980–2003	7.44	5.28	6.45
		2004–2020	9.63	7.61	8.77
		1980–2020	9.63	5.28	7.41
	Adjusted data	1980–2003	7.44	5.28	6.45
		2004–2020	8.56	6.30	7.49
		1980–2020	8.56	5.28	6.88
permafrost regions	Actual observed data	1980–2003	1.21	−1.29	−0.08
		2004–2020	4.98	0.78	4.10
		1980–2020	4.98	−1.29	1.65
	Adjusted data	1980–2003	1.21	−1.29	−0.08
		2004–2020	2.50	−0.26	1.18
		1980–2020	2.50	−1.29	0.44

After adjusting the GST data, as shown in Figure 2b, the mean annual ground surface temperature warming rates for Northeast China and permafrost regions from 1980 to 2020 are 0.53 °C/10a and 0.63 °C/10a, respectively.

3.2. TTOP and Spatial Distribution of Permafrost

Using the ANUSPLIN software for spatial interpolation of TTOP, Figure 3a shows the TTOP results obtained for Northeast China. Upon analysis of the spatial distribution of TTOP and classification of permafrost types, the following observations were made: The highest temperature in Northeast China rose from 9.306 °C to 10.164 °C, indicating a warming trend of 0.858 °C or an increase of 9.22%. The lowest temperature increased from −6.972 °C to −3.886 °C, showing a warming of 3.086 °C or an increase of 44.26%. Seasonally frozen ground areas were defined as regions with TTOP ≥ 0 °C, while permafrost areas were defined as regions with TTOP < 0 °C, Figure 3b. The permafrost areas in the 1980s, 1990s, 2000s, and 2010s were about 37.43×10^4 km^2, 24.87×10^4 km^2, 18.93×10^4 km^2, and 16.48×10^4 km^2, respectively. The largest stage of permafrost degradation happened between the 1980s and the 1990s, resulting in a decrease in the area of 12.57×10^4 km^2. Over the span of four decades, from the 1980s to the 2010s, the extent of permafrost decreased by about 20.95×10^4 km^2, accounting for a 55.98% decline. In contrast, the proportion of seasonally frozen ground areas increased considerably as the permafrost area percentage fell from 25.90% to 11.41%. Presently, the Da Xing'anling Mountains host a significant proportion of the remaining permafrost, raising alarm about its preservation.

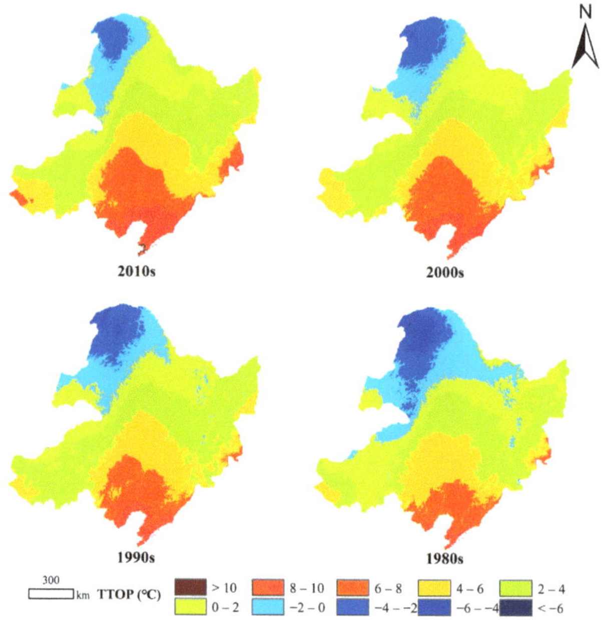

(**a**)

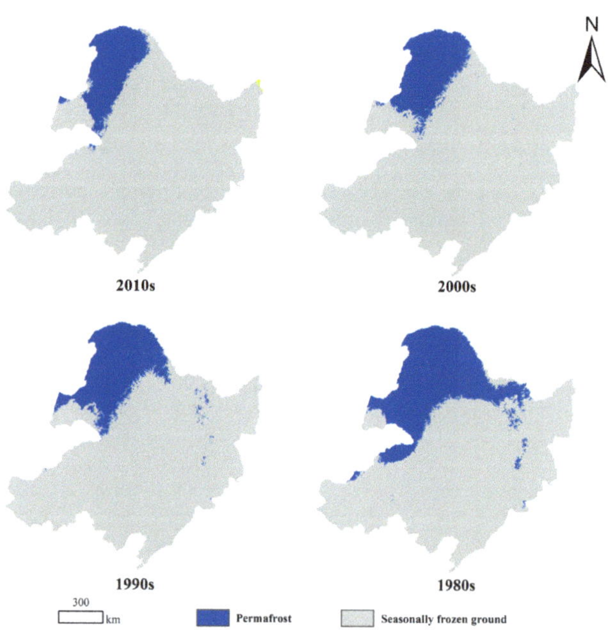

(**b**)

Figure 3. Mean annual top temperature of permafrost and spatial distribution of permafrost in Northeast China from the 1980s to the 2010s based on the TTOP model. (**a**) Mean annual top temperature of permafrost from 1980s to 2010s based on the TTOP model. (**b**) Permafrost distribution based on mean annual top temperature of permafrost division from 1980s to 2010s.

3.3. Changes in Soil Organic Carbon

As shown in Figure 4a, the regions in Northeast China that have the highest SOC density and SOC stocks are concentrated in the major forestland and grassland areas of the Da and Xiao Xing'anling Mountains, as well as the Songnen and Sanjiang Plains.

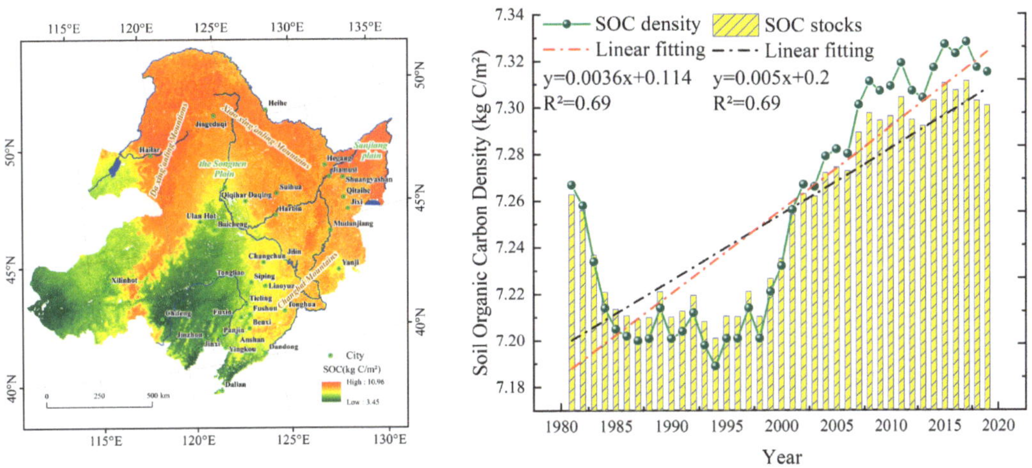

(a) (b)

Figure 4. (**a**) The spatial distribution of SOC density in Northeast China from 1981 to 2019. (**b**) The annual average variation trend of SOC density and stocks in Northeast China.

From the SOC data presented in Figure 4b, it can be observed that the overall trend of SOC density and SOC stocks in Northeast China experienced a rapid decrease followed by an increase. SOC density and stock increased at rates of 0.036 kg C/m^2 a and 5 Tg C/a, respectively. SOC density decreased from 7.22 kg C/m^2 in the 1980s to 7.20 kg C/m^2 in the 1990s, before rising to 7.28 kg C/m^2 in the 2000s and 7.32 kg C/m^2 in the 2010s. The SOC stock started at 10.15 Pg C in the 1980s, experienced a rapid decline of 24.18 Tg C to 10.13 Pg C in the 1990s, and then substantially rose to 10.23 Pg C in the 2000s, finally reaching 10.28 Pg C in the 2010s. The trend in SOC stocks corresponds to that of SOC density. According to Table 4, the corresponding total SOC stocks and proportions are displayed for the four periods based on various LULC types and the extent of permafrost.

Table 4. Proportion of SOC stocks (SOCS) corresponding to permafrost and different LULC types.

Year	LULC												Permafrost	
	Arable Land		Forestland		Grassland		Waters		Construction Land		Unused Land			
	SOCS (Pg C)	Proportion (%)	SOCS (Pg C)	Proportion (%)	SOCS (Pg C)	Proportion (%)	SOCS (Pg C)	Proportion (%)	SOCS (Pg C)	Proportion (%)	SOCS (Pg C)	Proportion (%)	SOCS (Pg C)	Proportion (%)
1980s	2.32	22.86	4.14	40.86	2.75	27.12	0.13	1.30	0.17	1.65	0.63	6.22	3.18	31.34
1990s	2.59	25.56	4.01	39.67	2.64	26.06	0.12	1.19	0.17	1.68	0.59	5.85	2.15	21.26
2000s	2.65	25.89	4.05	39.64	2.64	25.82	0.12	1.17	0.18	1.74	0.59	5.74	1.66	16.23
2010s	2.74	26.72	3.96	38.59	2.52	24.59	0.11	1.09	0.23	2.21	0.70	6.81	1.46	14.17

The area of arable land has steadily increased between 1980 and 2020, showing a total growth of 5.8×10^4 km^2. However, during the same period, the area covered by forestland and grasslands has consistently decreased, with a cumulative reduction of 7.0×10^4 km^2. Some of these lands have been converted into construction land, while others have become unused land. Table 5 displays the dynamics of LULC change in areas affected by permafrost degradation across various time periods. During the period from the 1980s to the 1990s, the increase in SOC stocks in arable land was the most significant in Northeast China, amounting to 267.94 Tg C. Conversely, forestland experienced the greatest loss in SOC stocks, totaling 129.71 Tg C, followed by grasslands. The region where permafrost is located witnessed a decrease in SOC stocks by 1026.97 Tg C. In the 1990s to 2000s, arable land contributed the most to the increase in SOC stocks, with an addition of 60.38 Tg C, followed by forestland, which had transitioned from losses to gains at this point, with an increase of 37.32 Tg C. Unused land experienced the greatest reduction in SOC stocks, amounting to 4.77 Tg C. In the region with permafrost, SOC stocks decreased by 494.52 Tg C. Between the 2000s and 2010s, unused land reversed its trend, registering the highest increase in soil SOC stocks, with an addition of 112.14 Tg C. Arable land followed closely behind. Both forestland and grasslands switched from gains to losses, with grasslands experiencing the most significant reduction in SOC stocks at 115.01 Tg C. The region with permafrost witnessed a decrease in SOC stocks of 202.45 Tg C.

Table 5. Dynamics of LULC change in areas affected by permafrost degradation across various time periods (%).

Year	LULC					
	Arable Land	Forestland	Grassland	Waters	Construction Land	Unused Land
1980–2000	45.23	−7.91	−1.76	19.82	17.33	−3.26
1990–2010	50.68	−1.25	−0.24	−8.55	27.45	−0.51
2000–2020	10.84	−1.86	−9.33	−8.47	72.97	184.07

3.4. Relationship between LULC and SOC Changes in Permafrost Degradation Areas

Permafrost degradation can cause alterations in hydrogeological conditions, which subsequently impact the succession of LULC types.

During the period from the 1980s to the 2000s, the largest increase in arable land area occurred, totaling 5201 square kilometers and resulting in a dynamic change rate of 5.23%. The highest reduction in land area during this time period occurred in forestland areas, decreasing by 4436 square kilometers and resulting in a dynamic change rate of −7.91%. Additionally, grassland also experienced a decrease of 834 square kilometers, resulting in a dynamic change rate of −1.76%. Permafrost degradation was identified as the most significant contributing factor to the decrease in forestland and grassland areas, ultimately making the land more suitable for cultivation. The changes in SOC are significantly impacted by the combined effect of permafrost degradation and human activity.

SOC changes can be divided into two parts: SOC increase (density ≥ 0) and SOC decrease (density < 0). Comparing the area occupied by various LULC types in the permafrost degradation zone with their corresponding contributions to SOC alterations, the following trends can be observed. The outcomes are displayed in Figure 5:

The data suggest that ongoing permafrost degradation leads to a significant increase in the proportion of forestland and grassland within the affected area, accounting for over 75% during each succession period. Additionally, their respective contributions to SOC increase or decrease show the most notable changes. Forestland dominates the increase in SOC, with a contribution rate that is consistent with its proportion of the total area. The contribution rates for forestland are 50%, 75%, and 72%, for the periods of 1980s–1990s, 1990s–2000s, and 2000s–2010s, respectively. A decrease in SOC mainly occurs in grasslands, with contribution rates of 50%, 83%, and 60% for the same periods. This relationship remains unchanged, even when the area of forestland exceeds that of grassland. Unlike the situation of SOC increase, the ratio of forestland area to grassland area displays a fluctuation in strength when associated with SOC decrease. This trend may be related to the different carbon sequestration capacities and sensitivities of the two land types.

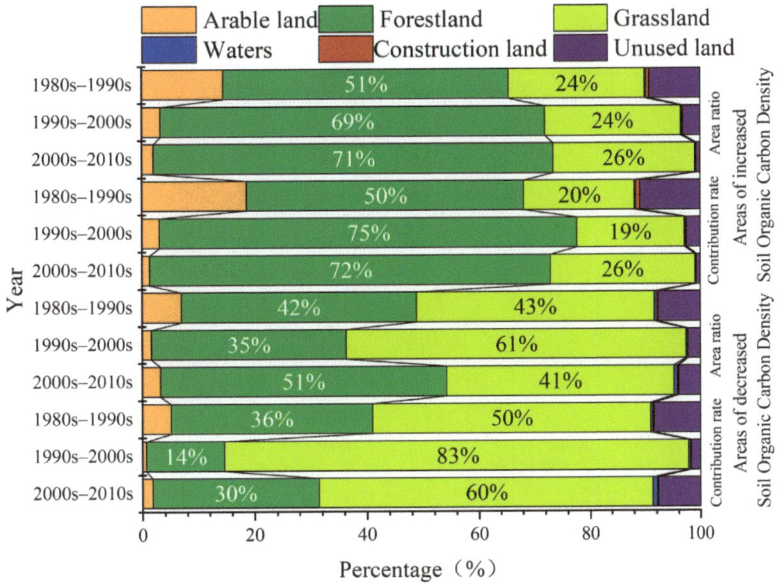

Figure 5. Proportion of LULC type area and contribution to increase/decrease in SOC stock in areas of permafrost degradation during different successional periods.

3.5. Relationship between Permafrost Change and SOC Change

When comparing the permafrost change status with the corresponding regional SOC stocks, the following observations can be made (Table 6).

The area of permafrost degradation between the 1980s and the 1990s was 12.52×10^4 km^2. Of this area, 3.77×10^4 km^2 was an area of SOC increase, with a total increase of SOC stocks of 2.05 Tg C, and 8.75×10^4 km^2 was an area of SOC decrease, with a total decrease of SOC stocks of 8.14 Tg C. Surprisingly, although the area of permafrost degradation in this period accounted for only 8.92% of the area of Northeast China, the total amount of the change in SOC stocks was 25.17% of the total change in SOC stocks in Northeast China, which is inconsistent with the correspondence between area and SOC and positively illustrates that the effect of permafrost degradation on SOC is significant and is especially pronounced when permafrost degradation are evident.

In the four periods of the 1980s, 1990s, 2000s, and 2010s, there was a region of permafrost between the Da Xing'anling Mountains, Genhe City, and Arctic Village with an area of 15.12×10^4 km^2. The SOC stocks in this region were 1.359 Pg C, 1.346 Pg C, 1.347 Pg C, and 1.349 Pg C for the respective periods. In the 1980s and 1990s, there was a significant increase in temperature, a rapid decrease in permafrost area, and a rapid decrease in SOC stocks. During this period, the longitudinal SOC stocks in this region also decreased rapidly. From the 1990s to the 2010s, there was a gradual increase, indicating that permafrost degradation not only affects SOC through area reduction, but also influences SOC changes in the active layer of permafrost.

In the permafrost growth area, the growth area is 12,610 km^2, the increase in SOC stocks is 0.24 Tg, the decrease is 0.47 Tg, and the total amount is a decrease of 0.23 Tg. Among them, the SOC growth area is 5,486 km^2, which accounts for 0.56% of SOC growth, and the SOC increment accounts for 0.32%; the SOC decrease area is 7,124 km^2, which accounts for 1.66%, and the SOC decrease accounts for 2.20%. The total change in SOC stocks had an impact of −0.42% on SOC stocks in Northeast China.

Table 6. Types of permafrost changes and corresponding types and proportions of SOC alterations.

Year	Subject	Areas of Permafrost Degradation	Northeast China	Proportion (%)	Subject	Areas of Permafrost Degradation	Northeast China	Proportion (%)
1980s–1990s	Areas of increased SOC (km^2)	3.77×10^4	61.66×10^4	6.11	Areas of reduced SOC (km^2)	8.75×10^4	78.60×10^4	11.13
	Total increase in SOC stocks (Tg C)	2.05	44.33	4.63	Total reduction in SOC stocks (Tg C)	8.14	68.52	11.88
	Total area (km^2)	12.52×10^4	140.26×10^4	8.92	The total amount of change in SOC stocks (Tg C)	−6.09	−24.18	25.17
1990s–2000s	Areas of increased SOC (km^2)	3.90×10^4	106.22×10^4	3.66	Areas of reduced SOC (km^2)	1.99×10^4	34.04×10^4	5.84
	Total increase in SOC stocks (Tg C)	3.91	121.10	3.23	Total reduction in SOC stocks (Tg C)	1.93	18.26	10.57
	Total area (km^2)	5.88×10^4	140.26×10^4	4.19	The total amount of change in SOC stocks (Tg C)	1.98	102.84	1.93
2000s–2010s	Areas of increased SOC (km^2)	2.50×10^4	97.36×10^4	2.57	Areas of reduced SOC (km^2)	1.17×10^4	42.90×10^4	2.72
	Total increase in SOC stocks (Tg C)	1.42	75.81	1.88	Total reduction in SOC stocks (Tg C)	0.66	21.42	3.07
	Total area (km^2)	3.67×10^4	140.26×10^4	2.62	The total amount of change in SOC stocks (Tg C)	0.77	54.39	1.41
	Subject	**Permafrost Growth Areas**	**Northeast China**	**Proportion (%)**	**Subject**	**Permafrost Growth Areas**	**Northeast China**	**Proportion (%)**
2000s–2010s	Areas of increased SOC (km^2)	0.55×10^4	97.36×10^4	0.56	Areas of reduced SOC (km^2)	0.71×10^4	42.90×10^4	1.66
	Total increase in SOC stocks (Tg C)	0.24	75.81	0.32	Total reduction in SOC stocks (Tg C)	0.47	21.42	2.20
	Total area (km^2)	1.26×10^4	140.26×10^4	0.90	The total amount of change in SOC stocks (Tg C)	−0.23	54.39	−0.42

3.6. Model Evaluation

The regression model is a machine learning technique that exhibits efficient and stable advantages when dealing with large datasets. The trend in MAGST from 1980 to 2002 is generally consistent with the 2003–2020 MAGST for the Northeast and permafrost regions, which has been adjusted by regression modeling. The model assesses the error outcomes (Figure 6) of a day-to-day surface air temperature simulation (N = 4748) from 1990 to 2002 with an RMSE of 1.289 °C, an MAE of 1.028 °C, and an R^2 of 0.99. Thus, the model is deemed to have an accurate fit.

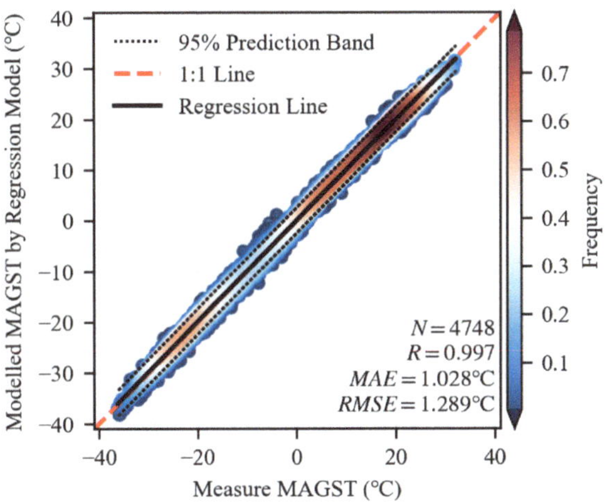

Figure 6. The regression model calculates MAGST versus measured MAGST and error analysis.

In this paper, 50 borehole data (MAGT) were collected from permafrost regions of Northeast China between 2000 and 2020 [51–56]. The data were divided into 25 groups each from the 2000s and 2010s and compared with the TTOP of the same period in the present study. The results are presented in Table 7:

Table 7. The average annual ground temperature (MAGT) measured by boreholes and the top temperature of permafrost (TTOP)simulated in this study. (When the borehole-measured MAGT is greater than 0 °C, that temperature represents the average temperature of the borehole and there is no permafrost at that borehole location).

Site	Longitude (°E)	Latitude (°N)	Borehole Measured MAGT (°C)	TTOP (°C)	TTOP (°C) by Obu, J. et al. (2019) [7]
CW1	123.98	53.33	−1.63	−2.11	0.74
CW3	124.58	52.99	−0.1	−1.87	1.86
CW4	124.51	52.74	1.9	−1.8	1.32
CW5	124.58	52.54	−1.8	−1.75	1.24
CW6	124.66	52.43	−0.8	−1.59	1.75
CW7	124.66	52.06	−0.7	−1.19	1.58
CW8	124.53	51.81	−0.5	−1.1	0.21
CW9	124.39	51.70	−1.3	−1.28	0.21
CW10	124.27	51.47	−2.75	−1.26	0.74
CW12	124.21	51.21	−1.3	−0.91	0.72
CW13	124.31	50.70	1.9	0.12	0.61

Table 7. *Cont.*

Site	Longitude (°E)	Latitude (°N)	Borehole Measured MAGT (°C)	TTOP (°C)	TTOP (°C) by Obu, J. et al. (2019) [7]
CW14	124.21	50.47	3	0.17	0.85
CW15	124.65	49.77	2.3	1.12	3.03
XL1	124.39	51.69	−1.13	−1.13	0.83
XL2	124.39	51.69	−1.71	−1.13	0.83
XL3	124.39	51.69	−1.43	−1.13	0.83
GH1	121.51	50.94	−0.24	−2.43	−0.03
GH2	121.51	50.94	−0.8	−2.43	−0.03
GH3	121.50	50.93	−3.87	−2.33	−0.20
GH4	121.50	50.93	−2.84	−2.33	−0.20
GH5	121.53	50.80	−0.59	−1.95	−0.72
GH6	121.50	50.80	−1.92	−2.19	0.01
GH9	121.51	50.94	−3.3	−2.43	−0.03
YT1	121.55	50.63	−2.18	−1.78	−0.22
YT2	121.55	50.63	−1.92	−1.78	−0.22
DW01	124.19	53.21	0.2	−0.57	1.38
DW02	124.20	53.19	−1.2	−0.56	1.95
DW04	124.42	53.10	0.5	−0.43	1.61
DW05	124.57	53.00	−1.4	−0.37	1.49
DW06	124.49	52.69	−0.5	−0.77	0.51
DW07	124.69	52.23	−0.6	0.51	1.40
DW08	124.66	52.03	−1.4	−0.64	1.53
DW09	124.34	51.59	−1.7	−0.58	0.40
DW10	124.33	50.90	0.9	0.8	0.16
DW11	124.34	50.74	1	1.14	0.19
DW12	124.29	50.60	1.9	1.28	1.15
MH1	122.02	53.03	−2.9	−2.35	0.28
MH2	122.02	53.02	−1.8	−2.36	0.83
NE2	125.14	51.13	−1.94	0.72	0.97
NE3	125.14	51.13	0.92	0.72	0.97
NE4	125.14	51.12	2.16	0.71	1.84
NE5	125.15	51.15	−1.32	0.69	1.91
YTLH1	121.55	50.63	−2.18	−1.71	−0.22
YTLH2	121.55	50.63	−1.93	−1.71	−0.22
MG	122.28	52.28	−2.31	−2.74	−0.70
ALS	121.90	51.89	−3.63	−2.61	−0.15
P1	125.14	51.13	−0.52	0.72	0.97
P2	125.14	51.13	−1.19	0.72	0.97
P3	125.14	51.13	0.17	0.72	0.97
P4	125.14	51.12	1.65	0.71	1.84
	Average Value		−0.81	−0.89	0.75

The average temperature for the 50 boreholes' data during both periods was −0.81 °C. In this study, the average TTOP was −0.89 °C, resulting in a difference of 0.09 °C. The root mean square error (RMSE) was 1.23 °C and the mean absolute error (MAE) was 0.94 °C.

This indicates that the map of the distribution of permafrost in Northeast China developed through this study has a high degree of confidence.

As shown in Figure 7, the correlation (R) between the model results obtained in this study and the borehole data was 0.65 ($p < 0.001$), which is higher than that reported by Obu et al. [7] (R = 0.50, N = 50). Therefore, we believe that after adjusting the GST for the years 2003–2020 through the regression model, using the TTOP model to simulate the distribution of permafrost in Northeast China region becomes more reliable.

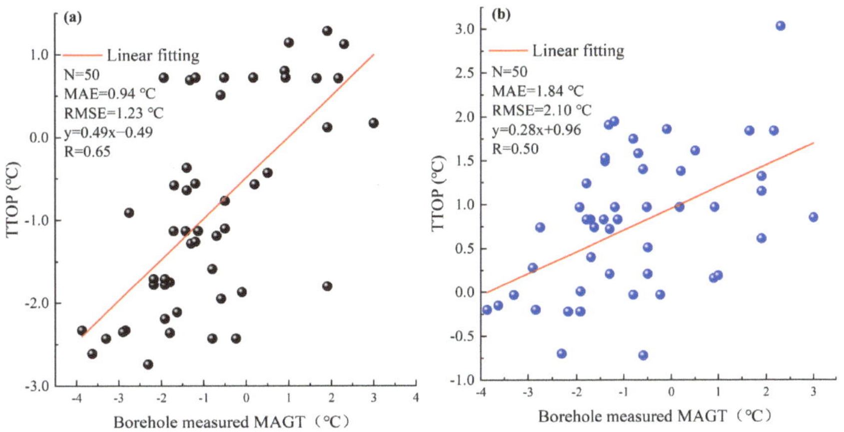

Figure 7. Comparison of correlation between modeled mean annual ground temperatures at the top temperature of permafrost (TTOP) and mean annual ground temperatures measured in boreholes (MAGT). (**a**) is the result of this research; (**b**) is from Obu et al. (2019) [7].

4. Discussion

4.1. Distribution and Degradation of Permafrost

Under the influence of global climate change, the degradation of permafrost in the northern hemisphere is an undeniable fact [57–60]. Different researchers have used different methods and data to assess the distribution of permafrost. In this study, the primary data source for permafrost distribution comes from ground meteorological station data. The data from 1980 to 2002 are consistent with the data used by other researchers. Therefore, in delineating permafrost regions, the areas for the 1980s (37.43×10^4 km^2) and 1990s (24.87×10^4 km^2) are basically consistent with the results of other studies [61,62]. Concerning the daily GST data, which were adjusted using regression models during the period of 2003–2020, it is crucial to note the substantial dissimilarities in GST measurements before and after a specific point. The discrepancy was observed in almost half of the data between 1980 and 2020, having a significant influence on the analysis of permafrost changes. Consequently, data adjustment is requisite to alleviate the irregularity in the information [63]. The study's findings align with Ran et al.'s permafrost distribution map for the Northern Hemisphere during 2000–2016(Figure 8), available at https://data.tpdc.ac.cn (accessed on 22 August 2023), with the exception of a slightly lower total permafrost area. The discrepancy could be attributed to the use of different time intervals for the simulation [64].

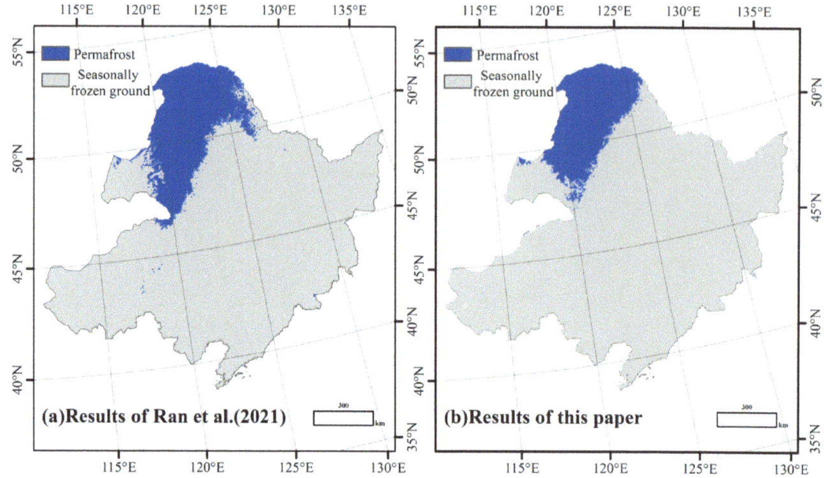

Figure 8. Comparison of spatial distribution of permafrost in Northeast China. (**a**) is based on the results of Ran et al. (2021) [64] from 2000 to 2016; (**b**) Results of this study from 2000 to 2019.

4.2. The Relationship between SOC and TTOP

The classification of permafrost is based on mean annual ground temperature at the top temperature of permafrost (TTOP), so changes in permafrost are fundamentally related to changes in TTOP. A comparison between TTOP and SOC in Northeast China from 1981 to 2019 is shown in Figure 9a:

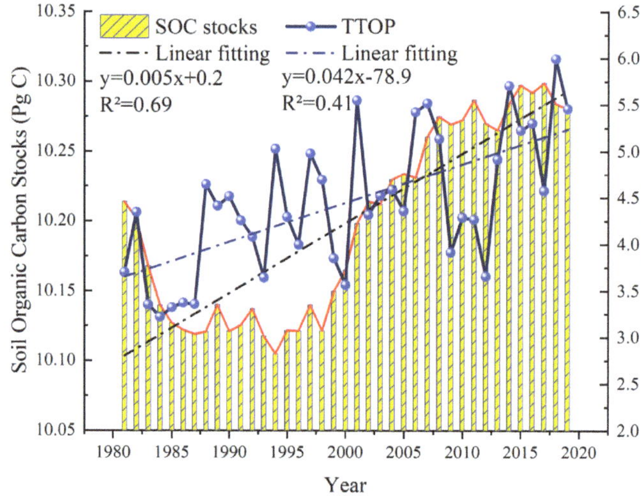

(**a**)

Figure 9. *Cont.*

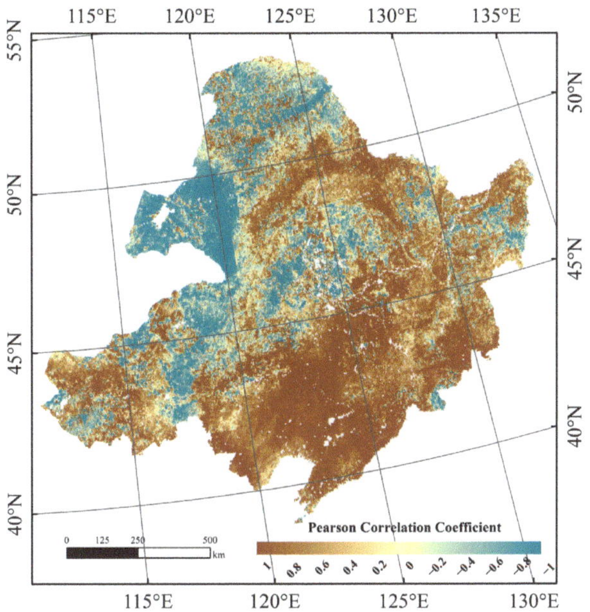

(b)

Figure 9. (**a**) is the average annual trend of SOC stocks and TTOP; (**b**) is the spatial correlation analysis of SOC density and TTOP.

Both TTOP and SOC stocks show an overall increasing trend, but TTOP shows a fluctuating pattern with significant variations every few years, which is influenced by the climatic conditions at that time. In contrast, SOC stocks showed a rapid decline from 1981 to 1985, followed by small fluctuations. Starting in 1998, SOC stocks began to increase rapidly and have maintained a stable upward trend since then. In the early period, the increase in TTOP led to the degradation and release of a significant amount of organic matter, resulting in a decrease in SOC. In the middle period, as TTOP continued to increase, it promoted vegetation growth and increased the carbon sequestration capacity of both vegetation and soil, which slowed and stabilized SOC changes. In the later period, both the carbon sequestration capacity and the carbon content of vegetation and soil began to increase, indicating that Northeast China was continuously adding carbon to SOC during this period. This not only compensated for earlier losses, but also contributed to further increases in SOC, resulting in a stable upward trend.

The spatial correlation analysis between TTOP and SOC density over about 40 years is shown in Figure 9b. The average absolute correlation coefficient between SOC density and TTOP in Northeast China over about 40 years is 0.51. In the southeastern part of the region, TTOP and SOC density mostly show a positive correlation, indicating that higher TTOP is associated with higher SOC density levels. In the northwestern part of the region, TTOP and SOC density are mostly negatively correlated, indicating that lower ground surface temperatures are associated with higher SOC density. As ground surface temperatures continue to rise, the positive correlation is expected to increase. The permafrost area of Northeast China may become an enhanced carbon sink, serving as one of the buffer zones against global warming.

4.3. SOC in Different LULC and Response to Changes in Permafrost

Permafrost degradation can impact hydrogeological conditions [65], leading to changes in land use rotation and succession. These modifications, in turn, affect the rates of input and decomposition of SOC [66]. The surface soil layer (0–30 cm) is the most biologically active part, requiring careful attention [67]. As a result, changes in LULC types under the influence of permafrost degradation can have a significant impact on SOC [68]. The change in land area corresponds directly to the change in SOC within different LULC types in areas undergoing permafrost degradation. Northeast China has the highest land areas devoted to forestland, grasslands, and arable lands, resulting in the highest SOC content. Forestland and grasslands are the predominant land types in areas with concentrated permafrost distribution. Additionally, as one travels further northward and up to higher altitudes, there is a higher proportion of forestland and grasslands. This trend aligns with the degradation of permafrost.

Vegetation has a beneficial impact on safeguarding permafrost, leading to slower rates of permafrost degradation and subsequently, a deceleration in SOC change [69]. Within the region of permafrost degradation, forestlands are the primary contributors to an increase in SOC, while grasslands are a factor in SOC reduction. This can be attributed to several factors. On one hand, the extensive coverage of forestland in the area leads to an increase in SOC. Forestland produce greater amounts of plant litter and organic matter due to their vast size. This organic matter, combined with tree growth, improves carbon sequestration in the soil, increasing the land's capacity for carbon storage and facilitating the growth of SOC. In contrast, grasslands have a weaker ability to sequester carbon compared to forestland. In large grassland areas, the release of SOC into the atmosphere through decomposition processes tends to exceed the rate of carbon fixation, resulting in a net decrease in SOC levels in regions dominated by grasslands. On the other hand, in the part of SOC growth, the contribution of forestland and grasslands to SOC growth is positively correlated with the ratio of their respective areas, with a ratio of approximately 1. In the part of SOC reduction, the loss of SOC from grasslands is intensified, with a contribution-to-area ratio ranging from 1.22 to 2.02. Conversely, the reduction in SOC from forestland is mitigated, with a contribution-to-area ratio ranging from 0.4 to 0.8. Even when forestland areas exceed those of grasslands, grasslands remain the primary contributor to SOC reduction. When there is a positive net change in SOC (1990s–2000s and 2000s–2010s), the ratio of the percentage of SOC growth to the percentage of multi-year permafrost area is approximately 0.46–0.53. However, when there is a negative net change in SOC (1980s–1990s), the ratio of the percentage of SOC decrease to the percentage of multi-year permafrost area is about 2.82. At the time of severe degradation of permafrost, the reduction in SOC was exacerbated by the enhancing effect of grassland compared to the inhibiting effect of forestland. This suggests that the decrease in SOC was greater, despite the fact that the area was dominated by forestland. Therefore, although the total area of permafrost in Northeast China is relatively small, permafrost degradation has a significant effect on the reduction in SOC.

The permafrost degradation trend in Northeast China undergoes an initial substantial decrease, followed by a gradual decline, resembling the SOC changes. The primary difference is that SOC exhibits a slow increase in the later stages. Rising air temperatures mainly contribute to permafrost degradation [70,71] which, combined with its degradation, induces significant SOC decomposition and release. Even small amounts of permafrost degradation can cause significant changes in SOC. Permafrost degradation upsets the equilibrium between the input of SOC and its decomposition, leading to a continuous adjustment of the environment to this dynamic. While SOC will initially decrease and then increase in the long term due to permafrost degradation, the prolonged impact on interrelationships between these factors under complex ecological circumstances may pose further complications.

5. Conclusions

In this study, we utilized a regression model and TTOP model to simulate the distribution of permafrost in Northeast China from 1980 to 2020. At the same time, we analyzed the changes in SOC over a forty-year period and examined the impacts of alterations in both permafrost and LULC on SOC changes. The study results demonstrate that climate warming has caused ongoing permafrost degradation in Northeast China, and the outlook for permafrost retention is not optimistic. SOC in Northeast China experienced a sharp decline, but has now recovered and is growing over the long term. In the short term, permafrost degradation has a negative impact on SOC growth, but over the long term, it has a positive effect. In the future, there is a need to enhance monitoring methods in Northeast China's permafrost regions to provide more accurate observations of permafrost conditions. Additionally, further research efforts should be directed towards refining models that describe the nuanced response of SOC to permafrost degradation, taking into account different carbon fractions. This holistic approach will contribute to a more nuanced comprehension of the impacts of permafrost thaw on SOC dynamics in the region.

Author Contributions: Conceptualization, Y.S. and Y.L.; methodology, Y.S., Y.L. and L.D.; software, Y.S., S.H. and Q.W.; validation, Y.S., S.H., H.Z. and Q.W.; formal analysis, Y.S., S.H., H.Z. and Q.W.; investigation, Y.S. and Y.L.; resources, L.D. and Y.L.; data curation, Y.S., S.H. and H.Z.; writing—original draft preparation, Y.S.; writing—review and editing, S.H., H.Z., L.D. and Y.L.; visualization, Y.S. and Q.W.; supervision, Y.L. and L.D.; project administration, L.D.; funding acquisition, Y.L. and L.D. All authors have read and agreed to the published version of the manuscript.

Funding: This research was funded by Natural Science Foundation of Heilongjiang Province (grant number LH2023D022).

Data Availability Statement: Publicly available datasets were analyzed in this study. This data can be found here: Meteorological Data, https://data.cma.cn, accessed on 20 May 2023; Digital Elevation Model (DEM) data, https://earthexplorer.usgs.gov, accessed on 20 May 2023; Land use and land cover data, https://www.resdc.cn, accessed on 20 May 2023; Soil Or-ganic Carbon (SOC) data, http://www.geodata.cn, accessed on 20 May 2023.

Conflicts of Interest: The authors declare no conflict of interest.

References

1. Ciais, P.; Sabine, C.; Bala, G.; Bopp, L.; Brovkin, V.; Canadell, J.; Chhabra, A.; DeFries, R.; Galloway, J.; Heimann, M. Carbon and other biogeochemical cycles. In *Climate Change 2013: The Physical Science Basis*; Contribution of Working Group I to the Fifth Assessment Report of the Intergovernmental Panel on Climate Change; Cambridge University Press: Cambridge, UK, 2014; pp. 465–570.
2. Scharlemann, J.P.; Tanner, E.V.; Hiederer, R.; Kapos, V. Global soil carbon: Understanding and managing the largest terrestrial carbon pool. *Carbon Manag.* **2014**, *5*, 81–91. [CrossRef]
3. Köchy, M.; Hiederer, R.; Freibauer, A. Global distribution of soil organic carbon–Part 1: Masses and frequency distributions of SOC stocks for the tropics, permafrost regions, wetlands, and the world. *Soil* **2015**, *1*, 351–365. [CrossRef]
4. Schrumpf, M.; Schulze, E.; Kaiser, K.; Schumacher, J. How accurately can soil organic carbon stocks and stocks changes be quantified by soil inventories? *Biogeosciences* **2011**, *8*, 1193–1212. [CrossRef]
5. Yigini, Y.; Panagos, P. Assessment of soil organic carbon stocks under future climate and land cover changes in Europe. *Sci. Total Environ.* **2016**, *557*, 838–850. [CrossRef] [PubMed]
6. Taylor, P.G.; Cleveland, C.C.; Wieder, W.R.; Sullivan, B.W.; Doughty, C.E.; Dobrowski, S.Z.; Townsend, A.R. Temperature and rainfall interact to control carbon cycling in tropical forests. *Ecol. Lett.* **2017**, *20*, 779–788. [CrossRef] [PubMed]
7. Obu, J.; Westermann, S.; Bartsch, A.; Berdnikov, N.; Christiansen, H.H.; Dashtseren, A.; Delaloye, R.; Elberling, B.; Etzelmüller, B.; Kholodov, A.; et al. Northern Hemisphere permafrost map based on TTOP modelling for 2000–2016 at 1 km^2 scale. *Earth-Sci. Rev.* **2019**, *193*, 299–316. [CrossRef]
8. Gruber, N.; Friedlingstein, P.; Field, C.B.; Valentini, R.; Heimann, M.; Richey, J.E.; Lankao, P.R.; Schulze, E.-D.; Chen, C.-T.A. The vulnerability of the carbon cycle in the 21st century: An assessment of carbon-climate-human interactions. *Scope-Sci. Comm. Probl. Environ. Int. Counc. Sci. Unions* **2004**, *62*, 45–76.
9. Hugelius, G.; Strauss, J.; Zubrzycki, S.; Harden, J.W.; Schuur, E.A.; Ping, C.-L.; Schirrmeister, L.; Grosse, G.; Michaelson, G.J.; Koven, C.D. Estimated stocks of circumpolar permafrost carbon with quantified uncertainty ranges and identified data gaps. *Biogeosciences* **2014**, *11*, 6573–6593. [CrossRef]
10. Ding, J.; Chen, L.; Ji, C.; Hugelius, G.; Li, Y.; Liu, L.; Qin, S.; Zhang, B.; Yang, G.; Li, F.; et al. Decadal soil carbon accumulation across Tibetan permafrost regions. *Nat. Geosci.* **2017**, *10*, 420–424. [CrossRef]
11. IPCC. *Climate Change 2021: The Physical Science Basis*; Masson-Delmotte, V., Zhai, P., Pirani, A., Connors, S.L., Péan, C., Berger, S., Caud, N., Chen, Y., Goldfarb, L., Gomis, M.I., et al., Eds.; Contribution of Working Group I to the Sixth Assessment Report of the Intergovernmental Panel on Climate Change; Cambridge University Press: Cambridge, UK; New York, NY, USA, 2021. [CrossRef]

12. Wei, D.; Qi, Y.; Ma, Y.; Wang, X.; Ma, W.; Gao, T.; Huang, L.; Zhao, H.; Zhang, J.; Wang, X. Plant uptake of CO_2 outpaces losses from permafrost and plant respiration on the Tibetan Plateau. *Proc. Natl. Acad. Sci. USA* **2021**, *118*, e2015283118. [CrossRef]
13. Jin, H.; Li, S.; Cheng, G.; Shaoling, W.; Li, X. Permafrost and climatic change in China. *Glob. Planet. Chang.* **2000**, *26*, 387–404 [CrossRef]
14. Jin, H.; Yu, Q.; Lü, L.; Guo, D.; He, R.; Yu, S.; Sun, G.; Li, Y. Degradation of permafrost in the Xing'anling Mountains, Northeastern China. *Permafr. Periglac. Process.* **2007**, *18*, 245–258. [CrossRef]
15. Jin, H.; Ma, Q. Impacts of Permafrost Degradation on Carbon Stocks and Emissions under a Warming Climate: A Review. *Atmosphere* **2021**, *12*, 1425. [CrossRef]
16. Jin, H.; He, R.; Cheng, G.; Wu, Q.; Wang, S.; Lü, L.; Chang, X. Changes in frozen ground in the Source Area of the Yellow River on the Qinghai–Tibet Plateau, China, and their eco-environmental impacts. *Environ. Res. Lett.* **2009**, *4*, 045206. [CrossRef]
17. Schuur, E.A.G.; McGuire, A.D.; Schädel, C.; Grosse, G.; Harden, J.W.; Hayes, D.J.; Hugelius, G.; Koven, C.D.; Kuhry, P.; Lawrence, D.M.; et al. Climate change and the permafrost carbon feedback. *Nature* **2015**, *520*, 171–179. [CrossRef] [PubMed]
18. Philipp, M.; Dietz, A.; Buchelt, S.; Kuenzer, C. Trends in satellite Earth observation for permafrost related analyses—A review. *Remote Sens.* **2021**, *13*, 1217. [CrossRef]
19. Nelson, F.E.; Outcalt, S.I. A computational method for prediction and regionalization of permafrost. *Arct. Alp. Res.* **1987**, *19*, 279–288. [CrossRef]
20. Cheng, G.-D. Problems on zonation of high-altitude permafrost. *Acta Geogr. Sin. Beijing* **1984**, *39*, 185–193.
21. Smith, M.; Riseborough, D. Permafrost monitoring and detection of climate change. *Permafr. Periglac. Process.* **1996**, *7*, 301–309. [CrossRef]
22. Che, L.; Zhang, H.; Wan, L. Spatial distribution of permafrost degradation and its impact on vegetation phenology from 2000 to 2020. *Sci. Total Environ.* **2023**, *877*, 162889. [CrossRef]
23. Ran, Y.; Li, X.; Cheng, G. Climate warming over the past half century has led to thermal degradation of permafrost on the Qinghai–Tibet Plateau. *Cryosphere* **2018**, *12*, 595–608. [CrossRef]
24. Lewkowicz, A.G.; Bonnaventure, P.P. Equivalent elevation: A new method to incorporate variable surface lapse rates into mountain permafrost modelling. *Permafr. Periglac. Process.* **2011**, *22*, 153–162. [CrossRef]
25. Gao, B.; Yang, D.; Qin, Y.; Wang, Y.; Li, H.; Zhang, Y.; Zhang, T. Change in frozen soils and its effect on regional hydrology, upper Heihe basin, northeastern Qinghai–Tibetan Plateau. *Cryosphere* **2018**, *12*, 657–673. [CrossRef]
26. Morrissey, L.A.; Strong, L.; Card, D. Mapping permafrost in the boreal forest with thematic mapper satellite data. *Photogramm. Eng. Remote Sens.* **1986**, *52*, 1510.
27. Ferreira, A.; Vieira, G.; Ramos, M.; Nieuwendam, A. Ground temperature and permafrost distribution in Hurd Peninsula (Livingston Island, Maritime Antarctic): An assessment using freezing indexes and TTOP modelling. *CATENA* **2017**, *149*, 560–571. [CrossRef]
28. Garibaldi, M.C.; Bonnaventure, P.P.; Lamoureux, S.F. Utilizing the TTOP model to understand spatial permafrost temperature variability in a High Arctic landscape, Cape Bounty, Nunavut, Canada. *Permafr. Periglac. Process.* **2021**, *32*, 19–34. [CrossRef]
29. Ni, J.; Wu, T.; Zhu, X.; Hu, G.; Zou, D.; Wu, X.; Li, R.; Xie, C.; Qiao, Y.; Pang, Q. Simulation of the present and future projection of permafrost on the Qinghai-Tibet Plateau with statistical and machine learning models. *J. Geophys. Res. Atmos.* **2021**, *126*, e2020JD033402. [CrossRef]
30. Huang, S.; Ding, Q.; Chen, K.; Hu, Z.; Liu, Y.; Zhang, X.; Gao, K.; Qiu, K.; Yang, Y.; Ding, L. Changes in near-surface permafrost temperature and active layer thickness in Northeast China in 1961–2020 based on GIPL model. *Cold Reg. Sci. Technol.* **2023**, *206*, 103709. [CrossRef]
31. Romanovsky, V.; Smith, S.; Isaksen, K.; Nyland, K.; Kholodov, A.; Shiklomanov, N.; Streletskiy, D.; Farquharson, L.; Drozdov, D.; Malkova, G. Terrestrial Permafrost [in "State of the Climate in 2019"]. *Bull. Am. Meteorol. Soc.* **2020**, *101*, S265–S271.
32. Romanovsky, V.; Isaksen, K.; Drozdov, D.; Anisimov, O.; Instanes, A.; Leibman, M.; McGuire, A.D.; Shiklomanov, N.; Smith, S.; Walker, D. *Changing Permafrost and Its Impacts, Snow, Water, Ice and Permafrost in the Arctic (SWIPA), 2017*; Arctic Monitoring and Assessment Programme (AMAP): Oslo, Norway, 2017; pp. 65–102.
33. Fu, Z.; Wu, Q.; Zhang, W.; He, H.; Wang, L. Water migration and segregated ice formation in frozen ground: Current advances and future perspectives. *Front. Earth Sci.* **2022**, *10*, 826961. [CrossRef]
34. Schuur, E.A.; Vogel, J.G.; Crummer, K.G.; Lee, H.; Sickman, J.O.; Osterkamp, T. The effect of permafrost thaw on old carbon release and net carbon exchange from tundra. *Nature* **2009**, *459*, 556–559. [CrossRef] [PubMed]
35. Abbott, B.W.; Brown, M.; Carey, J.C.; Ernakovich, J.; Frederick, J.M.; Guo, L.; Hugelius, G.; Lee, R.M.; Loranty, M.M.; Macdonald, R. We must stop fossil fuel emissions to protect permafrost ecosystems. *Front. Environ. Sci.* **2022**, *10*, 889428. [CrossRef]
36. Piao, S.; Nan, H.; Huntingford, C.; Ciais, P.; Friedlingstein, P.; Sitch, S.; Peng, S.; Ahlström, A.; Canadell, J.G.; Cong, N. Evidence for a weakening relationship between interannual temperature variability and northern vegetation activity. *Nat. Commun.* **2014**, *5*, 5018. [CrossRef] [PubMed]
37. Jiang, L.; Chen, H.; Zhu, Q.; Yang, Y.; Li, M.; Peng, C.; Zhu, D.; He, Y. Assessment of frozen ground organic carbon pool on the Qinghai-Tibet Plateau. *J. Soils Sediments* **2019**, *19*, 128–139. [CrossRef]
38. Ran, Y.; Li, X.; Cheng, G.; Zhang, T.; Wu, Q.; Jin, H.; Jin, R. Distribution of permafrost in China: An overview of existing permafrost maps. *Permafr. Periglac. Process.* **2012**, *23*, 322–333. [CrossRef]

39. Romanovsky, V.E.; Smith, S.L.; Christiansen, H.H. Permafrost thermal state in the polar Northern Hemisphere during the international polar year 2007–2009: A synthesis. *Permafr. Periglac. Process.* **2010**, *21*, 106–116. [CrossRef]
40. *Northeast Regional Climate Change Assessment Report: 2020 Writing Committee, Northeast Regional Climate Change Assessment Report: 2020 Summary for Decision Makers*; Meteorological Press: Beijing, China, 2021.
41. Shan, W.; Zhang, C.; Guo, Y.; Qiu, L.; Xu, Z.; Wang, Y. Spatial Distribution and Variation Characteristics of Permafrost Temperature in Northeast China. *Sustainability* **2022**, *14*, 8178. [CrossRef]
42. Li, X.; Jin, H.; Sun, L.; Wang, H.; Huang, Y.; He, R.; Chang, X.; Yu, S.; Zang, S. TTOP-model-based maps of permafrost distribution in Northeast China for 1961–2020. *Permafr. Periglac. Process.* **2022**, *33*, 425–435. [CrossRef]
43. Yang, D.; Zhan, D.; Li, M.; Zang, S. Factors Influencing the Spatiotemporal Changes of Permafrost in Northeast China from 1982 to 2020. *Land* **2023**, *12*, 350. [CrossRef]
44. Jin, H.; Sun, G.; Yu, S.; Jin, R.; He, R. Symbiosis of marshes and permafrost in Da and Xiao Hinggan Mountains in northeastern China. *Chin. Geogr. Sci.* **2008**, *18*, 62–69. [CrossRef]
45. Chen, S.; Zang, S.; Sun, L. Characteristics of permafrost degradation in Northeast China and its ecological effects: A review. *Sci. Cold Arid. Reg.* **2020**, *12*, 1–11.
46. Cantarello, E.; Newton, A.C.; Hill, R.A. Potential effects of future land-use change on regional carbon stocks in the UK. *Environ. Sci. Policy* **2011**, *14*, 40–52. [CrossRef]
47. Xu, W.; Sun, C.; Zuo, J.; Ma, Z.; Li, W.; Yang, S. Homogenization of Monthly Ground Surface Temperature in China during 1961–2016 and Performances of GLDAS Reanalysis Products. *J. Clim.* **2019**, *32*, 1121–1135. [CrossRef]
48. Darajeh, N.; Idris, A.; Fard Masoumi, H.R.; Nourani, A.; Truong, P.; Sairi, N.A. Modeling BOD and COD removal from Palm Oil Mill Secondary Effluent in floating wetland by *Chrysopogon zizanioides* (L.) using response surface methodology. *J. Environ. Manag.* **2016**, *181*, 343–352. [CrossRef] [PubMed]
49. Nguyen, X.C.; Nguyen, T.P.; Lam, V.S.; Le, P.-C.; Vo, T.D.H.; Hoang, T.-H.T.; Chung, W.J.; Chang, S.W.; Nguyen, D.D. Estimating ammonium changes in pilot and full-scale constructed wetlands using kinetic model, linear regression, and machine learning. *Sci. Total Environ.* **2024**, *907*, 168142. [CrossRef] [PubMed]
50. Luo, D.; Jin, H.; Jin, R.; Yang, X.; Lü, L. Spatiotemporal variations of climate warming in northern Northeast China as indicated by freezing and thawing indices. *Quat. Int.* **2014**, *349*, 187–195. [CrossRef]
51. Hong, H. Study on the Impact of China-Russia Oil Transmission Project on Surrounding Permafrost. Master's Thesis, Heilongjiang University, Harbin, China, 2018. (In Chinese).
52. Chang, X.; Jin, H.; He, R.; Jing, H.; Li, G.; Wang, Y.; Luo, D.; Yu, S.; Sun, H. Progress of perennial permafrost monitoring in the northern DaXingAnLing. *Glacial Permafr.* **2013**, *35*, 93–100. (In Chinese)
53. Li, X.; Jin, H.; Wang, H.; Jin, X.; Bense, V.F.; Marchenko, S.S.; He, R.; Huang, Y.; Luo, D. Effects of fire history on thermal regimes of permafrost in the northern Da Xing'anling Mountains, NE China. *Geoderma* **2022**, *410*, 115670. [CrossRef]
54. He, R.; Jin, H.; Luo, D.; Huang, Y.; Ma, F.; Li, X.; Wang, H.; Li, Y.; Jia, N.; Li, X. Changes in the permafrost environment under dual impacts of climate change and human activities in the Hola basin, northern Da Xing'anling Mountains, Northeast China. *Land Degrad. Dev.* **2022**, *33*, 1219–1234. [CrossRef]
55. He, R.-X.; Jin, H.-J.; Luo, D.-L.; Li, X.-Y.; Zhou, C.-F.; Jia, N.; Jin, X.-Y.; Li, X.-Y.; Che, T.; Yang, X.; et al. Permafrost changes in the Nanwenghe Wetlands Reserve on the southern slope of the Da Xing'anling–Yile'huli mountains, Northeast China. *Adv. Clim. Change Res.* **2021**, *12*, 696–709. [CrossRef]
56. He, R.; Jin, H.; Hao, J.; Chang, X.; Luo, D. Evaluation of environmental impacts of permafrost along the Sino-Russian oil pipeline (Mohe-Urqi section). *Civ. Eng. Environ. Eng.* **2011**, *33* (Suppl. S2), 128–134. (In Chinese)
57. Wang, J.; Liu, D. Vegetation green-up date is more sensitive to permafrost degradation than climate change in spring across the northern permafrost region. *Glob. Change Biol.* **2022**, *28*, 1569–1582. [CrossRef] [PubMed]
58. Zhang, Z.-Q.; Wu, Q.-B.; Hou, M.-T.; Tai, B.-W.; An, Y.-K. Permafrost change in Northeast China in the 1950s–2010s. *Adv. Clim. Change Res.* **2021**, *12*, 18–28. [CrossRef]
59. Liu, L.; Zhao, D.; Wei, J.; Zhuang, Q.; Gao, X.; Zhu, Y.; Zhang, J.; Guo, C.; Du, Z. Permafrost sensitivity to global warming of 1.5 °C and 2 °C in the Northern Hemisphere. *Environ. Res. Lett.* **2021**, *16*, 034038. [CrossRef]
60. Hu, G.; Zhao, L.; Wu, T.; Wu, X.; Park, H.; Fedorov, A.; Wei, Y.; Li, R.; Zhu, X.; Sun, Z.; et al. Spatiotemporal variations and regional differences in air temperature in the permafrost regions in the Northern Hemisphere during 1980–2018. *Sci. Total Environ.* **2021**, *791*, 148358. [CrossRef] [PubMed]
61. Desyatkin, R. Climate Change and Dynamics of Permafrost Ecosystems of the Center of the Continental Cryolithozone of the Northern Hemisphere. *Her. Russ. Acad. Sci.* **2018**, *88*, 494–501. [CrossRef]
62. Zhongqiong, Z.; Qingbai, W.; Xueyi, X.; Yuncheng, L. Spatial distribution and changes of Xing'an permafrost in China over the past three decades. *Quat. Int.* **2019**, *523*, 16–24. [CrossRef]
63. Zhou, C.; Wang, K.; Ma, Q. Evaluation of Eight Current Reanalyses in Simulating Land Surface Temperature from 1979 to 2003 in China. *J. Clim.* **2017**, *30*, 7379–7398. [CrossRef]
64. Ran, Y.; Li, X.; Cheng, G.; Nan, Z.; Che, J.; Sheng, Y.; Wu, Q.; Jin, H.; Luo, D.; Tang, Z.; et al. Mapping the permafrost stability on the Tibetan Plateau for 2005–2015. *Sci. China Earth Sci.* **2021**, *64*, 62–79. [CrossRef]
65. Carpino, O.A.; Berg, A.A.; Quinton, W.L.; Adams, J.R. Climate change and permafrost thaw-induced boreal forest loss in northwestern Canada. *Environ. Res. Lett.* **2018**, *13*, 084018. [CrossRef]

66. Xu, M.; Li, X.; Cai, X.; Gai, J.; Li, X.; Christie, P.; Zhang, J. Soil microbial community structure and activity along a montane elevational gradient on the Tibetan Plateau. *Eur. J. Soil Biol.* **2014**, *64*, 6–14. [CrossRef]
67. Buraka, T.; Elias, E.; Lelago, A. Soil organic carbon and its' stock potential in different land-use types along slope position in Coka watershed, Southern Ethiopia. *Heliyon* **2022**, *8*, e10261. [CrossRef] [PubMed]
68. Tellen, V.A.; Yerima, B.P.K. Effects of land use change on soil physicochemical properties in selected areas in the North West region of Cameroon. *Environ. Syst. Res.* **2018**, *7*, 3. [CrossRef]
69. Jia, Y.; Chen, S.; Wu, M.; Gu, Y.; Wei, P.; Wu, T.; Shang, Z.; Wang, S.; Yu, H. Improved permafrost stability by revegetation in extremely degraded grassland of the Qinghai-Tibetan Plateau. *Geoderma* **2023**, *430*, 116350. [CrossRef]
70. Biskaborn, B.K.; Smith, S.L.; Noetzli, J.; Matthes, H.; Vieira, G.; Streletskiy, D.A.; Schoeneich, P.; Romanovsky, V.E.; Lewkowicz, A.G.; Abramov, A.; et al. Permafrost is warming at a global scale. *Nat. Commun.* **2019**, *10*, 264. [CrossRef]
71. Zhang, G.; Nan, Z.; Zhao, L.; Liang, Y.; Cheng, G. Qinghai-Tibet Plateau wetting reduces permafrost thermal responses to climate warming. *Earth Planet. Sci. Lett.* **2021**, *562*, 116858. [CrossRef]

forests

Article

The Different Factors Driving SOC Stability under Different N Addition Durations in a *Phyllostachys edulis* Forest

Yue Wu [1,2,3], Quanxin Zeng [1,2,3], Xianchu Su [1,2,3], Wei Zheng [1,2,3], Qiufang Zhang [1,2,3] and Yuehmin Chen [1,2,3,*]

1 School of Geographical Sciences, Fujian Normal University, Fuzhou 350007, China
2 State Key Laboratory for Subtropical Mountain Ecology of the Ministry of Science and Technology and Fujian Province, Fujian Normal University, Fuzhou 350007, China
3 Fujian Provincial Key Laboratory for Subtropical Resources and Environment, Fujian Normal University, Fuzhou 350007, China
* Correspondence: ymchen@fjnu.edu.cn

Abstract: As one of the most widespread driving forces in the world, atmospheric nitrogen (N) deposition can significantly alter the carbon cycling of ecosystems. In order to understand how N deposition regulates soil organic carbon (SOC) dynamics and its underlying mechanisms, a 7-year N addition experiment was set in a *Phyllostachys edulis* forest with three N addition levels (+0, +20, and +80 kg N hm^{-2} year^{-1}) to evaluate the effects of N addition on the concentration and stability of SOC fractions in the third, fifth, and seventh years. The results are as follows: (1) short-term (third year) N addition markedly increased SOC stability by decreasing the concentration of particulate organic carbon (POC) and increasing the mineral-associated organic carbon (MAOC); longer duration of N addition (5 and 7 years) had an insignificant effect on SOC stability and fractions, suggesting that the effects of N deposition on the SOC stability varied under different duration regimes; (2) N addition did not significantly affect microbial community composition while increasing the ratio of fungi to bacteria (F:B) in the seventh year, and microbial biomass carbon (MBC) and carbon use efficiency (CUE) were significantly increased in the short-term (third year) high N addition regime and enzyme activity was significantly increased in the seventh years' high N addition regime; (3) variation partitioning analysis and multiple regression analysis showed that SOC fractions are mainly regulated by CUE and MBC under short-term N addition, while enzyme activity was mainly regulated under the longer duration of N addition. Our results show that SOC stability was more sensitive in the short term, and the role of microbial characteristics varied under different N addition durations in the *P. edulis* forests. Overall, our findings provide a new perspective for the responses of the SOC pool to N deposition and contribute to predicting SOC dynamics in terrestrial ecosystems under future climate change.

Keywords: soil organic carbon fractions; nitrogen deposition; time scales; microbial enzyme activity; microbial carbon use efficiency

Citation: Wu, Y.; Zeng, Q.; Su, X.; Zheng, W.; Zhang, Q.; Chen, Y. The Different Factors Driving SOC Stability under Different N Addition Durations in a *Phyllostachys edulis* Forest. *Forests* **2023**, *14*, 1890. https://doi.org/10.3390/f14091890

Academic Editor: Thomas H. DeLuca

Received: 7 August 2023
Revised: 11 September 2023
Accepted: 16 September 2023
Published: 17 September 2023

1. Introduction

Nitrogen (N) deposition increased recently and will remain at high levels in the future [1]. N deposition extensively affects the structure and function of terrestrial ecosystems [2,3]. Soil has the largest carbon (C) pool in terrestrial ecosystems, with global soil organic carbon (SOC) storage of approximately 2270–2770 Pg (1 Pg = 10^{15} g) [4,5]. Due to the close coupling relationship of C and N cycles in terrestrial ecosystems [6], it has a significant meaning in exploring the SOC dynamics under N deposition.

It is well known that SOC is complex. In recent years, SOC was often divided into particulate organic carbon (POC) and mineral-associated organic carbon (MAOC) based on the physical methods, which provides a new way to further study the dynamics and response of SOC to N deposition [7–9]. In detail, POC (labile and fast-turnover, >53 μm) is unattached to mineral particles and vulnerable to disturbance, which consists of partially decomposed plant materials and byproducts [7,10]. On the contrary, MAOC (slow turnover, <53 μm) is adsorbed on the mineral surfaces by physicochemical mechanisms and normally inaccessible to microbes [7]. Given that the turnover rates and stability mechanisms vary among SOC fractions, the effects of the N deposition on them may be markedly different, which in turn may affect the stability of SOC [1]. For example, a 3-year study by Duan et al. [11] found that N addition significantly increased the concentration of POC, while decreasing that of MAOC. However, a 7-year study showed that N addition significantly reduced the POC concentration while increasing the MAOC concentration [12]. In summary, the fractions of SOC are affected not only by the levels of N addition, but also likely by the duration. Therefore, clarifying the responses of POC and MAOC under different N deposition durations is the key to understanding the changes in SOC dynamics and stability under N deposition.

POC promotes microbial growth, the products in the process of which, such as microbial exudates and necromass, are important precursors of MAOC [13–15]; thus, microbial communities contribute significantly to the processes of SOC [16]. Microbial groups and their produced extracellular enzymes participate in the decomposition of soil organic matter in different ways [17–19]. Moreover, microbial C use efficiency (CUE) is a key indicator to represent microbial physiological metabolism and plays an important role in understanding the stability of SOC, which was defined as the ratio of C for microbes growth to C uptake by themselves [19,20]. N addition may affect microbial activity and microbial CUE, thus affecting the decomposition of POC and the accumulation of MAOC [11,21] based on previous studies. For example, higher CUE may accelerate the decomposition of POC, in which the process of the production of microbial residues is conducive to the formation of MAOC [6,22]. In addition, CUE varied among different N addition durations, for example, a previous study by Stone et al. [23] showed that short-term N addition could improve the availability of N by providing NH_4^+-N and NO_3^--N, thus enhancing microbial CUE. In contrast, soil acidification caused by long-term N addition will reduce microbial activity and decrease microbial CUE [24,25]. Our previous studies at this site also found that the response of microbial characteristics to N addition may be mediated by the addition duration [26]. Therefore, we suggest that the different responses of microbial characteristics to N addition under different durations may lead to differences in POC and MAOC, which have great significance in understanding the response of SOC dynamics to long-term N deposition in the future.

The subtropical zone in China is one of the areas with the most severe N deposition in the world [27,28], and there is an abundant *P. edulis* forest resource of approximately 4.68 million ha^{-1}, which is an important part of China's subtropical forest resources [29]. In our study, a 7-year N addition experiment was set in a *P. edulis* forest in southeastern China to (1) determine the effects of different N addition durations on POC and MAOC and (2) explore the main factors driving changes in SOC stability. Based on our previous studies at this site, we hypothesized that: (1) a 3-year (short-term) N addition increase in N availability accelerates the decomposition of POC and increases microbial CUE and MBC, thereby promoting the accumulation of MAOC, and (2) a longer duration of N addition N addition inhibits microbial CUE by soil acidification, resulting in the little effect on POC and MAOC.

2. Materials and Methods

2.1. Study Sites and Long-Term Experimental Design

In 2014, the N addition experiment was launched at the *P. edulis* forest in Daiyun Mountain National Nature Reserve (25°38′07″ to 25°43′40″ N, 118°05′22″ to 118°20′15″ E), Fujian province, China. The soil at this site is classified as an Ultisol with a pH of about 4.7. The experimental station has a subtropical marine monsoon climate with a mean annual temperature of 15.6 to 19.5 °C and precipitation of 1700 to 2000 mm. In addition, the *P. edulis* in this experiment was mainly 5 to 6 years, and there were almost no other trees or shrubs in the forest with sparse understory vegetation. The basic information on understory vegetation is shown in Table S1 [26,27].

A completely randomized block design with three replicates was implemented in the experiment. Based on the N deposition background values in this area, three N addition treatments were set in with ammonium nitrate [30]: (i) no N addition (N0) [+0 kg N hm^{-2} a^{-1}], (ii) low N addition (N20) [+20 kg N hm^{-2} a^{-1}], and (iii) high N addition (N80) [+80 kg N hm^{-2} a^{-1}]. A total of nine plots with a size of 3 m×10 m were set, with an interval of 2 m between each plot as a buffer zone to prevent mutual interference. The ammonium nitrate was dissolved in 20 L of Milli-Q water and applied every year in the growing season (March to September).

2.2. Sample Collection and Analysis

Soil samples were collected from every plot with a depth of 0 to 20 cm at the end of July in 2017, 2019, and 2020. Five soil cores were collected and the samples from the same treatment plot were then mixed. Thus, nine samples in total were returned to the laboratory for each sampling time. Then, gravel, debris, and roots were removed, and the soils were passed through a 2 mm sieve, and the soil physicochemical properties, microbial biomass, enzyme activity, and SOC fractions were determined.

2.3. Determination of Soil Physicochemical Properties

Soil pH was measured at a soil/water ratio of 1:2.5 (weight/volume) by a pH meter (STARTER 300, OHAUS, Parsippany, NJ, USA). Soil total carbon (STC) and total nitrogen (STN) were measured using an elemental analyzer (Elementar Vario EL III; Langenselbod, Germany). The results of soil carbonate determination by adding acid indicate that carbonates were not present, so the STC concentrations were equated to SOC concentrations. Dissolved organic carbon (DOC) and dissolved organic nitrogen (DON) concentrations were extracted by a soil/water ratio of 1:4 (weight/volume), filtered through a 0.45 μm Millipore filter and determined by a SHIMADZU TOC-VCPH/CPN analyzer (Shimadzu, Kyoto, Japan) and a continuous flow analysis system (Skalar san++; Skalar, Breda, Netherlands), respectively. Soil-available phosphorus (AP) and total phosphorus (TP) were extracted by Mehlich III and digested by $HClO_4$-H_2SO_4, respectively, and determined by a continuous flow analyzer. Soil inorganic N was analyzed for ammonia (NH_4^+-N) and nitrate (NO_3^--N) by using a continuous flow analysis system after extraction with a 2 mol L^{-1} KCl solution [26,27]. The results of physicochemical properties are shown in Table S2.

2.4. Determination of Soil Organic Carbon Fractions

The POC (size > 53 μm) and MAOC (size < 53 μm) were determined according to Robertson et al. [31]. Specifically, 5 g of air-dried soil passed through a 2 mm sieve and 50 mL of sodium poly tungstate (SPT) were shaken for 18 h at 95 rpm to disperse the sample. After this, the suspension was poured onto a 53 μm sieve and washed thoroughly to separate the two SOC fractions. The soil passed through the sieve was identified as MAOC, while the soil on the sieve was identified as POC. All fractions were oven dried at 60 °C and weighed [19], and the STC of each was analyzed by an elemental analyzer.

2.5. Determination of Soil Microbial Biomass and Composition

Microbial biomass carbon (MBC), and microbial biomass nitrogen (MBN) were determined by Vance et al. [32]. MBC was determined by a TOC analyzer. MBN and MBP were determined by the continuous flow analyzer. The calculation equations are as follows:

$$MBC = \Delta E_C / K_C \quad (1)$$

$$MBN = \Delta E_N / K_N \quad (2)$$

$$MBP = \Delta E_P / K_P / K_b \quad (3)$$

where ΔE_X is the difference between fumigated and unfumigated total carbon, nitrogen and phosphorus; and K_C, K_N, and K_P are the extraction coefficients of MBC, MBN, and MBP for 0.45, 0.54, and 0.40, respectively. K_b is the recovery rate.

According to White et al. [33] and Bardgett et al. [34], the phospholipid fatty acid (PLFA) analysis was used to determine the microbial community composition. In detail lipids were extracted from 3 g of freeze-dried soil by a mixture of chloroform–methanol-phosphate buffer with a ratio of 1:2:0.8 (volume/volume/volume). The extracted phospholipids were identified on a gas chromatograph (Agilent 6890 N, Santa Clara, CA, USA after methylated to form fatty acid methyl esters (FAMEs). According to Joergensen [35] we identified the total PLFAs of Gram-positive bacteria (GP; i14:0, i15:0, i16:0, i17:0, i18 ai15:0, ai16:0, ai17:0, ai18:0, ai19:0, 10Me16:0, 10Me17:0, and 10Me18:0), Gram-negative bacteria (GN, cy17:0, cy19:0, 16:1ω7, 16:1ω9, 17:1ω8, and 18:1ω7), Actinobacteria (ACT 10Me16:0, 10Me17:0, and 10Me18:0), fungi (18:1ω9c, 18:2ω6c, 16:1ω5c, and 18:3ω6,9,12) and arbuscular mycorrhiza fungi (AMF, 16:1ω5c). Total bacteria were calculated as the sum of GP and GN. The ratio of fungal to bacterial PLFAs (F:B) was used to estimate the change in the microbial community composition.

2.6. Determination of Soil Enzyme Activities and Microbial Carbon Use Efficiency

The soil enzyme activities, including β-glucosidase (βG), nacetyl glucosaminidase (NAG), and acid phosphatase (ACP), were determined according to Saiya-Cork et al. [36] and all enzyme activities were measured by a SpectraMax M5 Microplate reader (MDS Analytical Technologies, Sunnyvale, CA, USA). The microbial CUE was calculated as follows [37]:

$$CUE = CUE_{max} \times \{(S_{C:N} \times S_{C:P}) \div [(K_{C:N} + S_{C:N}) \times (K_{C:P} + S_{C:P})]\}^{0.5} \quad (4)$$

$$S_{C:N} = B_{C:N}/L_{C:N} \times 1/EEA_{C:N} \quad (5)$$

$$S_{C:P} = B_{C:P}/L_{C:P} \times 1/EEA_{C:P} \quad (6)$$

where CUE_{max} is the ceiling for microbial growth efficiency, which was set to 0.6 [37], and $K_{C:N}$ and $K_{C:P}$ are set to 0.5 fot the half-saturation constants. $EEA_{C:N}$ and $EEA_{C:P}$ are βG/NAG and βG/ACP ratios, respectively. $L_{C:N}$ and $L_{C:P}$ are the SOC/STN and SOC/TP ratios, respectively. $B_{C:N}$ and $B_{C:P}$ were the ratios of MBC/MBN and MBC/MBP.

2.7. Statistical Analysis

A two-way ANOVA test was implemented to examine the N addition duration, levels, and their interaction. Tukey's test was used to analyze the significant differences in the soil physicochemical properties, microbial biomass, SOC fractions, and microbial CUE under different N additions ($p < 0.05$). Before this, the normality and homogeneity of variance of the residuals were tested by Shapiro–Wilk and Levene' tests, respectively. Box–Cox was used to transform the data when they were not normally distributed or did not have the same variance as the observed results. We calculated the response ratios (RRs) by the ratio of the variable in the N addition treatment (N20 and N80) to that in the corresponding control (N0). An RR value larger than 1 indicates a positive effect, whereas a value less than 1 indicates a negative effect.

In order to elucidate the influence exerted by soil physiochemical and microbial attributes on soil SOC fractions, a multiple linear regression analysis that followed was conducted by a variation partitioning analysis (VPA) [38], as follows: (1) all predictors and response variables were standardized by Z-scores to interpret parameter estimates on a comparable scale; (2) principal component analysis (PCA) was conducted, respectively, for soil physicochemical and microbial properties, and the PC1 and PC2 were subsequently utilized in the next multiple linear regression analysis; (3) the relative importance of physicochemical and microbial properties on SOC fractions was assessed through VPA, and the variable with a higher R^2 was selected for further analysis; (4) the potential collinearity of relationships among model covariates were evaluated by the calculation of variance–inflation factors (VIFs) in "vif" function from the package "car" in R, and the variables exhibiting a VIF > 10 were omitted [39,40]; (5) the optimal model for selecting the best explanatory variable was identified based on the Akaike's information criterion (AICc; AICc < 4) by the "dredge" function from the package "MuMIn" [41,42]. Model averaging was taken relying on the AICc weights when several models were selected. All statistical analyses were executed in R (3.4.1), and all images were generated using Origin 2023 and Adobe Illustrator 2021.

3. Results

3.1. On the Response of SOC Fractions to Different N Addition Durations

The two-way analysis showed that N addition duration, N addition levels, and their interactions had significant effects on SOC and its fractions ($p < 0.05$). N addition had a negative effect on POC, POC/SOC, and POC/MAOC in the third year of N addition. In contrast, in the fifth year of N addition, POC, POC/SOC, MAOC/SOC, and POC/MAOC did not change significantly under different N addition treatments. In the seventh year of N addition, SOC, MAOC, POC/SOC, MAOC/SOC, and POC/MAOC did not change significantly under different N addition treatments. The RR-MAOC and RR-MAOC/SOC increased and then decreased with N addition duration. The MAOC was significantly decreased under N20 in the fifth year. SOC was significantly increased by N addition in the third year ($p < 0.05$, Figure 1b,e). Compared with the N0, N addition had an insignificant effect on SOC (Figure 1c), while the POC/MAOC was markedly decreased in the third year.

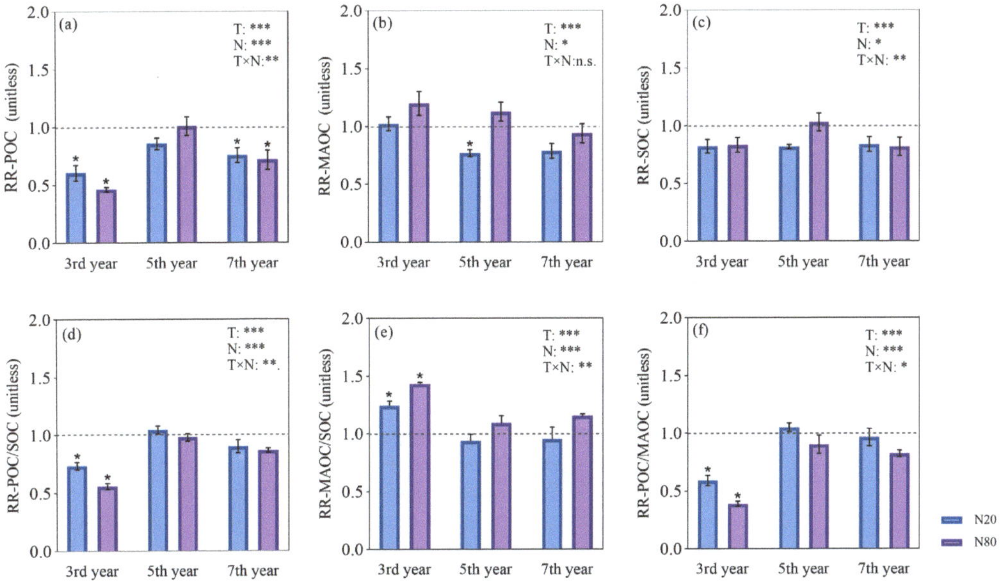

Figure 1. The response ratio (RR) of SOC fractions compared to the control treatment under different nitrogen addition durations (mean ± standard deviation). Asterisks indicate significant differences compared to the N0. The horizontal dotted line (RR = 1) indicates an insignificant effect of N addition on SOC fractions. RR > 1.0 indicates that the treatment has a positive effect, and RR < 1.0 indicates that the processing has a negative effect. (**a**) POC: particulate organic carbon, (**b**) MAOC: mineral-associated organic carbon, (**c**) SOC: soil organic carbon, (**d**) POC/SOC: the ratio of particulate and soil organic carbon, (**e**) MAOC/SOC: the ratio of mineral-associated and soil organic carbon, (**f**) POC/MAOC: the ratio of particulate and mineral-associated organic carbon. T: time scale, N: nitrogen addition, and T×N: time scale and nitrogen addition interaction. * $p < 0.05$, ** $p < 0.01$, *** $p < 0.001$, n.s. non-significant.

3.2. Effects of Nitrogen Addition on Microbial Characteristics

Compared to N0, high N addition increased the MBC in the third year, increased the MBN in the seventh year, and decreased the MBC and MBP in the seventh year (Figure 2). N addition had an insignificant effect on the proportion of PLFAs except AMF in the seventh year (Figure 3). Specifically, the proportion of AMF was increased by N addition, and it increased with the increasing addition. In addition, GP:GN and F:B were not affected by N addition (>0.05).

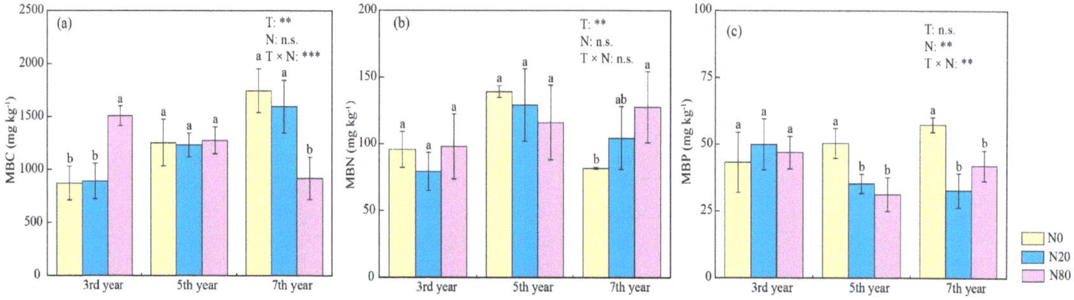

Figure 2. The soil microbial biomass under different nitrogen addition durations (mean ± standard deviation). (**a**) MBC: microbial biomass carbon, (**b**) MBN: microbial biomass nitrogen, and (**c**) MBP: microbial biomass phosphorus. Lowercase letters indicate significant differences in the variables at the three nitrogen levels per year. T: time scale, N: nitrogen addition, and T×N: time scale and nitrogen addition interaction. ** $p < 0.01$, *** $p < 0.001$, n.s. non-significant.

The enzyme activity and CUE were affected by N addition (Figure 4). βG and ACP were increased by N addition except βG under N80 in the third year ($p < 0.05$, Figure 4a, 4c). The NAG activity was, respectively, increased under N20 in the third and seventh years, and under N80 in the seventh year. Distinctively, the NAG activity was inhibited by N addition in the fifth year ($p < 0.05$, Figure 4b). In addition, CUE was markedly increased under N80 compared to N0 in the third year ($p < 0.05$, Figure 4d).

3.3. The Driving Factors of SOC Fractions under Different Durations

Soil physicochemical properties and the microbial community composition accounted for 74.16%, 83.35%, and 85.40% of the variation in SOC fractions in the third, fifth, and seventh years of N addition, respectively. Variance partitioning showed that the proportion of soil physicochemical properties decreased with N addition duration, accounting for 17.17%, 2.50%, and 6.11% of the variation in SOC fractions, respectively, for each duration, and the proportion of microbial properties increased, accounting for 34.48%, 67.48%, and 52.83%, respectively. The interaction of the two driving factors accounted for 22.51%, 13.38.00%, and 26.47% of the variation, respectively (Figure 5a–c).

As a whole, microbial properties dominated the variation in SOC fractions; therefore, the microbial biomass, composition, and function (enzyme activity and CUE) were selected in the model selection process to screen the main factors driving the variations in SOC fractions, as shown in Figure 5. In detail, POC was negatively related to CUE, while MAOC was positively related to MBC, and ACP, MBN, and MBP also explanatory variables for SOC and its fractions in the third year of N addition (Figure 5d,g,j). In the fifth year of N addition, CUE and ACP were positively and negatively related to POC, MAOC, and SOC, respectively. MBP and βG are also the explanatory variables for SOC and its fractions in the fifth year (Figure 5e,h,k). As shown in Figure 5f,i,l, NAG and βG were negatively related to POC in the seventh year, and ACP and MBP CUE were explanatory variables for MAOC and SOC, respectively.

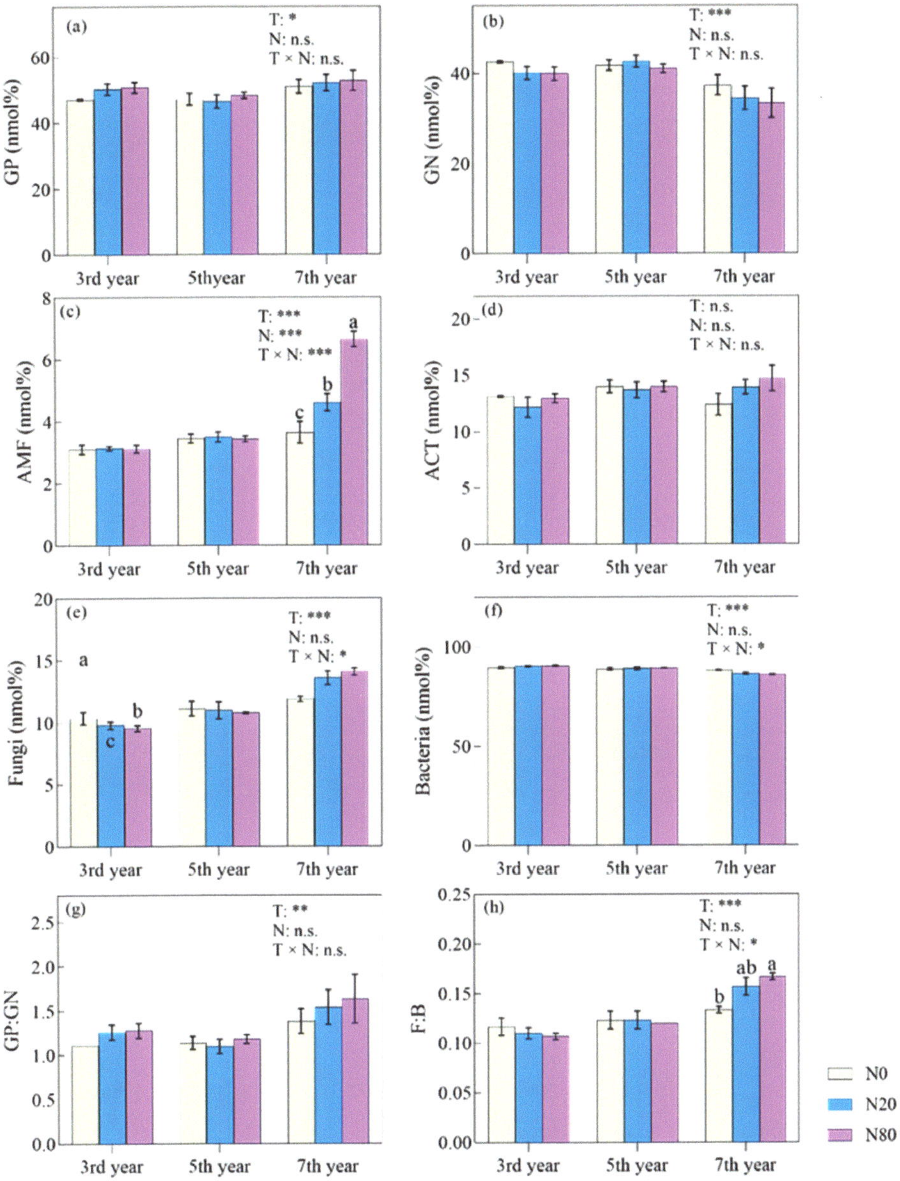

Figure 3. The proportional nmol% of phospholipid fatty acid biomarkers under different nitrogen addition durations (mean ± standard deviation). Lowercase letters indicate significant differences in the variables at the three nitrogen levels per year. (**a**) GP: Gram-positive, (**b**) GN: Gram-negative, (**c**) AMF: arbuscular mycorrhizal fungi, (**d**) ACT: actinomycetes, (**e**) fungi, (**f**) bacteria, (**g**) GP:GN: the ratio of Gram-positive and Gram-negative, (**h**) the ratio of fungi and bacteria. T: Time scale, N: Nitrogen addition, T×N: time scale and nitrogen addition interaction. * $p < 0.05$, ** $p < 0.01$, *** $p < 0.001$, n.s. non-significant.

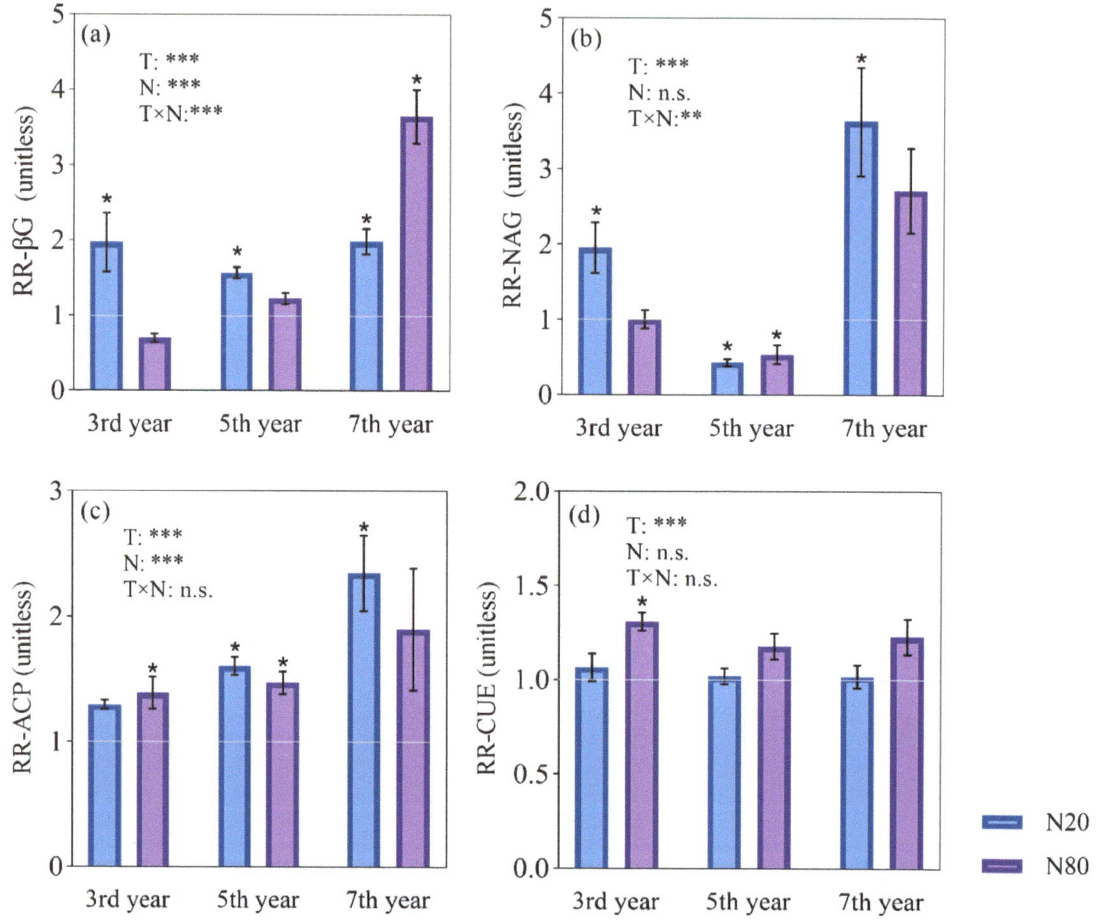

Figure 4. The response ratio (RR) of enzyme activity and carbon use efficiency compared to the control treatment under different nitrogen addition durations (mean ± standard deviation). (**a**) βG: β-glucosidase, (**b**) NAG: nacetyl glucosaminidase, (**c**) ACP: acid phosphatase, and (**d**) CUE: carbon use efficiency. Asterisks indicate significant differences compared to the N0. The gray horizontal lines (RR = 1) indicate no effect of N addition on SOC fractions. RR > 1.0 indicates that the treatment has a positive effect, and RR < 1.0 indicates that the processing has a negative effect. T: time scale, N: nitrogen addition, and T×N: time scale and nitrogen addition interaction. * $p < 0.05$, ** $p < 0.01$, *** $p < 0.001$, n.s. non-significant.

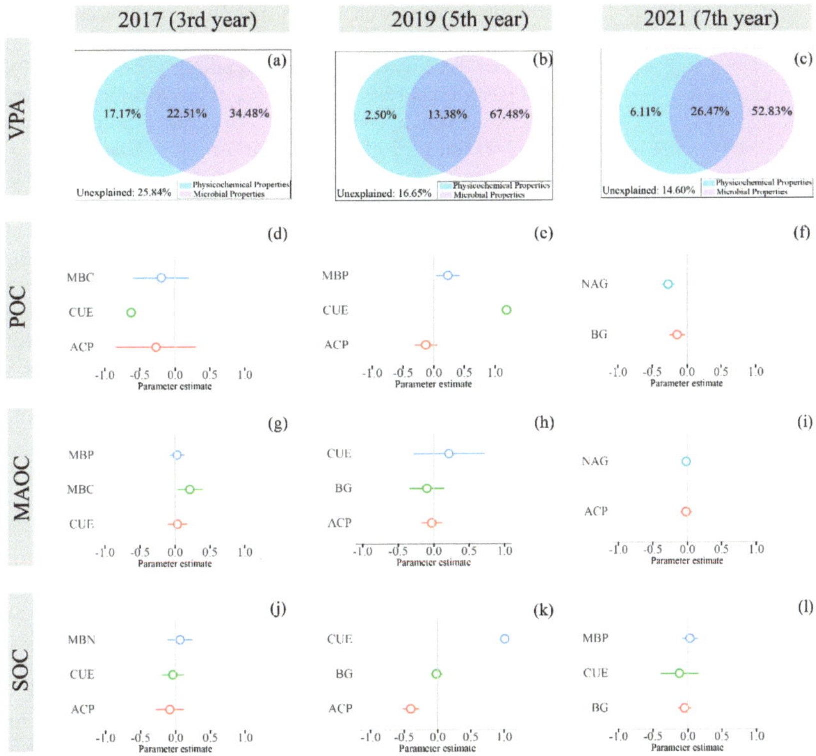

Figure 5. Variation partition analysis (**a–c**) in soil organic carbon fractions under different nitrogen addition duration in two sets of explanatory variables (soil physicochemical properties and microbial properties, values denote the proportion of variance accounted for by each of the explanatory variables), and the model-averaged effect size of the predictors (**d–l**) on SOC fractions under different nitrogen addition durations.

4. Discussion

4.1. SOC Fractions Responded Differently under Different N Addition Durations

Our study showed that N addition had an insignificant effect on the SOC concentration (Figure 1, Table S3). SOC storage depends on the balance between input and output [43,44]. Increased N deposition generally increases C inputs to soil, as elevated N can facilitate plant biomass production [12,45,46] while increasing microbial activity and accelerating the mineralization of SOC, resulting in the relative balance of the SOC stock. The RR of POC and MAOC under N addition was opposite among N addition durations, suggesting that SOC fractions were significantly changed by N addition, and the effects varied with different N addition levels and durations.

The POC concentration decreased significantly under 3 years N addition, consistent with hypothesis 1, which is contrary to what was concluded in some studies that found that N induced increases in plant input and soil acidification, which inhibited the microbial decomposition of litter and led to the accumulation of POC [1,12,47,48]. However, we did not find soil acidification, and microbial composition was not significantly affected by N addition (Figure 3). In contrast, the addition of N increased the concentration of soil-available N, which may reduce the belowground C allocation to obtain N. Considering that plant-derived C is the main source of POC [49–51], this may be the reason for the decrease in POC under short-term N addition. In addition, relevant studies also pointed out that the increase in N availability will increase microbial activity and CUE, making the loss of POC by decomposition much larger than its input [21], which is similar to our findings.

Consistent with part of our hypothesis 1, MAOC/SOC increased under N addition in the third year of N addition, suggesting that short-term N addition significantly increased SOC stability. MAOC is considered a recalcitrant SOC with a relatively long turnover time [7,52], and is difficult for microbes to utilize [44]. Therefore, the increased microbial activity by N addition might not affect the persistence of MAOC. Moreover, it was also pointed out that microbial residues contribute to MAOC [15], and the accelerated microbial turnover rate and the microbe biomass determine the formation of new MAOC [14,53]. In our study, MBC significantly increased under short-term N addition in the third year (Figure 2a), which might lead to an increase in MAOC. In the fifth year of N addition, the MAOC of the N20 treatment was significantly lower than that of N0, but there was an insignificant difference between N80 and the control, suggesting that lower N addition in the fifth year might lead to a decline in SOC stability. Our study indicated that short-term N addition significantly increased MAOC/SOC, while a longer duration of N addition had an insignificant effect on MAOC/SOC (Figure 1), suggesting that the response of SOC stability to short-term N addition was more intense, and the short-term N addition significantly increased SOC stability, while longer duration of N addition had an insignificant effect on it.

4.2. Effects of N Addition on Microbial Properties under Different Durations

N addition had an insignificant effect on the microbial composition except for an increase in the proportion of AMF in the seventh year. AMF obtains carbohydrates from plant roots, which are used to mineralize SOM to provide nutrients for plant growth [54,55]. Our study showed that AMF was increased under long-term (>7 years) N addition, which is inconsistent with some previous studies. Han et al. [55] pointed out in a meta-analysis that pH change (soil acidification) rather than available N was more important in reducing AMF. Moreover, when phosphorus is limited in soil, there is a possibility that N addition will have a positive effect on AMF [55]. Our study area, located in the subtropical zone, is considered a relatively P-limited area [56], and the ACP enzyme activity was enhanced compared to N0 in our study, which shows our finding of relative P deficiency. In addition, Zhou et al. [57] also found that N addition improved GP:GN and decreased F:B, which is different from what we observed in our study in that N addition did not affect GP:GN and F:B ratios. The inconsistency may be due to the high utilization efficiency of soil-available N by *P. edulis* [49], which weakens the effect of N addition on soil microbes. Moreover, N addition increased BG and ACP activity at different periods except for NAG in the fifth year (Figure 4). NAG is the main extracellular enzyme for microbes to obtain N, and N addition reduces the microbial demand for N and thus NAG activity. Relevant studies also found that N promotes the synthesis of extracellular enzymes rich in N by microbes, and thus improves enzyme activity, which supports our results [58,59].

4.3. The Relationship between Microbial Communities and SOC Fractions under Different Durations

CUE is a function of both substrate quality, which is defined as the proportion of assimilated substrates that are used for growth and enzyme production, and being mineralized or respired [20]. It is announced that microbial CUE plays a key role in the formation and turnover of SOC fractions under N addition [13,19]. POC lacks a protection mechanism compared with MAOC [44] and might be easily utilized by microbes. Therefore, short-term N addition might improve microbial activity [23,44,60], and thus facilitate the utilization of labile fractions (POC). Consistently, we found that N addition increased the MBC, βG, and ACP while decreasing the POC concentration. Additionally, microbial utilization of POC might promote an increase in CUE, which explains the negative correlation between CUE and POC in our study (Figure 5d).

In addition, a positive correlation between MAOC and MBC under short-term N (third year) addition was found (Figure 5g), suggesting a positive impact of increased MBC on the formation of MAOC. Duan et al. [11] found that improved soil N availability by short-term N addition can increase MBC and microbial residues, and further promote the accumulation of MAOC, which is in agreement with the results of our study. Greater microbial CUE and/or turnover will lead to more MBC and microbial residue accumulation, thereby increasing the MAOC pool [11,61]. Therefore, the change in MAOC concentration may be caused by the change in MBC in response to N addition [62–64]. These findings support our hypothesis that increasing MBC promotes the accumulation of MAOC.

There was an insignificant change in CUE in the fifth and seventh years (Figure 4d), which might not be the most important driving factor for change in SOC and its fractions under longer duration of N addition. The longer duration did not change the soil pH in our study (Table S2), which is inconsistent with our hypothesis that microbial CUE is inhibited by soil acidification. However, N addition had greater positive impacts on βG and NAG activities in the seventh year than in to the third and fifth years (Figure 4). Moreover, these enzymes were the best predictors of the POC concentration under N addition for 7 years (Figure 5). Indeed, previous studies emphasized the importance of βG and NAG in the decomposition of SOC [65,66]. Therefore, we posit that long-term N addition (>7 years) will decrease the POC concentration by promoting enzyme activity. Notably, we did not find changes in POC/SOC and MAOC/POC in the fifth and seventh years as in the third year; thus, we still believe that a longer duration of N addition will not affect the SOC stability, although the RR-POC was decreased in the seventh year.

Moreover, some studies pointed out that increased available N in soil under long-term N addition might lead to a significant decrease in soil C/N, resulting in relative C limitation for microbes [67,68], and subsequently causing the microbial biomass to be reduced (Figure 2a,c) [69]. When soil microbes are relatively limited by a certain element, this will promote them to synthesize corresponding element acquisition enzymes [65,66], which is supported by the rapid increase in enzyme activity in our study (Figure 4). Additionally, the promotion of enzyme activity means that more energy or C may be allocated for microbial metabolism rather than growth, and thus explains the minor effect of N addition on CUE and MBC in the fifth and seventh years. Therefore, the fifth and seventh years of N addition did not increase the proportion of MAOC, probably because the longer durations of N addition reduce the contribution of CUE and MBC to MAOC formation. Collectively, our findings suggest that CUE and enzyme activity are important drivers of SOC fractions in response to different durations of N addition. Moreover, we found that microbial properties contributed more to SOC fractions in the fifth and seventh years than in the third year of N addition. Therefore, it is necessary to further quantify the effects of N addition on SOC sequestration and its microbial mechanisms over long durations. However, the duration of N addition was 3 to 7 years in our study, and in order to understand the response of SOC fractions under different N addition durations more clearly, it is necessary to extend the time scale to observe the long-term effects of N addition in the future research.

5. Conclusions

Although the dynamic response of SOC fractions to N deposition was extensively studied, whether and how the N deposition of different durations affects these responses remains poorly understood. SOC stability was more sensitive to short-term N addition than to a longer duration of N addition. Short-term N addition significantly changed the SOC fractions, decreasing the POC and increasing the MAOC, and affecting the stability of SOC, increasing the MAOC/SOC. However, there was an insignificant change in SOC fractions under the longer duration of N addition. Moreover, N addition had no effect on microbial composition, but regulated SOC fractions by increasing CUE and MBC in the short term, and by accelerating enzyme activity over a longer duration. Our results highlight the key role of different microbial characteristics in regulating SOC stability under different durations of N addition in *P. edulis* forests, and the effects of N addition over long durations on SOC and its microbial mechanisms should be paid more attention to in further research. Our study contributed to a better understanding of the effects of different durations of N addition on SOC stability in subtropical *Phyllostachys edulis* forests, and provided a theoretical basis for improving the artificial management and protection of forest soil under the background of N deposition.

Supplementary Materials: The following supporting information can be downloaded at: https://www.mdpi.com/article/10.3390/f14091890/s1; Table S1: Basic vegetation information of soil in our study. Table S2: Physical and chemical properties of soil in our study (mean ± Standard deviation). Table S3: The soil organic carbon fractions under different nitrogen addition durations (Mean ± Standard deviation). Table S4: The carbon use efficiency under different nitrogen addition durations (Mean ± Standard deviation).

Author Contributions: Conceptualization, Y.C.; Software, Q.Z. (Quanxin Zeng) and W.Z.; Formal analysis, Q.Z. (Quanxin Zeng) and W.Z.; Investigation, X.S.; Data curation, X.S.; Writing—original draft, Y.W.; Writing—review & editing, Y.W., Q.Z. (Qiufang Zhang) and Y.C.; Visualization, Y.W., Q.Z. (Quanxin Zeng), X.S. and W.Z.; Supervision, Q.Z. (Qiufang Zhang); Funding acquisition, Y.C. All authors have read and agreed to the published version of the manuscript.

Funding: This research was funded by Fujian Provincial Department of Science and Technology, grant number 2023R1002002.

Data Availability Statement: The data presented in this study are available on request from the corresponding author. The data are not publicly available due to privacy restrictions.

Conflicts of Interest: The authors declare no conflict of interest.

References

1. Chen, J.G.; Xiao, W.; Zheng, C.Y.; Zhu, B. Nitrogen addition has contrasting effects on particulate and mineral-associated soil organic carbon in a subtropical forest. *Soil Biol. Biochem.* **2020**, *142*, 107708. [CrossRef]
2. Clark, C.M.; Tilman, D. Loss of plant species after chronic low-level nitrogen deposition to prairie grasslands. *Nature* **2008**, *451*, 712–715. [CrossRef] [PubMed]
3. Liu, X.J.; Zhang, Y.; Han, W.X.; Tang, A.H.; Shen, J.L.; Cui, Z.L.; Vitousek, P.; Erisman, J.M.; Goulding, K.; Christie, P.; et al. Enhanced nitrogen deposition over China. *Nature* **2013**, *494*, 459–462. [CrossRef] [PubMed]
4. Bradford, M.A.; Wieder, W.R.; Bonan, G.B.; Fierer, N.; Raymond, P.A.; Crowther, T.W. Managing uncertainty in soil carbon feedbacks to climate change. *Nat. Clim. Chang.* **2016**, *6*, 751–758. [CrossRef]
5. Jackson, R.B.; Lajtha, K.; Crow, S.E.; Hugelius, G.; Kramer, M.G.; Piñeiro, G. The ecology of soil carbon: Pools, vulnerabilities, and biotic and abiotic controls. *Annu. Rev. Ecol. Evol. Syst.* **2017**, *48*, 419–445. [CrossRef]
6. Guenet, B.; Gabrielle, B.; Chenu, C.; Arrouays, D.; Balesdent, J.; Bernoux, M.; Bruni, E.; Caliman, J.P.; Cardinael, R.; Chen, S.C.; et al. Can N_2O emissions offset the benefits from soil organic carbon storage? *Glob. Chang. Biol.* **2021**, *27*, 237–256. [CrossRef]
7. Lavallee, J.M.; Soong, J.L.; Cotrufo, M.F. Conceptualizing soil organic matter into particulate and mineral-associated forms to address global change in the 21st century. *Glob. Chang. Biol.* **2020**, *26*, 261–273. [CrossRef]
8. Lugato, E.; Lavallee, J.M.; Haddix, M.L.; Panagos, P.; Cotrufo, M.F. Different climate sensitivity of particulate and mineral associated soil organic matter. *Nat. Geosci.* **2021**, *14*, 295–300. [CrossRef]
9. Cotrufo, M.F.; Ranalli, M.G.; Haddix, M.L.; Six, J.; Lugato, E. Soil carbon storage informed by particulate and mineral-associated organic matter. *Nat. Geosci.* **2019**, *12*, 989–994. [CrossRef]
10. Yu, W.J.; Huang, W.J.; Weintraub-Leff, S.R.; Hall, S.J. Where and why do particulate organic matter (POM) and mineral-associated organic matter (MAOM) differ among diverse soils? *Soil Biol. Biochem.* **2022**, *172*, 108756. [CrossRef]
11. Duan, P.P.; Wang, K.L.; Li, D.J. Nitrogen addition effects on soil mineral-associated carbon differ between the valley and slope in a subtropical karst forest. *Geoderma* **2023**, 116357. [CrossRef]
12. Zhu, X.M.; Zhang, Z.L.; Liu, D.Y.; Kou, Y.P.; Zheng, Q.; Liu, M.; Xiao, J.; Liu, Q.; Yin, H.J. Differential impacts of nitrogen addition on rhizosphere and bulk-soil carbon sequestration in an alpine shrubland. *J. Ecol.* **2020**, *108*, 2309–2320. [CrossRef]
13. Cotrufo, M.F.; Soong, J.L.; Horton, A.J.; Campbell, E.E.; Haddix, M.L.; Wall, D.H.; Parton, A.J. Formation of soil organic matter via biochemical and physical pathways of litter mass loss. *Nat. Geosci.* **2015**, *8*, 776–779. [CrossRef]
14. Kallenbach, C.M.; Frey, S.D.; Grandy, A.S. Direct evidence for microbial-derived soil organic matter formation and its ecophysiological controls. *Nat. Commun.* **2016**, *7*, 13630. [CrossRef] [PubMed]
15. Liang, C.; Schimel, J.P.; Jastrow, J.D. The importance of anabolism in microbial control over soil carbon storage. *Nat. Microbiol.* **2017**, *2*, 17105. [CrossRef] [PubMed]
16. Luo, R.Y.; Luo, J.F.; Fan, J.L.; Liu, D.Y.; He, J.S.; Perveen, N.; Ding, W.X. Responses of soil microbial communities and functions associated with organic carbon mineralization to nitrogen addition in a Tibetan grassland. *Pedosphere* **2020**, *30*, 214–225. [CrossRef]
17. Zumsteg, A.; Bååth, E.; Stierli, B.; Zeyer, J.; Frey, B. Bacterial and fungal community responses to reciprocal soil transfer along a temperature and soil moisture gradient in a glacier forefield. *Soil Biol. Biochem.* **2013**, *61*, 121–132. [CrossRef]

18. Cenini, V.L.; Fornara, D.A.; McMullan, G.; Ternan, N.; Carolan, R.; Crawley, M.J.; Clément, J.; Lavorel, S. Linkages between extracellular enzyme activities and the carbon and nitrogen content of grassland soils. *Soil Biol. Biochem* **2016**, *96*, 198–206. [CrossRef]
19. Feng, X.H.; Qin, S.Q.; Zhang, D.Y.; Chen, P.D.; Hu, J.; Wang, G.Q.; Liu, Y.; Wei, B.; Li, Q.L.; Yang, Y.H.; et al. Nitrogen input enhances microbial carbon use efficiency by altering plant–microbe–mineral interactions. *Glob. Chang. Biol.* **2022**, *28*, 4845–4860. [CrossRef]
20. Cotrufo, M.F.; Wallenstein, M.D.; Boot, C.M.; Denef, K.; Paul, E. The Microbial Efficiency-Matrix Stabilization (MEMS) framework integrates plant litter decomposition with soil organic matter stabilization: Do labile plant inputs form stable soil organic matter? *Glob. Chang. Biol.* **2013**, *19*, 988–995. [CrossRef]
21. Chen, Q.Y.; Hu, Y.L.; Hu, A.; Niu, B. Shifts in the dynamic mechanisms of soil organic matter transformation with nitrogen addition: From a soil carbon/nitrogen-driven mechanism to a microbe driven mechanism. *Soil Biol. Biochem.* **2021**, *160*, 108355. [CrossRef]
22. Averill, C.; Waring, B. Nitrogen limitation of decomposition and decay: How can it occur? *Glob. Chang. Biol.* **2018**, *24*, 1417–1427. [CrossRef] [PubMed]
23. Stone, B.W.; Li, J.H.; Koch, B.J.; Blazewicz, S.J.; Dijkstra, P.; Hayer, M.; Hofmockel, K.S.; Liu, X.J.A.; Mau, R.L.; Morrissey, E.M.; et al. Nutrients cause consolidation of soil carbon flux to small proportion of bacterial community. *Nat. Commun.* **2021**, *12*, 3381. [CrossRef] [PubMed]
24. Mueller, N.D.; Gerber, J.S.; Johnston, M.; Ray, D.K.; Ramankutty, N.; Foley, J.A. Closing yield gaps through nutrient and water management. *Nature* **2012**, *490*, 254–257. [CrossRef] [PubMed]
25. Horn, E.L.; Cooledge, E.C.; Jones, D.L.; Hoyle, F.C.; Brailsford, F.L.; Murphy, D.V. Addition of base cations increases microbial carbon use efficiency and biomass in acidic soils. *Soil Biol. Biochem.* **2021**, *161*, 108392. [CrossRef]
26. Zeng, Q.X.; Yuan, X.C.; Zhou, J.C.; Wu, J.M. Effects of nitrogen addition on the kinetic parameters of soil acid phosphomonoesterase in a Moso bamboo forest. *Chin. J. Appl. Ecol.* **2022**, *33*, 2178–2186. (In Chinese)
27. Cheng, L.; Lin, K.M.; Zhou, J.C.; Zhang, Q.F.; Zeng, X.M.; Ji, Y.H.; Zheng, Y.; Xu, J.G.; Chen, Y.M. Effects of nitrogen deposition on the concentration and spectral characteristics of dissolved organic matter in soil in Moso bamboo plantations. *Chin. J. Appl. Ecol.* **2019**, *30*, 1754–1762. (In Chinese)
28. Chen, J.; Luo, T.S.; Zhou, Z.; Xu, H.; Chen, D.X.; Li, Y.D.; Jiang, Z.Y. Research advances in nitrogen deposition effects on microbial processes involved in soil nitrogen cycling in tropical and subtropical forests. *Acta Ecol. Sin.* **2020**, *40*, 8528–8538. (In Chinese)
29. Yu, G.R.; Jia, Y.L.; He, N.P.; Zhu, J.X.; Chen, Z.; Wang, Q.F.; Piao, S.L.; Liu, X.J.; He, H.L.; Guo, X.B.; et al. Stabilization of atmospheric nitrogen deposition in China over the past decade. *Nat. Geosci.* **2019**, *12*, 424–429. [CrossRef]
30. Jia, Y.L.; Yu, G.R.; Gao, Y.N.; He, N.P.; Wang, Q.F.; Jiao, C.C.; Zuo, Y. Global inorganic nitrogen dry deposition inferred from ground- and space-based measurements. *Sci. Rep.* **2016**, *6*, 19810. [CrossRef]
31. Robertson, A.D.; Paustian, K.; Ogle, S.; Wallenstein, M.D.; Lugato, E.; Cotrufo, M.F. Unifying soil organic matter formation and persistence frameworks: The MEMS model. *Biogeosciences* **2019**, *16*, 1225–1248. [CrossRef]
32. Vance, E.D.; Brookes, P.C.; Jenkinson, D.S. An extraction method for measuring soil microbial biomass C. *Soil Biol. Biochem.* **1987**, *19*, 703–707. [CrossRef]
33. White, D.C.; Davis, W.M.; Nickels, J.S.; King, J.D.; Bobbie, R.J. Determination of the sedimentary microbial biomass by extractible lipid phosphate. *Oecologia* **1979**, *40*, 51–62. [CrossRef] [PubMed]
34. Bardgett, R.D.; Hobbs, P.J.; Frostegård, Å. Changes in soil fungal: Bacterial biomass ratios following reductions in the intensity of management of an upland grassland. *Biol. Fertil. Soils* **1996**, *22*, 261–264. [CrossRef]
35. Joergensen, R.G. Phospholipid fatty acids in soil-drawbacks and future prospects. *Biol. Fertil. Soils* **2022**, *58*, 1–6. [CrossRef]
36. Saiya-Cork, K.R.; Sinsabaugh, R.L.; Zak, D.R. The effects of long term nitrogen deposition on extracellular enzyme activity in an Acer saccharum forest soil. *Soil Biol. Biochem.* **2002**, *34*, 1309–1315. [CrossRef]
37. Sinsabaugh, R.L.; Turner, B.L.; Talbot, J.M.; Waring, B.G.; Powers, J.S.; Kuske, C.R.; Moorhead, D.L.; Shah, J.J.F. Stoichiometry of microbial carbon use efficiency in soils. *Ecol. Monogr.* **2016**, *86*, 172–189. [CrossRef]
38. Borcard, D.; Legendre, P.; Drapeau, P. Partialling out the spatial component of ecological variation. *Ecology* **1992**, *73*, 1045–1055. [CrossRef]
39. Schmidt-Nielsen, K. *Scaling: Why Is Animal Size So Important?* Cambridge University Press: New York, NY, USA, 1984.
40. Dormann, C.F.; Elith, J.; Bacher, S.; Buchmann, C.; Carl, G.; Carré, G.; Marquéz, J.R.G.; Gruber, B.; Lafourcade, B.; Leitão, P.J.; et al. Collinearity: A review of methods to deal with it and a simulation study evaluating their performance. *Ecography* **2013**, *36*, 27–46. [CrossRef]
41. Burnham, K.; Anderson, D. *Model Selection and Multimodel Inference: A Practical Information-Theoretic Approach*; Springer: New York, NY, USA, 2002.
42. Bartoń, K. MuMIn: Multi-Model Inference. R Package v. 1.43.17. 2020. Available online: https://CRAN.R-project.org/package=MuMIn (accessed on 15 May 2023).
43. Shi, L.; Dech, J.P.; Yao, H.; Zhao, P.; Shu, Y.; Zhou, M. The effects of nitrogen addition on dissolved carbon in boreal forest soils of northeastern China. *Sci. Rep.* **2019**, *9*, 8274. [CrossRef]
44. Tang, B.; Rocci, K.S.; Lehmann, A.; Rillig, M.C. Nitrogen increases soil organic carbon accrual and alters its functionality. *Glob. Chang. Biol.* **2023**, *29*, 1971–1983. [CrossRef] [PubMed]

45. Chen, H.; Li, D.; Feng, W.; Niu, S.; Plante, A.; Luo, Y.; Wang, K. Different responses of soil organic carbon fractions to additions of nitrogen. *Eur. J. Soil Sci.* **2018**, *69*, 1098–1104. [CrossRef]
46. Wan, D.; Ma, M.K.; Peng, N.; Luo, X.S.; Chen, W.L.; Cai, P.; Wu, L.H.; Pan, H.B.; Chen, J.B.; Yu, G.H.; et al. Effects of long-term fertilization on calcium-associated soil organic carbon: Implications for C sequestration in agricultural soils. *Sci. Total Environ* **2021**, *772*, 145037. [CrossRef]
47. Zak, D.R.; Freedman, Z.B.; Upchurch, R.A.; Steffens, M.; Kögel-Knabner, I. Anthropogenic N deposition increases soil organic matter accumulation without altering its biochemical composition. *Glob. Chang. Biol.* **2017**, *23*, 933–944. [CrossRef] [PubMed]
48. Argiroff, W.A.; Zak, D.R.; Upchurch, R.A.; Salley, S.O.; Grandy, A.S. Anthropogenic N deposition alters soil organic matter biochemistry and microbial communities on decaying fine roots. *Glob. Chang. Biol.* **2019**, *25*, 4369–4382. [CrossRef]
49. Lai, Y.N.; Zhong, Q.L.; Li, B.Y.; Cheng, D.L.; Yu, H.; Xu, C.B.; Xue, X.H.; Zheng, W.T. Effect of nitrogen addition on the growth of *Phyllostachys edulis* and evergreen broad-leaved trees in different Vegetation Types. *Prot. For. Sci. Technol.* **2021**, *2*, 1–7. (In Chinese)
50. Niu, L.M.; Qin, H.; Xu, Q.F.; Wu, Q.F.; Miao, J.J.; Peng, D.C. Effects of Long-term plantation of *Phyllostachys edulis* on evolution of arbuscular mycorrhizal fungus community in soil. *Acta Pet. Sin.* **2017**, *54*, 722–734. (In Chinese)
51. Ye, C.L.; Chen, D.M.; Hall, S.J.; Pan, S.; Yan, X.B.; Bai, T.S.; Guo, H.; Zhang, Y.; Bai, Y.F.; Hu, S.J. Reconciling multiple impacts of nitrogen enrichment on soil carbon: Plant, microbial and geochemical controls. *Ecol. Lett.* **2018**, *21*, 1162–1173. [CrossRef]
52. Schmidt, M.W.I.; Torn, M.S.; Abiven, S.; Dittmar, T.; Guggenberger, G.; Janssens, I.A.; Kleber, M.; Kögel-Knabner, I.; Lehmann, J.; Manning, D.A.C.; et al. Persistence of soil organic matter as an ecosystem property. *Nature* **2011**, *478*, 49–56. [CrossRef]
53. Liang, C.; Amelung, W.; Lehmann, J.; Kästner, M. Quantitative assessment of microbial necromass contribution to soil organic matter. *Glob. Chang. Biol.* **2019**, *25*, 3578–3590. [CrossRef]
54. Treseder, K.K. A meta-analysis of mycorrhizal responses to nitrogen, phosphorus, and atmospheric CO_2 in field studies. *New Phytol.* **2004**, *164*, 347–355. [CrossRef] [PubMed]
55. Han, Y.; Feng, J.; Han, M.; Zhu, B. Responses of arbuscular mycorrhizal fungi to nitrogen addition: A meta-analysis. *Glob. Chang. Biol.* **2020**, *26*, 7229–7241. [CrossRef] [PubMed]
56. Fan, Y.X.; Zhong, X.J.; Lin, F.; Liu, C.C.; Yang, L.M.; Wang, M.H.; Chen, G.S.; Chen, Y.M.; Yang, Y.S. Responses of soil phosphorus fractions after nitrogen addition in a subtropical forest ecosystem: Insights from decreased Fe and Al oxides and increased plant roots. *Geoderma* **2019**, *337*, 246–255. [CrossRef]
57. Zhou, Z.H.; Wang, C.K.; Zheng, M.H.; Jiang, L.F.; Luo, Y.Q. Patterns and mechanisms of responses by soil microbial communities to nitrogen addition. *Soil Biol. Biochem.* **2017**, *115*, 433–441. [CrossRef]
58. Allison, S.D.; Czimczik, C.I.; Treseder, K.K. Microbial activity and soil respiration under nitrogen addition in Alaskan boreal forest. *Glob. Chang. Biol.* **2008**, *14*, 1156–1168. [CrossRef]
59. Sistla, S.A.; Schimel, J.P. Seasonal patterns of microbial extracellular enzyme activities in an arctic tundra soil: Identifying direct and indirect effects of long-term summer warming. *Soil Biol. Biochem.* **2013**, *66*, 119–129. [CrossRef]
60. Manzoni, S.; Taylor, P.; Richter, A.; Porporato, A.; Ågren, G.I. Environmental and stoichiometric controls on microbial carbon-use efficiency in soils. *New Phytol.* **2012**, *196*, 79–91. [CrossRef]
61. Wang, C.; Qu, L.R.; Yang, L.M.; Liu, D.W.; Morrissey, E.; Miao, R.H.; Liu, Z.P.; Wang, Q.K.; Fang, Y.T.; Bai, E. Large-scale importance of microbial carbon use efficiency and necromass to soil organic carbon. *Glob. Chang. Biol.* **2021**, *27*, 2039–2048. [CrossRef]
62. Samson, M.-È.; Chantigny, M.H.; Vanasse, A.; Menasseri-Aubry, S.; Royer, I.; Angers, D.A. Management practices differently affect particulate and mineral-associated organic matter and their precursors in arable soils. *Soil Biol. Biochem.* **2022**, *148*, 107867 [CrossRef]
63. Almeida, L.F.J.; Souza, I.F.; Hurtarte, L.C.C.; Teixeira, P.P.D.C.; Inagaki, T.M.; Silva, I.; Mueller, C.M. Forest litter constraints on the pathways controlling soil organic matter formation. *Soil Biol. Biochem.* **2021**, *163*, 108447. [CrossRef]
64. Tian, J.; Zong, N.; Hartley, I.P.; He, N.P.; Zhang, J.J.; Powlson, D.; Zhou, J.Z.; Kuzyakov, Y.; Zhang, F.S.; Yu, G.R.; et al. Microbial metabolic response to winter warming stabilizes soil carbon. *Glob. Chang. Biol.* **2021**, *27*, 2011–2028. [CrossRef] [PubMed]
65. DeForest, J.L.; Moorhead, D.L. Effects of elevated pH and phosphorus fertilizer on soil C, N and P enzyme stoichiometry in an acidic mixed mesophytic deciduous forest. *Soil Biol. Biochem.* **2020**, *150*, 107996. [CrossRef]
66. Fanin, N.; Mooshammer, M.; Sauvadet, M.; Meng, C.; Alvarez, G.; Bernard, L.; Bertrand, I.; Blagodatskaya, E.; Bon, L.; Fontaine, S.; et al. Soil enzymes in response to climate warming: Mechanisms and feedbacks. *Funct. Ecol.* **2022**, *36*, 1378–1395. [CrossRef]
67. Poeplau, C.; Helfrich, M.; Dechow, R.; Szoboszlay, M.; Tebbe, C.C.; Don, A.; Greiner, B.; Zopf, D.; Thumm, U.; Korevaar, H.; et al. Increased microbial anabolism contributes to soil carbon sequestration by mineral fertilization in temperate grasslands. *Soil Biol. Biochem.* **2019**, *130*, 167–176. [CrossRef]
68. Soares, M.; Rousk, J. Microbial growth and carbon use efficiency in soil: Links to fungal-bacterial dominance, SOC-quality, and stoichiometry. *Soil Biol. Biochem.* **2019**, *131*, 195–205. [CrossRef]
69. Tian, D.; Jiang, L.; Ma, S.H.; Fang, W.J.; Schmid, B.; Xu, L.C.; Zhu, J.X.; Li, P.; Losapio, G.; Jing, X.; et al. Effects of nitrogen deposition on soil microbial communities in temperate and subtropical forests in China. *Sci. Total Environ.* **2017**, *607–608*, 1367–1375. [CrossRef]

forests

MDPI

Article

Effect of In Situ Large Soil Column Translocation on CO_2 and CH_4 Fluxes under Two Temperate Forests of Northeastern China

Xingkai Xu [1,2,*], Tingting Xu [1,2] and Jin Yue [1]

1 State Key Laboratory of Atmospheric Boundary Layer Physics and Atmospheric Chemistry, Institute of Atmospheric Physics, Chinese Academy of Sciences, Beijing 100029, China; xutt@tea.ac.cn (T.X.); yuejin@mail.iap.ac.cn (J.Y.)

2 Department of Atmospheric Chemistry and Environmental Science, College of Earth and Planetary Sciences, University of Chinese Academy of Sciences, Beijing 100049, China

* Correspondence: xingkai_xu@mail.iap.ac.cn

Abstract: Global warming has a significant impact on soil carbon dioxide (CO_2) and methane (CH_4) fluxes in temperate forests. However, due to a lack of field observations, limited information is available about the responses of soil CO_2 and CH_4 fluxes to changes in temperature during the non-growing season and throughout the year in temperate forests. The broadleaf and Korean pine mixed mature forest (MF) and adjacent secondary white birch forest (BF) at different succession stages in the Changbai mountain region in northeastern China were selected, to study the effect of in situ soil column translocation on CO_2 and CH_4 fluxes in temperate forests. On average, the air temperature and soil temperature at 5 cm depth under BF stands from October 2018 to October 2022 increased by 0.64 and 0.42 °C during the non-growing season and by 0.49 and 0.43 °C throughout the year, respectively, compared with those under MF stands. Based on multi-year measurements in field experiments, it was shown that during the non-growing season, fluxes of CO_2 and CH_4 from soil columns under MF and BF stands ranged from 0.004 to 1.175 and from 0.015 to 1.401 (averages of 0.321 and 0.387) μmol CO_2 m^{-2} s^{-1}, and from −1.003 to 0.048 and from −1.037 to −0.013 (averages of −0.179 and −0.250) nmol CH_4 m^{-2} s^{-1}, respectively, accounting for approximately 20.8% and 25.3%, and 48.8% and 69.1% of the corresponding average fluxes during the growing season. When undisturbed soil columns of MF were transferred to a BF stand, to simulate warming, the cumulative soil CO_2 emissions and CH_4 uptake increased by 23.5% and 15.3% during the non-growing season, and by 9.5% and 16.3% across the year, respectively. However, when soil columns of BF were transferred to a MF stand, to simulate cooling, the cumulative soil CO_2 emissions decreased by 16.9% and 0.1% during the non-growing season and across the year, respectively. Upon cooling, the cumulative soil CH_4 uptake decreased by 21.8% during the non-growing season but increased by 15.4% across the year. The soil temperature and moisture at 5 cm depth in soil columns could explain 84–86% of the variability in CO_2 fluxes and 16–51% of the variability in CH4 fluxes under the two forest stands throughout the field measurement period. The results of the in situ soil column translocation experiments highlight that a small climate warming in nature can increase soil CO_2 emissions and CH_4 uptake in the temperate forests of northeastern China, particularly during the non-growing season, which should be considered when predicting soil C fluxes in the temperate forests of northeastern China under global warming scenarios.

Keywords: carbon dioxide emission; climate change; heterotrophic respiration; methane flux; soil carbon flux; soil moisture

check for updates

Citation: Xu, X.; Xu, T.; Yue, J. Effect of In Situ Large Soil Column Translocation on CO_2 and CH_4 Fluxes under Two Temperate Forests of Northeastern China. *Forests* **2023**, *14*, 1531. https://doi.org/10.3390/f14081531

Academic Editor: Luca Belelli Marchesini

Received: 23 June 2023
Revised: 16 July 2023
Accepted: 21 July 2023
Published: 27 July 2023

1. Introduction

Climate change is expected to increase global surface temperatures by 1.4 to 4.4 °C above pre-industrial levels by the end of the 21st century, mainly due to increased atmospheric greenhouse gas (GHG) concentrations resulting from human activities [1]. Rising air temperatures have been shown to affect the growth of vegetation and the activity of soil microorganisms, probably influencing carbon (C) cycling in soil and the C fluxes between soil and the atmosphere [2–5]. As a vital process of soil C cycling, soil respiration can account for more than 60% of forest ecosystem respiration [6,7], and a small change in soil respiration due to climate change may lead to a significant change in forest ecosystem respiration. Based on an updated global soil respiration database, it was documented that the ratios of soil heterotrophic respiration to total soil respiration around the globe increased significantly from 0.54 to 0.63 between 1990 and 2014, indicating that climate-driven losses of soil C are currently occurring in many terrestrial ecosystems around the globe [4]. Apart from soil C loss as CO_2, well-drained forest soils are a major sink for atmospheric CH_4 and can play an important role in modulating the global CH_4 budget [8,9]. Climate warming and its attendant changes in precipitation are known to remarkably affect soil CO_2 emissions and CH_4 uptake in forest ecosystems across different climate zones, especially in temperate forests [2,4,5,8,10–14]. The interaction between soil temperature and soil moisture can make the responses of soil CO_2 and CH_4 fluxes to climate change more complicated [5,12,15,16]. Therefore, it is of great importance to study the effects of climate warming on soil CO_2 and CH_4 fluxes in temperate forest ecosystems.

In recent decades, soil warming experiments conducted in forest ecosystems around the globe have shown that warming generally results in a considerable short-term soil C loss [3,10,13,17–20]. However, some experimental observations showed that soil warming did not affect soil CO_2 emissions under boreal-temperate forests [15] and a subtropical planation forest [21]. Compared with the response of soil CO_2 emissions to climate warming, relatively less attention has so far been paid to the CH_4 uptake by soils in forest ecosystems under warming [15,16,22]. In recent meta-analyses, Liu et al. [23] and Yan et al. [5] reported that soil warming could increase soil CH_4 uptake across different upland ecosystems by 7.5% and 15%, respectively. Due to minimal changes in soil moisture and NH_4-N contents, no effects of warming on soil CH_4 fluxes were observed for boreal-temperate forests [15,22]. During the decadal soil warming of a temperate mountain forest, Heinzle et al. [16] reported that soil warming by 4 °C reduced the CH_4 uptake by 19.5% during the initial years of warming, whereas no warming effects occurred during the later years. Taken together, the different responses of CO_2 and CH_4 fluxes in forest soils to warming during the growing season are mainly ascribed to various changes in the duration and intensity of soil warming, climate types, as well as soil physicochemical and microbial properties associated with soil warming at different experimental sites [5,10,13,16,22,24–26]. More importantly, to date, few studies have been paid attention to the C fluxes between soils and the atmosphere in temperate forests, considering the background of global warming, especially during the non-growing season [12,16,27].

In a recent meta-analysis, all in situ soil warming experiments were characterized by different levels of warming intensity (from 0.5 to 9 °C) and mostly by relatively short durations during the growing season [13], which implies some limitations for simulating global climate changes in nature, i.e., the gradual increase in both soil and air temperature over decadal time scales. The relatively high magnitude of changes in temperature relative to the smaller climate changes in nature, and the differences in soil properties and annual precipitation across soil warming experiments, can to some extent lead to different responses of soil CO_2 and CH_4 fluxes to climate changes [5,16,22]. Unlike infrared or cable heating manipulations, translocation experiments can result in the air and soil being warmed under natural conditions, by transferring soil columns from high-latitude to low-latitude zones. Among the soil translocation experiments previously reported, most studies focused on the measurement of C fluxes between soils and the atmosphere within and across years at different sites with varying annual precipitation, with less attention given to the soil CH_4 flux [24,28–30]. The coupled relationship between annual average soil temperature and annual precipitation across different soil-translocation experimental sites can, in turn, make the response of annual soil CO_2 and CH_4 fluxes to climate change more complicated [5,16,24,29]. Furthermore, to date, there has been a lack of knowledge about soil CO_2 and CH_4 fluxes in response to small temperature changes in nature, which is associated with precisely evaluating the responses of annual soil CO_2 and CH_4 fluxes in temperate forests to future climate changes.

Translocation experiments with a small gradient in micrometeorological factors (e.g., temperature and rainfall) can provide an opportunity to simulate climate change in nature and, thus, to evaluate the long-term effects of climate change within a natural environment [19,28–30]. Northeastern China is located on the northeastern edge of the global monsoon climate and is affected by the continental climate, with long cold winters and short summers; this region, which recorded an annual increase of 0.04 °C per year from 1961 to 2004, is currently becoming sensitive to climate change [31]. It is likely that C fluxes between soils and the atmosphere in the temperate forest ecosystems of northeastern China would be sensitive to future climate changes. The forest coverage in northeastern China accounts for approximately 27% of national forest coverage [32]. Broadleaf and Korean pine mixed mature forest (MF) represents the main zonal forest ecosystem in northeastern China, and the near-surface air temperature under MF is lower than under the adjacent secondary white birch forest (BF) across the year, particularly in the non-growing season (from November to April), due to differences in vegetation cover and understory light availability [33]. Furthermore, the properties of soil and aboveground litter in MF stands differ from those in BF stands [34,35]. The differences in annual average near-surface air temperature and soil properties under MF and BF stands can thus provide a good platform for studying the effects of warming and cooling in nature on soil CO_2 and CH_4 fluxes in temperate forests over the year, using reciprocal translocation experiments of large in situ soil columns between the two forest stands.

We predicted that changes in air temperature in nature would affect the CO_2 and CH_4 fluxes from temperate forest soil during the non-growing season and throughout the year and that the C fluxes between the soil and the atmosphere during the non-growing season would be more sensitive to climate change. To reduce the effects of other environmental factors, as much as possible, in situ large soil column reciprocal translocation experiments were used, as an effective method for studying the effects of changes in air temperature and/or precipitation on C fluxes between the soil and the atmosphere in forest and grassland ecosystems [19,24,28–30]. Based on multi-year field measurements of soil column reciprocal translocation experiments under MF and BF stands, the objectives of this study were to explore how climate warming and cooling in nature can affect CO_2 and CH_4 fluxes from temperate forest soils during the non-growing season and throughout the year. The results can provide a basis for understanding the responses of C fluxes between soils and the atmosphere in the temperate forests of northeastern China to future climate change.

2. Materials and Methods

2.1. Site Description and Collection of Soil Columns

The soil column reciprocal translocation experiments were carried out under MF and BF stands near the National Research Station of Changbai Mountain Forest Ecosystems (42°24′ N and 128°6′ E), at the foot of Changbai mountains, northeastern China. The Changbai mountain region is located on the northeastern edge of the global monsoon climate and is influenced by the continental climate, with long cold winters and short summers. MF represents the main zonal forest ecosystem in northeastern China, and its aboveground litter and soil properties are different from those of BF [34,35]. The dominant trees of the MF include Korean pine (*Pinus koraiensis* Sieb. et Zucc.), Tuan linden (*Tilia amurensis* Rupr.), Mongolian oak (*Quercus mongolica* Fisch. ex Turcz), and Manchurian ash (*Fraxinus mandshurica* Rupr.), mostly >200 years old, and their stand densities are 98.9, 117.1, 37.1, and 27.2 trees ha^{-1}, respectively [36]. The BF is dominated by white birch (*Betula platyphylla* Suk.) and mountain poplar (*Populus davidiana* Dode), mostly >70 years old, and its stand density is 1402 stem ha^{-1}; white birch and mountain polar account for approximately 51.5% and 23.1% of the basal area of tree species, respectively [33]. Due to the differences in vegetation cover and photophilic tree species, the air temperature and understory photosynthetically active radiation transmittances for BF in the non-growing season are normally higher than for MF [33]. The temperate forest soil is classified as an Andosol, with a 3–5 cm organic horizon and approximately 10 cm depth of A horizon. The main soil properties and the properties of the aboveground litter for the MF and BF stands were reported in detail by Xu et al. [37,38] and Wu et al. [34,35].

Eight 2 m × 2 m small plots with well-preserved aboveground litter layers were selected under MF and BF stands in October 2017, prior to autumn freeze–thaw events, to collect large undisturbed soil columns at approximately 30 cm depth. At the center of each small plot, a stainless steel cylindrical grinding tool (20 cm in diameter and 40 cm in height) was first driven into the soil to approximately 30 cm depth and then replaced by thickened UPVC (unplasticized polyvinyl chloride) tubes (20 cm in diameter and 40 cm in height), and the soil column was finally taken out as one undisturbed large soil column, with the aboveground litter. In total, there were eight undisturbed large soil columns for each forest stand for the layout of the soil translocation experiments in the field.

2.2. Setup of the In Situ Large Soil Column Translocation Experiments

As mentioned above, the air temperature under BF stands is normally higher than under MF stands, especially in the non-growing season, due to differences in vegetation cover and understory light availability [33], which provides a good platform for simulating climate change in the field. To study changes in soil CO_2 and CH_4 fluxes in temperate forests with warming and cooling throughout the year, half of the undisturbed large soil columns (n = 4) from the MF stand were translocated to the BF stand, to simulate warming, and half of the undisturbed large soil columns (n = 4) from the BF stand were translocated to the MF stand, to simulate cooling; the other soil columns (n = 4) were separately incubated in situ under the MF and BF stands as controls. To eliminate the effects of fine root decomposition on the soil CO_2 and CH_4 fluxes as much as possible, all undisturbed large soil columns were placed into MF and BF stands for one year, following the reciprocal translocation of soil columns, considering that the turnover rates of fine roots in the 0–10 cm soil depth ranged from 0.49 year^{-1} to 1.0 year^{-1} for five forest types in northeastern China [39]. Starting from October 2018, the CO_2 and CH_4 fluxes from the soil columns under MF and BF stands were determined on non-rainy days, mostly between 9:00 and 11:30 in the morning, using a portable greenhouse gas analyzer (915-0011, Los Research Inc., Fremont, CA, USA) coupled to a smart respiratory chamber (SC-11, Beijing LICA United Technology Limited, Beijing, China). From December to February each year, the soil CO_2 and CH_4 fluxes were generally measured once each month, because of low fluxes; in the other months of the year, the soil CO_2 and CH_4 fluxes were generally measured twice to three times each month. From January to September 2020, the measurement of soil CO_2 and CH_4 fluxes was not performed, due to equipment problems. During the non-freezing period of each year, the soil temperature and volumetric water contents (v/v, %) at 5 cm depth in the soil columns were determined using a temperature and moisture device attached to a portable greenhouse gas analyzer (915-0011, Los Research Inc., Fremont, CA, USA). During the period from October 2017 to October 2022, the air temperature and soil temperature at 5 cm depth under the MF and BF stands were recorded using an ONSET HOBO temperature logger (HOBO U23-003, Onset Computer Corporation, Bourne, MA, USA) at 30 min intervals.

2.3. Data Calculation and Statistical Analysis

Soil CO_2 and CH_4 fluxes per unit area (m^{-2}) were calculated, only concerning the surface area of the soil columns. The mean and standard error of four replicates were calculated in each experimental treatment. Cumulative soil CO_2 and CH_4 fluxes during the non-growing and growing seasons and for the year were calculated. All variables for air temperature, soil temperature, and moisture at 5 cm depth, as well as CO_2 and CH_4 fluxes, were first tested for normality using a Shapiro–Wilk test and homogeneity of variance using Levene's test, and log-transformed where necessary. The monthly and seasonal CO_2 and

CH_4 fluxes, as well as soil temperature and volumetric water contents at 5 cm depth in all treatments were described with box and box normal plots using OriginPro 2021 (OriginLab Corporation, Northampton, MA, USA). The monthly and seasonal changes in air temperature and soil temperature at 5 cm depth under MF and BF stands were also described with box and box normal plots, respectively, using OriginPro 2021 (OriginLab Corporation, Northampton, MA, USA). Pair-wise comparisons of significant effects were conducted using Tukey's HSD post hoc test, with significant differences identified at $p < 0.05$. The relationships among the soil CO_2 and CH_4 fluxes, soil temperature, and volumetric water contents at 5 cm depth in soil columns were fitted with nonlinear regressions using OriginPro 2021 (OriginLab Corporation, Northampton, MA, USA). The temperature sensitivity of the soil CO_2 fluxes during the non-growing season and throughout the year were calculated. All statistical analyses were conducted using SPSS for Windows software (version 19.0) (IBM Corp., New York, NY, USA).

3. Results

3.1. Changes in Air Temperature and Soil Temperature at 5 cm Depth under MF and BF Stands

On average, the monthly air temperature and soil temperature at 5 cm depth under the BF stand from October 2018 to October 2022 were higher than those under the MF stand ($p < 0.01$), with the exception of the soil temperature in November (Figure 1a,b). Compared with the MF stand, the air temperature and soil temperature at 5 cm depth under the BF stand, on average, increased by 0.64 °C and 0.42 °C during the non-growing season ($p < 0.001$) and by 0.33 °C and 0.45 °C during the growing season ($p < 0.001$), respectively (Figure 1c,d). The persistent air and soil temperature gradients under MF and BF stands that occurred with a short distance of less than 500 m, enabled the effects of small climate changes in nature (e.g., warming and cooling) on soil CO_2 and CH_4 fluxes to be studied, using the means of the large soil column reciprocal translocation experiments between the two forest stands.

3.2. Effects of Soil Column Translocation on Soil CO_2 and CH_4 Fluxes under MF and BF Stands

Prior to the field experiments, the CO_2 and CH_4 fluxes from the soil columns under MF and BF stands ($n = 8$, for each forest type) varied little, indicating that the spatial variability of the background conditions across soil columns would be negligible in the subsequent analysis of CO_2 and CH_4 fluxes in the soil column reciprocal translocation experiment.

During the period from October 2018 to October 2022, the monthly average soil temperature at 5 cm depth in soil columns during the non-growing season was higher under the BF stand than under the MF stand ($p < 0.05$) (Figure 2a,b). However, due to the relatively high volumetric water contents of the soil columns under the BF stand during the growing season (Figure 2c,d), the monthly average soil temperature at 5 cm depth for the MF and BF soil columns under the BF stand was smaller than under the MF stand in May and June ($p < 0.001$) (Figure 2a,b), and the same trend for the BF soil columns occurred in September ($p < 0.01$) (Figure 2b). The average monthly CO_2 emissions from soil columns under the MF and BF stands were the maximum in July and the minimum in January (Figure 2e,f). Upon the translocation of soil columns from the MF stand to the BF stand, a significant increase in monthly average soil CO_2 emissions occurred in December, April, and May, due to warming ($p < 0.05$) (Figure 2e), and the monthly average CH_4 fluxes decreased significantly in April and May ($p < 0.001$) (Figure 2g). When the BF soil columns were transferred to the MF stand, the monthly average soil CO_2 emissions decreased significantly in December ($p < 0.05$), due to cooling, and increased significantly in June ($p < 0.01$) (Figure 2f), partially due to the relatively high volumetric water contents of the BF soil columns incubated in situ in June (Figure 2d); the monthly average soil CH_4 fluxes decreased significantly in May, June, and July following the translocation of the BF soil columns to the MF stand ($p < 0.05$) (Figure 2h), which paralleled the decrease in monthly soil volumetric water contents (Figure 2d). The monthly average CH_4 fluxes from soil columns under the MF and BF stands remained lowest in October and November

(Figure 2g,h), suggesting the maximum seasonal soil CH_4 uptake over the year in this study region.

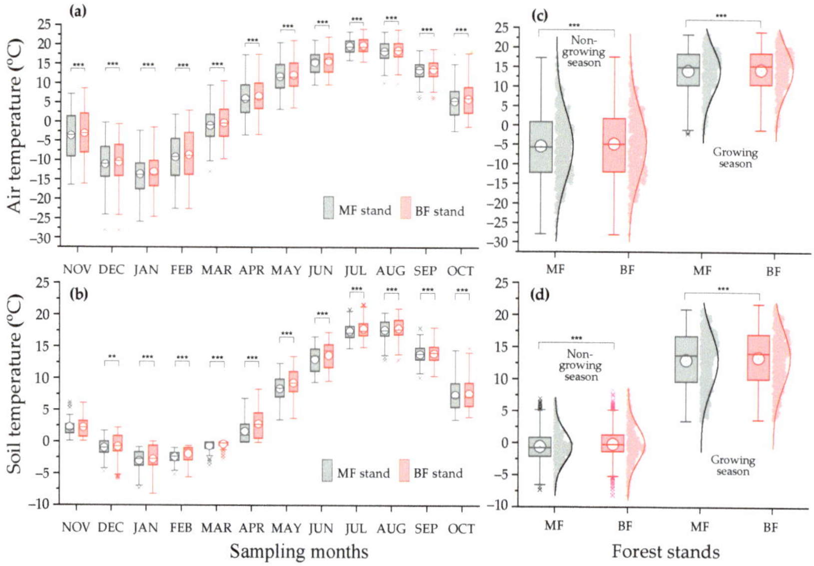

Figure 1. Overall pattern of air temperature and soil temperature at 5 cm depth under MF and BF stands from October 2018 to October 2022: (**a**,**b**) box plots of monthly air temperature and monthly soil temperature, (**c**,**d**) box normal plots of air temperature and soil temperature during the non-growing and growing seasons. Boxes represent interquartile ranges (IQRs), and horizontal lines and circles within boxes indicate median and mean values, respectively. Upper and lower whiskers (x) show 75 percentiles plus 1.5 IQR and 25 percentiles minus 1.5 IQR, respectively. ** $p < 0.01$; *** $p < 0.001$.

On average, the daily average soil temperature at 5 cm depth across MF and BF soil columns during the non-growing season were higher under the BF stand than under the MF stand ($p < 0.001$) (Figure 3a), which paralleled the changes in average soil CO_2 emissions during the non-growing season (Figure 4a). The relatively high average soil volumetric water contents under the BF stand during the growing season (Figure 3d) contributed to the lower average soil temperature at 5 cm depth in the soil columns under the BF stand than under the MF stand ($p < 0.01$) (Figure 3b). This phenomenon did not lead to any significant differences in the average CO_2 emissions from the MF and BF soil columns between BF and MF stands during the growing season (Figure 4b). The average CH_4 fluxes from the soil columns during the non-growing season appeared to be smaller under the BF stand than under the MF stand, due to the relatively low soil moisture (Figure 3c), but the differences were not significant (Figure 4c). During the growing season, CH_4 fluxes from the MF soil columns under the BF stand were smaller than under the MF stand ($p < 0.05$), whereas the CH_4 fluxes from the BF soil columns under the MF stand were smaller than under the BF stand ($p < 0.05$) (Figure 4d).

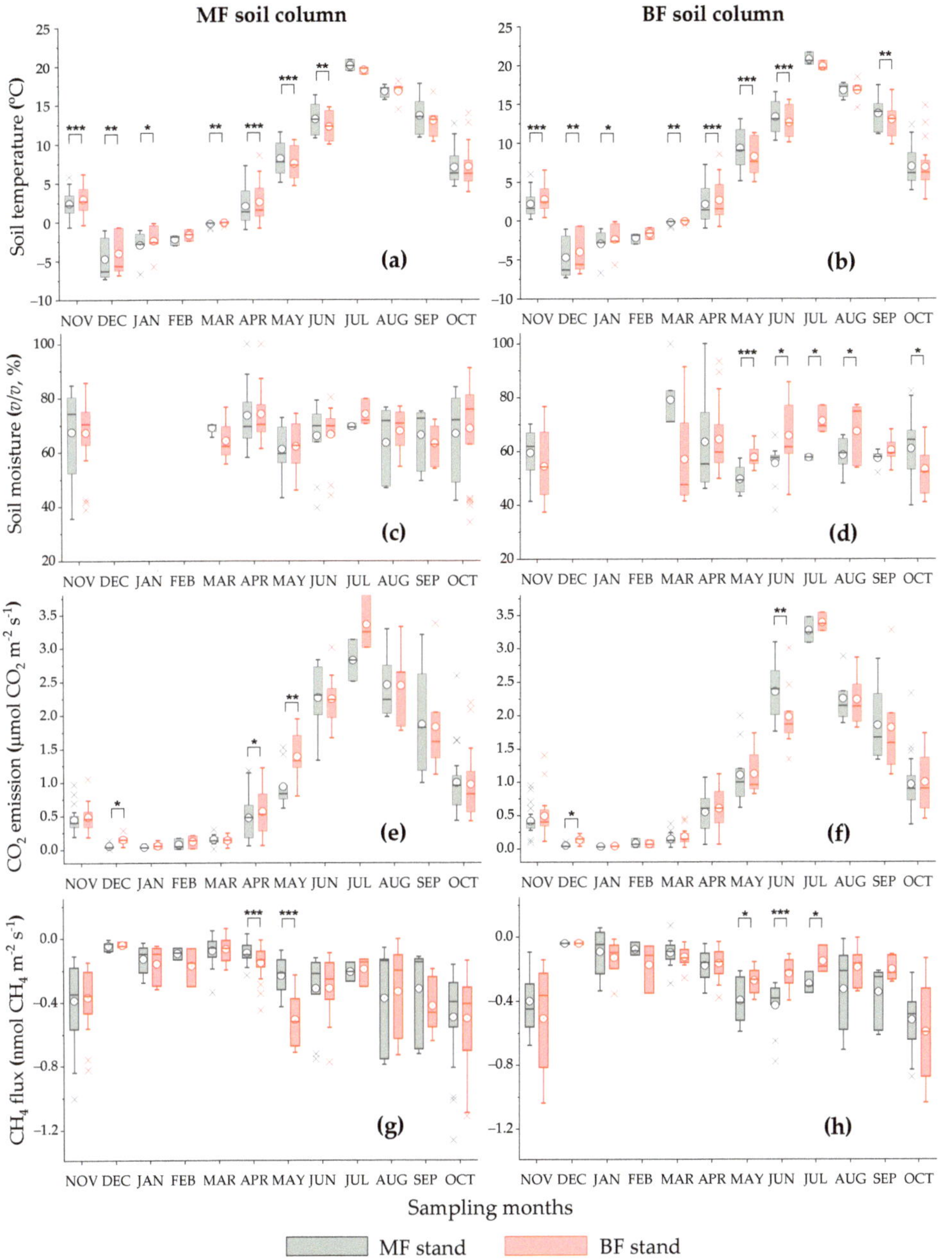

Figure 2. Box plots of monthly soil temperature (**a**,**b**) and moisture (**c**,**d**) at 5 cm depth, CO_2 emission (**e**,**f**), and CH_4 flux (**g**,**h**) from the MF and BF soil columns under MF and BF stands from October 2018 to October 2022. * $p < 0.05$; ** $p < 0.01$; *** $p < 0.001$. Descriptions of the box plots are shown in the caption of Figure 1.

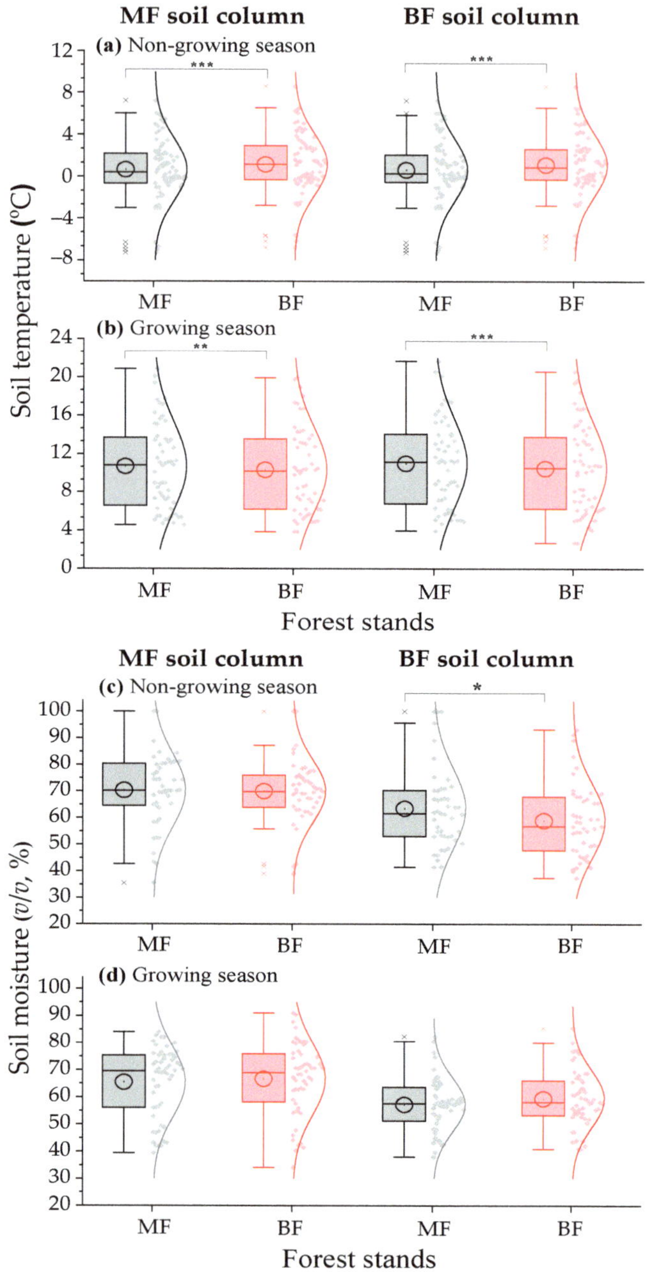

Figure 3. Box normal plots of soil temperature (**a**,**b**) and moisture (**c**,**d**) at 5 cm depth in the MF and BF soil columns under MF and BF stands during the non-growing and growing seasons from October 2018 to October 2022. * $p < 0.05$; ** $p < 0.01$; *** $p < 0.001$. Descriptions of the box plots are shown in the caption of Figure 1.

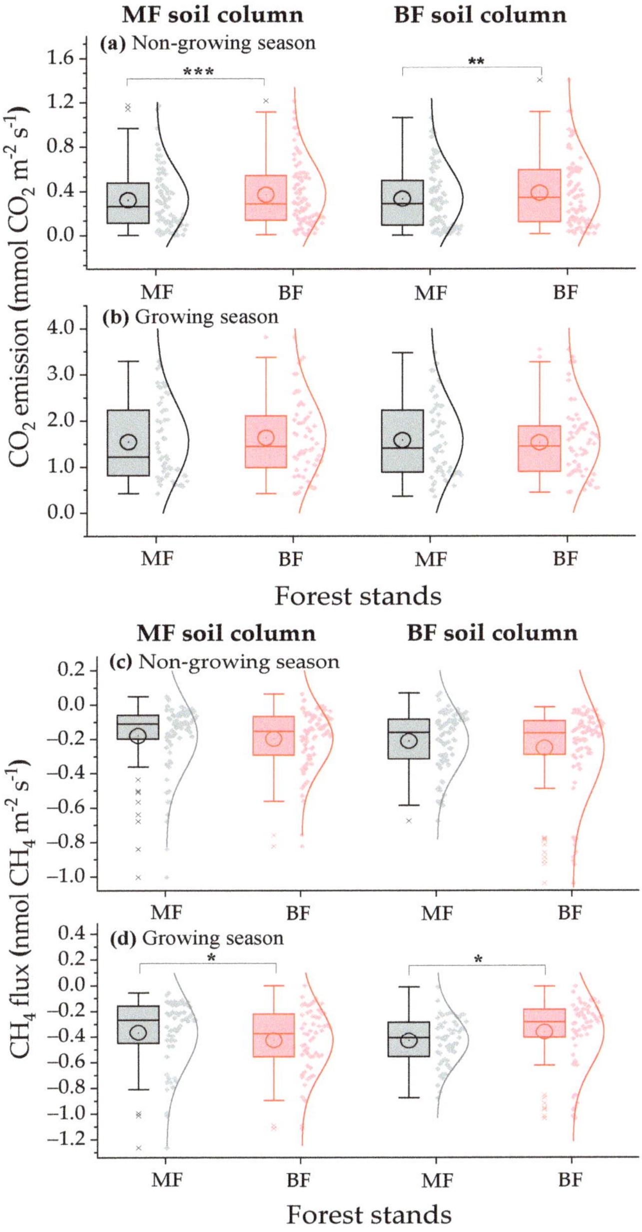

Figure 4. Box normal plots of CO_2 emissions (**a**,**b**) and CH_4 fluxes (**c**,**d**) from the MF and BF soil columns under MF and BF stands during the non-growing and growing seasons from October 2018 to October 2022. * $p < 0.05$; ** $p < 0.01$; *** $p < 0.001$. Descriptions of the box plots are shown in the caption of Figure 1.

3.3. Effects of Soil Temperature and Moisture on Soil CO_2 and CH_4 Fluxes

The combined effects of soil temperature and moisture on the daily CO_2 and CH_4 fluxes from the MF and BF soil columns under MF and BF stands are shown in Figures 5 and 6. The 84%–86% variation in soil CO_2 emissions and 16%–51% variation in soil CH_4 fluxes were explained by the soil temperature and moisture at 5 cm depth, respectively. The maximum daily CH_4 uptake occurred at soil temperature at 5 cm depth ranging from 7.5 to 14 °C, with soil volumetric water contents at 5 cm depth ranging from 38% to 48% (v/v) (Figure 6). The daily CO_2 emission increased significantly with the increase in soil temperature at 5 cm depth, and it reached a maximum at soil volumetric water contents ranging from 55% to 64% (v/v) under the experimental conditions (Figure 5).

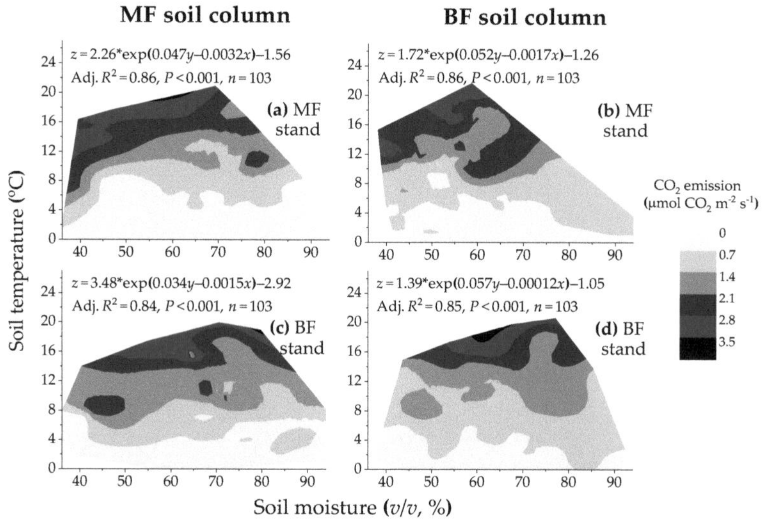

Figure 5. Combined effects of soil temperature and moisture at 5 cm depth on the daily CO_2 emissions from the MF (**a**,**c**) and BF (**b**,**d**) soil columns under MF and BF stands from October 2018 to October 2022. For winter (from late November to early March), soil moisture is not included, due to soil freezing. The combined effects of soil temperature and soil moisture on daily CO_2 emissions were fitted with multivariate regression curves. Legends with different gray scales represent the different ranges of soil CO_2 emissions.

On average, the temperature sensitivity (Q_{10}) of CO_2 emissions from MF and BF soil columns over the year from October 2018 to October 2022 ranged from 2.86 to 2.96 under MF and BF stands, which was not significantly affected by the soil column translocation (Figure 7c,d). The Q_{10} values of CO_2 emissions over the year were smaller than those during the non-growing season ($p < 0.05$) (Figure 7a,b). When the MF soil columns were translocated to the BF stand, the Q_{10} values of the soil CO_2 emissions decreased, on average, from 8.73 to 7.22 during the non-growing season (Figure 7a), which differed from the changes in the Q_{10} values of the soil CO_2 emissions (5.85 to 7.46) following the translocation of BF soil columns to the MF stand (Figure 7b). Due to the large annual variability, differences in the Q_{10} values of CO_2 emissions during the non-growing season across all soil columns were not significant (Figure 7a,b).

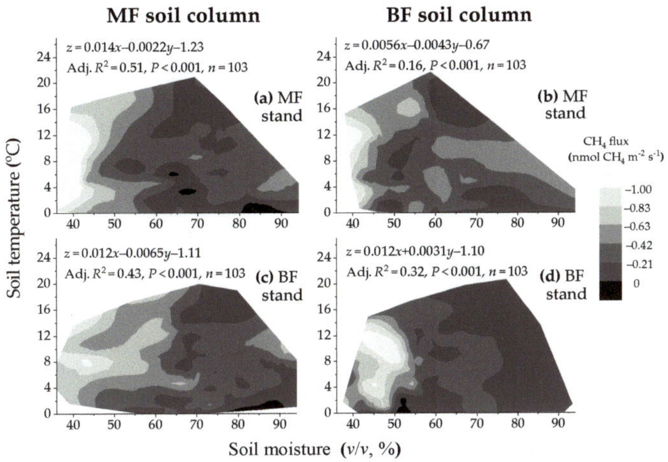

Figure 6. Combined effects of soil temperature and moisture at 5 cm depth on daily CH_4 fluxes from the MF (**a**,**c**) and BF (**b**,**d**) soil columns under MF and BF stands from October 2018 to October 2022. For winter (from late November to early March), soil moisture is not included, due to soil freezing. The combined effects of soil temperature and soil moisture on daily CH_4 fluxes were fitted with multivariate linear regressions. Legends with different gray scales represent different ranges of soil CH_4 fluxes.

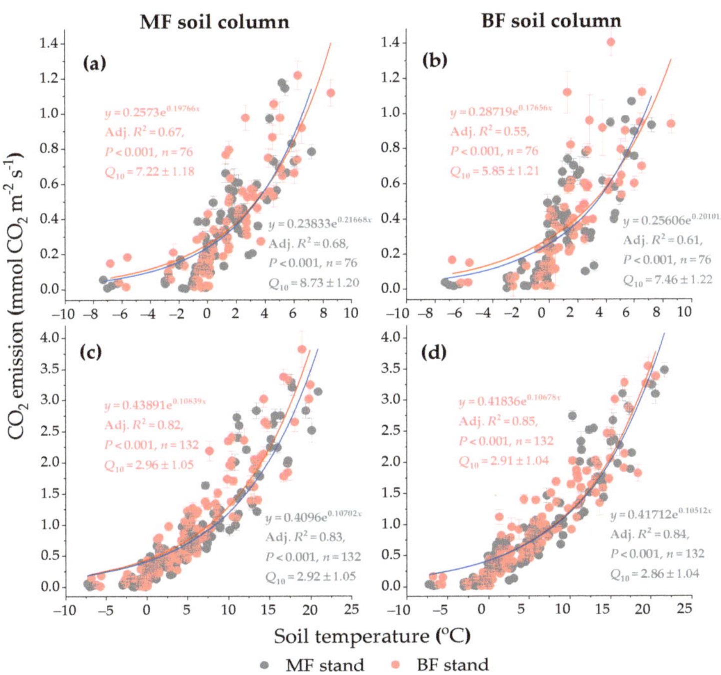

Figure 7. Relationships of soil temperature at 5 cm depth and daily CO_2 emissions from the MF and BF soil columns under MF and BF stands during the non-growing season (**a**,**b**) and for the whole year (**c**,**d**) from October 2018 to October 2022. In winter (from late November to early March), soil temperature in soil columns for the measurement of CO_2 emissions was collected with automatic monitoring of the soil temperature at 5 cm depth under the two forest stands.

3.4. Effects of Warming and Cooling on Cumulative Soil CO_2 and CH_4 Fluxes in Temperate Forests

Due to differences in the vegetation cover and understory light availability for th MF and BF stands, the translocation of the MF soil columns to BF stand simulated climat warming, with an increase of 0.64 °C and 0.33 °C in air temperature during the non-growin and growing seasons, respectively, and with a climate cooling phenomenon occurrin following the translocation of the BF soil column to the MF stand (Figure 1). Cumulativ CO_2 emissions from the MF soil columns during the non-growing season were significantl increased upon climate warming ($p < 0.05$), and the warming appeared to enhance th emissions during the growing season and throughout the year, but no significant difference were observed (Figure 8a). When the BF soil columns were translocated to the MF stanc to simulate climate cooling, the cumulative soil CO_2 emissions during the non-growin season were significantly reduced ($p < 0.05$) (Figure 8b). However, there were no difference in the cumulative CO_2 emissions from the BF soil column under the growing season anc across the year with cooling (Figure 8b). Following the translocation of MF soil columns t the BF stand, the cumulative soil CH_4 fluxes appeared to decrease during the non-growin and growing seasons and across the year, due to climate warming, but no significan differences were observed (Figure 8c). Upon the translocation of BF soil columns to the M stand, the cumulative soil CH_4 fluxes significantly decreased during the growing seasor due to climate cooling ($p < 0.05$) but appeared to increase during the non-growing seasor (Figure 8d). Taken together, these results indicate that the effects of warming and coolin on cumulative soil CO_2 and CH_4 fluxes in the temperate forests of the study region variec between the non-growing and growing seasons.

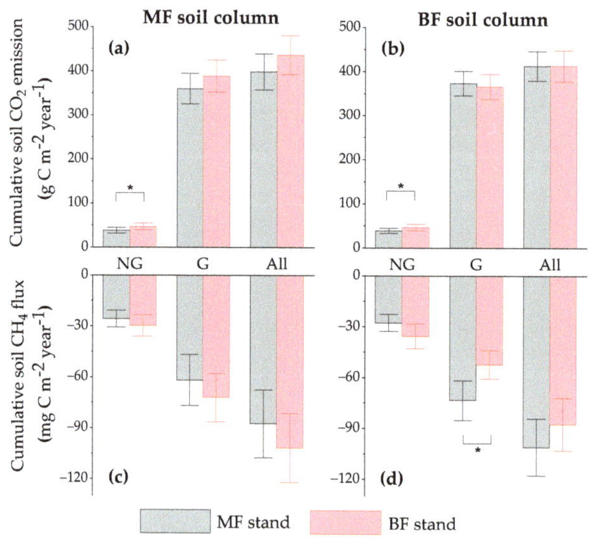

Figure 8. Changes in average cumulative soil CO_2 emissions (**a**,**b**) and CH_4 fluxes (**c**,**d**) from the MF and BF soil columns under MF and BF stands during the non-growing and growing seasons and fo the whole year from October 2018 to October 2022. NG and G represent non-growing and growing seasons, respectively. All signifies the whole year. Bars represent standard errors ($n = 4$). * $p < 0.05$.

Based on multi-year measurements of field experiments, on average, a slight warming and cooling of approximately 0.5 °C in annual mean air temperature (Figure 1) increased the cumulative CO_2 emissions from MF soil columns by 23.5% during the non-growing season ($p < 0.05$) and by 9.6% across the year and decreased the cumulative CO_2 emissions from the BF soil columns by 16.8% during the non-growing season ($p < 0.05$) and by 0.1%

across the year, respectively (Figure 9a,b). Compared with those during the growing season (Figure 8a,b), the cumulative soil CO_2 emissions in the temperate forests in the study region became more sensitive to small climate changes during the non-growing season (Figure 9a). The small climate warming appeared to increase the cumulative CH_4 uptake of the MF soil columns by 15.3% during the non-growing season and by 16.3% across the year (Figure 9a,b), suggesting that even such a slight warming of the climate can promote soil CH_4 uptake in temperate forests in the study region. With such a slight climate cooling, the cumulative CH_4 uptake of the BF soil columns during the non-growing season decreased by 21.8% (Figure 9a), whereas it significantly increased, by 40.7%, during the growing season ($p < 0.05$) (Figure 8d). This led to a 15.4% increase in the annual cumulative CH_4 uptake in the BF soil columns due to the cooling (Figure 9b).

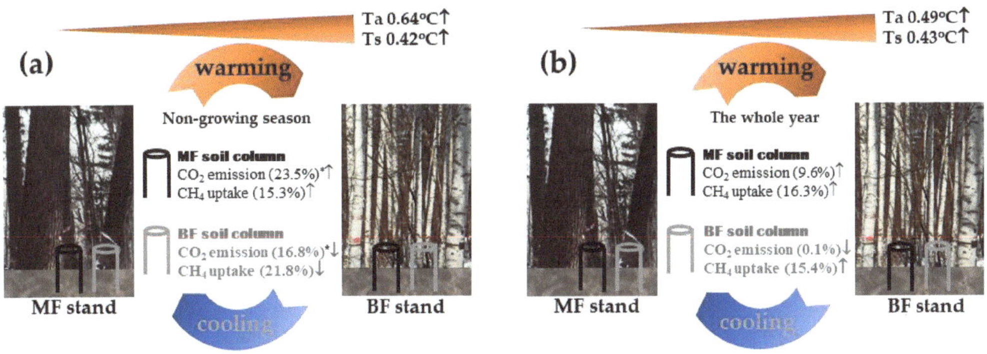

Figure 9. Conceptual model illustrating the effects of warming and cooling on soil cumulative CO_2 emissions and CH_4 uptake under MF and BF stands during the non-growing season (**a**) and across the year (**b**). Ta and Ts represent the average air temperature and soil temperature at 5 cm depth under MF and BF stands during the non-growing season and across the year from October 2018 to October 2022, respectively. Data in brackets indicate the percentile changes in soil cumulative CO_2 emissions and CH_4 uptake following the soil column translocation, relative to the soil columns incubated in situ. Upward and downward arrows represent the increase and decrease in cumulative CO_2 emissions and CH_4 uptake, respectively. * $p < 0.05$.

4. Discussion

4.1. Effects of Warming and Cooling on Soil CO_2 Emissions under Temperate Forest Stands

Under the experimental conditions, the cumulative CO_2 emissions from the soil columns under the MF and BF stands in the non-growing season ranged from 38.31 to 39.31 and from 47.26 to 47.35 g C per m^2, respectively, which accounted for 9.6% to 11.4% of the annual average soil CO_2 emissions (Figure 8a,b). The soil cumulative CO_2 emissions during the non-growing season and their contributions to annual soil CO_2 emissions for the two temperate forests were within the range reported by previous studies under temperate forests in Northeast Asia region [40–42].

In this study, an increase of approximately 0.5 °C in the annual mean near-surface air temperature (Figure 9b) resulted in a 9.6% increase in annual CO_2 emissions from the MF soil columns, which was much smaller than the values reported by many previous soil column translocation experiments [19,28,29]. Hart [19] reported a 120% increase in soil CO_2 emissions in a spruce–fir forest in the American Southwest as the result of an increase of 2.5 °C in soil temperature. Based on a reciprocal translocation experiment, Luan et al. [29] reported a 44% increase in soil CO_2 emissions under a warm-temperate oak forest in China, with a 13-month warming of soil temperatures by approximately 3.3 °C. The relatively low response of soil CO_2 emissions in this study may have resulted from the small magnitude of soil warming (approximately 0.4 °C across the year) and differences in soil physiochemical and microbial properties (e.g., soil C:N ratio, C substrate quality and quantity, nutrient and moisture availability, soil texture, and shifts in the microbial community) at the different experimental sites [5,20,24,30,43]. Additionally used to reduce the effects of the decomposition of fine roots remaining inside the soil columns on the soil CO_2 emissions, as much as possible, the delay of one year following the soil column reciprocal translocation experiments between the two forest stands was partially responsible for the relatively low responses to the slight climate warming, because the initial large stimulating effect of climate warming on the soil CO_2 emissions was excluded [28]. Nonetheless, the cumulative CO_2 emissions from the MF soil columns during the non-growing season with such warming significantly increased by 23.5% ($p < 0.05$) and vice versa (a decrease of 16.8%, $p < 0.05$) for those from the BF soil columns under cooling conditions (Figure 9a). Hence, the results indicated that slight soil warming can lead to a remarkable increase in soil CO_2 emissions in temperate forests in the study region, particularly in the non-growing season. An earlier meta-analysis indicated that experimental warming can increase soil respiration by 7%–11% across different terrestrial ecosystems [2], which is close to the 9.6% increase in annual soil CO_2 emissions with such small climate warming in this study (Figure 9b). The magnitude of the increase in annual soil heterotrophic respiration (9.6%) upon soil warming was also close to the results reported by Yan et al. [5] in a recent meta-analysis, who summarized a 9.9% increase in soil CO_2 emissions with experimental warming across multiple terrestrial ecosystems around the globe.

Our study showed that, under warming or cooling conditions, soil temperature and moisture at 5 cm depth could account for 84%–86% of the variation in daily soil CO_2 emissions during the period from October 2018 to October 2022 (Figure 5). Furthermore, the maximum daily soil CO_2 emission occurred at soil volumetric water contents ranging from 55% to 64% (v/v) under the experimental conditions (Figure 5), suggesting that soil CO_2 emissions decrease when soil moisture becomes either relatively wetter or drier [44–47]. This was obviously different from the results observed by Luan et al. [29] for a warm temperate forest, who reported that the responses of soil CO_2 emission to a soil column translocation experiment were mainly caused by changes in soil temperature, rather than soil moisture. This negligible effect of soil moisture could be ascribed to a small range of soil volumetric water contents (0.15 to 0.35 cm^{-3} cm^{-3}), which was much less than the soil moisture observed in our study (Figure 2c,d).

The surface soil at 10 cm depth under the BF stand was characterized by a relatively small sand content and relatively large silt content compared with surface soils of the MF stand [35], and the annual throughfall for the BF was generally larger than for the MF stand during the growing season (data not shown), due to the forest canopy vegetation [33]. These phenomena may have partially contributed to the relatively large volumetric water contents in the BF soil columns incubated in situ during the growing season (Figure 2d). The changes in soil moisture during the growing season could have weakened, to a greater extent, the negative responses of the cumulative CO_2 emissions from the BF soil columns to climate cooling, compared with those during the non-growing season (Figure 8b), thus resulting in a decrease of approximately 0.1% in the annual cumulative CO_2 emissions from the BF soil columns upon climate cooling (Figure 9b). These results indicate that changes in soil moisture induced by throughfall can, to some extent, affect the response to climate change in soil CO_2 emissions under temperate forests [11,15]. Based on various recent meta-analyses, soil CO_2 emissions in temperate forests are becoming more sensitive to climate changes, such as warming and precipitation, compared to those in boreal forests [13,14]. Hence, more field experiments on warming should be carried out at various specific sites in temperate forests, to precisely characterize the response of soil CO_2 emissions, especially soil heterotrophic respiration, to climate change.

4.2. Effects of Warming and Cooling on Soil CH_4 Flux under Temperate Forest Stands

Under the experimental conditions, the cumulative soil CH_4 fluxes in the non-growing season under the MF and BF stands ranged from –27.72 to –25.73 and from –35.43 to

−29.67 mg C per m^2, which accounted for 27.4% to 40.5% of the annual average soil CH_4 fluxes (Figure 8c,d). The contribution of soil CH_4 fluxes during the non-growing season to the annual soil CH_4 fluxes in this study is in agreement with the results reported by Heinzle et al. [16] for control and warmed plots in a temperate mountain forest in Austria (from 28% to 35% and from 24% to 38%, respectively). When MF soil columns were transferred to BF stands to simulate soil warming in nature, the annual average cumulative soil CH_4 uptake throughout the experimental period increased from 0.88 ± 0.20 to 1.02 ± 0.20 Kg C per hectare, due to soil warming of 0.43 °C (Figure 8c,d). The magnitude of the increase in the soil CH_4 uptake with soil warming is in agreement with the results described in recent meta-analyses [5,23]. However, Heinzle et al. [16] reported that soil warming by 4 °C reduced the soil CH_4 uptake by 19.5% during the initial years of warming in a temperate mountain forest in Austria with high annual precipitation, whereas no warming effects occurred during the later years. This difference could be mainly ascribed to changes in the soil moisture induced by annual precipitation across the different field experimental sites [8,11,16], because soil moisture is normally considered the most important factor in regulating the diffusion of atmospheric CH_4 into the soil in upland ecosystems [18,48].

Under soil warming conditions, soil moisture can, to some extent, decrease, which then strengthens the CH_4 uptake in soils by microbes via increasing gas diffusion [49]. However, soil warming can affect the availability of soil nutrients (e.g., NH_4-N) [22,25,50], the quantity and quality of substances (e.g., amount and biodegradation of dissolved organic C) [50–53], soil pH [50,54], and shifts in the population and activity of methanotrophic bacteria [25,54], thus influencing the soil CH_4 uptake [11–13,26,55]. In this study, an increase of approximately 0.5 °C in annual mean near-surface air temperature (Figure 9b) led to a 16.3% increase in the annual soil CH_4 uptake in temperate forests in northeastern China, which was higher than the results described in several recent meta-analyses [5,23], which documented that, on average, a 7.5% or 15% increase in the soil CH_4 uptake occurred in upland ecosystems under warming conditions. Furthermore, Hart [19] reported a 90% increase in soil CH_4 consumption in a spruce-fir forest in the American Southwest, as the result of an increase of 2.5 °C in soil temperature, which paralleled a 80% increase in net N mineralization and nitrification in transferred soil cores compared with in situ soil cores. The results of Hart [19] show that small increases in mean annual air temperature can have large impacts on soil nitrogen cycling and CH_4 consumption, even in ecosystems where water availability is a limited resource. However, due to minimal changes in soil moisture and NH_4-N contents, soil warming in a Swiss alpine tree line did not affect CH_4 fluxes from the forest soils [22]. Taken together, these different effects of climate warming on CH_4 fluxes in forest ecosystems during the growing season can be mainly ascribed to various changes in soil properties (e.g., soil water and nutrient availability [12,48,50], pH [50,54,55], quantity and quality of substances [12,50,51], and the population and activity of methanotrophic bacteria [5,22,25]) associated with soil warming at different experimental sites.

To date, limited information has been available about the responses of soil CH_4 fluxes to climate changes in the non-growing season in forest ecosystems at a global scale [16,27]. The results of this study showed that soil CH_4 fluxes appeared to be more sensitive to climate changes during the non-growing season than across the year in temperate forests in northeastern China (Figure 9a,b), probably due to the impacts of changes in soil moisture induced by throughfall during the growing season under MF and BF stands (Figure 2c,d). In a previous warming experiment in a northern hardwood forest in the USA, McHale et al. [18] reported that soil water content exhibited the most influence on soil CH_4 fluxes in the second year. Under warming or cooling conditions, soil temperature and soil moisture at 5 cm depth explained 16%–51% of the variation in soil CH_4 fluxes throughout the experimental period, and a maximum daily CH_4 uptake was observed at soil temperature ranging from 7.5 to 14 °C, along with soil volumetric water contents ranging from 38% to 48% (v/v) (Figure 6). Although there is a maximum CH_4 uptake capacity in temperate forest surface soils at 25 °C under optimum soil moisture conditions [55], the decrease in the soil CH_4 uptake at high soil temperatures in this study may have been, to a large extent, caused by the limitation of high soil moisture on CH_4 diffusivity during the growing season, particularly inside the BF soil columns incubated in situ (Figure 2d,h). This would partly explain the relatively large soil CH_4 uptake during the growing season when the BF soil columns were transferred to MF stands (Figures 2h and 8d). When the soil moisture is not limited, the CH_4 uptake by microbes in soils can increase linearly with increasing soil temperature up to 35 °C [56]. In mesic environments with soil moisture levels typically above optimum, soil warming normally increases the CH_4 uptake, which is opposite to the effect of warming on CH_4 uptake when soils are dry [49]. Nonetheless, contrary experimental results were reported for the warming effects of CH_4 fluxes between two alpine grasslands with a contrasting soil water status [25] and in a temperate mountain forest with high annual precipitation [16]. Together with the responses of soil CO_2 emissions to climate warming, the results of this study indicate that a remarkable response to climate warming of C fluxes between soils and the atmosphere is currently occurring in temperate forests of northeastern China, especially during the non-growing season. Considering that climate warming-induced soil C fluxes across different terrestrial ecosystems are dependent on the magnitude of changes in soil moisture, the intensity and duration of soil warming, and changes in other associated soil properties [3,5,18,25,27,50,51,57], long-term field observations of warming across different terrestrial ecosystems of northeastern China are needed, to evaluate the response and feedback of soil C fluxes in northeastern China to future climate changes.

5. Conclusions and Future Perspectives

The results from multi-year field measurements indicated that following the translocation of MF soil columns to a BF stand, the increase of approximately 0.5 °C in annual average near-surface air temperature, on average, increased the annual soil CO_2 emission by 9.6% and CH_4 uptake by 16.3% in temperate forests in the study region. These increases amounted to 760 kg ha^{-1} year^{-1} of soil C loss and 0.286 kg ha^{-1} year^{-1} of CH_4 uptake per 1 °C increase in the temperate forests of the study region. The combined effects of soil temperature and moisture could account for 84%–86% of the variability in soil CO_2 emissions and 16%–51% of the variability in soil CH_4 fluxes under the experimental conditions, respectively. Assuming a similar magnitude of response of soil C fluxes to climate warming across different temperate forests in northeastern China and an estimated forest coverage for this region of 58.6 million hectare [32], our results further illustrate that a moderate warming of 1 °C would increase the annual soil heterotrophic respiration and annual CH_4 uptake in the temperate forests of northeastern China by a magnitude of 44.536 Tg CO_2-C (1 Tg = 10^{12} g) and 0.0168 Tg CH_4-C, respectively. Provided that CH_4 from fossil sources has the ability to trap 29.8 times as much thermal radiation as CO_2 on a molar basis over a 100-year timescale [1], an increase in C fluxes between soils and the atmosphere from 1 °C warming across temperate forests in northeastern China was estimated at approximately 44.04 Tg CO_2-C-equivalent year^{-1}, indicating the remarkable positive feedback of forest soil C fluxes to future climate change in northeastern China. Considering an annual air temperature increase rate of 0.04 °C per year from 1961 to 2004 in northeastern China [31] this finding highlights that the temperate forests in northeastern China may trigger an overall positive feedback of soil C fluxes to climate warming, which should be effectively incorporated into soil C cycling-related models, to precisely predict soil C fluxes in the temperate forests of northeastern China under global warming scenarios.

Author Contributions: Conceptualization, methodology, investigation, formal analysis, resources, investigation, writing—original draft preparation, writing—review and editing, supervision, funding acquisition, X.X.; formal analysis, investigation, writing—review and editing, T.X.; formal analysis, writing—review and editing, J.Y. All authors have read and agreed to the published version of the manuscript.

Funding: This research was funded by the National Natural Science Foundation of China (grant number: 41175133, 41775163, 41975121 and 42275130).

Data Availability Statement: Not applicable.

Conflicts of Interest: The authors declare no conflict of interest.

References

1. IPCC. The Physical Science Basis. In *Contribution of Working Group I to the Sixth Assessment Report of the Intergovernmental Panel on Climate Change*; Cambridge University Press: Cambridge, UK, 2021.
2. Lu, M.; Zhou, X.H.; Yang, Q.; Li, H.; Luo, Y.Q.; Fang, C.M.; Chen, J.K.; Yang, X.; Li, B. Responses of ecosystem carbon cycle to experimental warming: A meta-analysis. *Ecology* **2013**, *94*, 726–738. [PubMed]
3. Melillo, J.M.; Frey, S.D.; DeAngelis, K.M.; Werner, W.J.; Bernard, M.J.; Bowles, F.P.; Pold, G.; Knorr, M.A.; Grandy, A.S. Long-term pattern and magnitude of soil carbon feedback to the climate system in a warming world. *Science* **2017**, *358*, 101–104. [PubMed]
4. Bond-Lamberty, B.; Bailey, V.L.; Chen, M.; Gough, C.M.; Vargas, R. Globally rising soil heterotrophic respiration over recent decades. *Nature* **2018**, *560*, 80–83. [PubMed]
5. Yan, W.M.; Zhong, Y.; Yang, J.Y.; Shangguan, Z.P.; Torn, M.S. Response of soil greenhouse gas fluxes to warming: A global meta-analysis of field studies. *Geoderma* **2022**, *419*, 115865.
6. Law, B.E.; Ryan, M.G.; Anthoni, P.M. Seasonal and annual respiration of a ponderosa pine ecosystem. *Glob. Chang. Biol.* **1999**, *5*, 169–182.
7. Bolstad, P.V.; Davis, K.J.; Martin, J.; Cook, B.D.; Wang, W. Component and whole-system respiration fluxes in northern deciduous forests. *Tree Physiol.* **2004**, *24*, 493–504.
8. Ni, X.Y.; Groffman, P.M. Declines in methane uptake in forest soils. *Proc. Natl. Acad. Sci. USA* **2018**, *115*, 8587–8590.
9. Yu, L.J.; Huang, Y.; Zhang, W.; Li, T.T.; Sun, W.J. Methane uptake in global forest and grassland soils from 1981 to 2010. *Sci. Total Environ.* **2017**, *607*, 1163–1172.
10. Romero-Olivares, A.L.; Allison, S.D.; Treseder, K.K. Soil microbes and their response to experimental warming over time: A meta-analysis of field studies. *Soil Biol. Biochem.* **2017**, *107*, 32–40.
11. Xu, X.K.; Duan, C.T.; Wu, H.H.; Luo, X.B.; Han, L. Effects of changes in throughfall on soil GHG fluxes under a mature temperate forest, northeastern China. *J. Environ. Manag.* **2021**, *294*, 112950.
12. Xu, X.K. Effect of freeze-thaw disturbance on soil C and N dynamics and GHG fluxes of East Asia forests: Review and future perspectives. *Soil Sci. Plant Nutr.* **2022**, *68*, 15–26. [CrossRef]
13. Wang, X.; Hu, H.B.; Zheng, X.; Deng, W.B.; Chen, J.Y.; Zhang, S.; Cheng, C. Will climate warming of terrestrial ecosystem contribute to increase soil greenhouse gas fluxes in plot experiment? A global meta-analysis. *Sci. Total Environ.* **2022**, *827*, 154114. [PubMed]
14. Xu, X.K. Effect of changes in throughfall on soil respiration in global forest ecosystems: A meta-analysis. *Forests* **2023**, *14*, 1037. [CrossRef]
15. Martins, C.S.C.; Nazaries, L.; Delgado-Baquerizo, M.; Macdonald, C.A.; Anderson, I.C.; Hobbie, S.E.; Venterea, R.T.; Reich, P.B.; Singh, B.K. Identifying environmental drivers of greenhouse gas emissions under warming and reduced rainfall in borealtemperate forests. *Funct. Ecol.* **2017**, *31*, 2356–2368. [CrossRef]
16. Heinzle, J.; Kitzler, B.; Zechmeister-Boltenstern, S.; Tian, Y.; Kengdo, S.K.; Wanek, W.; Borken, W.; Schindlbacher, A. Soil CH_4 and N_2O response diminishes during decadal soil warming in a temperate mountain forest. *Agri. For. Meteorol.* **2023**, *329*, 109287. [CrossRef]
17. Rustad, L.E.; Fernandez, I.J. Experimental soil warming effects on CO_2 and CH_4 flux from a low elevation spruce-fir forest soil in Maine, USA. *Glob. Chang. Biol.* **1998**, *4*, 597–605. [CrossRef]
18. McHale, P.J.; Mitchell, M.J.; Bowles, F.P. Soil warming in a northern hardwood forest: Trace gas fluxes and leaf litter decomposition. *Can. J. For. Res.* **1998**, *28*, 1365–1372. [CrossRef]
19. Hart, S.C. Potential impacts of climate change on nitrogen transformations and greenhouse gas fluxes in forests: A soil transfer study. *Glob. Chang. Biol.* **2006**, *12*, 1032–1046. [CrossRef]
20. Melillo, J.M.; Butler, S.; Johnson, J.; Mohan, J.; Steudler, P.; Lux, H.; Burrows, E.; Bowles, F.; Smith, R.; Scott, L.; et al. Soil warming, carbon-nitrogen interactions, and forest carbon budgets. *Proc. Natl. Acad. Sci. USA* **2011**, *108*, 9508–9512. [CrossRef]
21. Wang, H.; Liu, S.R.; Wang, J.X.; Li, D.J.; Shi, Z.M.; Liu, Y.C.; Xu, J.; Hong, P.Z.; Yu, H.L.; Zhao, Z.; et al. Contrasting responses of heterotrophic and root dependent respiration to soil warming in a subtropical plantation. *Agric. For. Meteorol.* **2017**, *247*, 221–228. [CrossRef]

22. Karbin, S.; Hagedorn, F.; Dawes, M.A.; Niklaus, P.A. Treeline soil warming does not affect soil methane fluxes and the spatial micro-distribution of methanotrophic bacteria. *Soil Biol. Biochem.* **2015**, *86*, 164–171. [CrossRef]
23. Liu, S.; Zheng, Y.; Ma, R.; Yu, K.; Han, Z.; Xiao, S.; Li, Z.; Wu, S.; Li, S.; Wang, J. Increased soil release of greenhouse gases shrinks terrestrial carbon uptake enhancement under warming. *Glob. Chang. Biol.* **2020**, *26*, 4601–4613. [CrossRef]
24. Zimmermann, M.; Meir, P.; Bird, M.I.; Malhi, Y.; Ccahuana, A.J.Q. Climate dependence of heterotrophic soil respiration from a soil-translocation experiment along a 3000 m tropical forest altitudinal gradient. *Eur. J. Soil Sci.* **2009**, *60*, 895–906. [CrossRef]
25. Li, F.; Yang, G.B.; Peng, Y.F.; Wang, G.Q.; Qin, S.Q.; Song, Y.T.; Fang, K.; Wang, J.; Yu, J.C.; Liu, L.; et al. Warming effects on methane fluxes differ between two alpine grasslands with contrasting soil water status. *Agri. For. Meteorol.* **2020**, *290*, 107988. [CrossRef]
26. Wu, H.H.; Xu, X.K.; Cheng, W.G.; Han, L. Dissolved organic matter and inorganic N jointly regulate greenhouse gases fluxes from forest soils with different moistures during a freeze-thaw period. *Soil Sci. Plant Nutr.* **2020**, *66*, 163–176. [CrossRef]
27. Wang, P.Y.; Wang, J.S.; Elberling, B.; Yang, L.; Chen, W.N.; Song, L.; Yan, Y.J.; Wang, S.; Pan, J.X.; He, Y.L.; et al. Increased annual methane uptake driven by warmer winters in an alpine meadow. *Glob. Chang. Biol.* **2022**, *28*, 3246–3259. [CrossRef]
28. Rey, M.; Guntiñas, E.; Gil-Sotres, F.; Leirós, M.C.; Trasar-Cepeda, C. Translocation of soils to stimulate climate change: CO_2 emissions and modifications to soil organic matter. *Eur. J. Soil Sci.* **2007**, *58*, 1233–1243. [CrossRef]
29. Luan, J.W.; Liu, S.R.; Chang, S.X.; Wang, J.X.; Zhu, X.L.; Liu, K.; Wu, J.H. Different effects of warming and cooling on the decomposition of soil organic matter in warm-temperate oak forests: A reciprocal translocation experiment. *Biogeochemistry* **2014**, *121*, 551–564. [CrossRef]
30. Nottingham, A.T.; Whitaker, J.; Ostle, N.J.; Bardgett, R.D.; McNamara, N.P.; Fierer, N.; Salinas, N.; Ccahuana, A.J.Q.; Turner, B.L.; Meir, P. Microbial responses to warming enhance soil carbon loss following translcoation across a typical forest elevation gradient. *Ecol. Lett.* **2019**, *22*, 1889–1899. [CrossRef]
31. Zhao, D.S.; Zheng, D.; Wu, S.H.; Wu, Z.F. Climate changes in northeastern China during last four decades. *Chin. Geogr. Sci.* **2007**, *17*, 317–324. [CrossRef]
32. Zhang, J.B.; Sun, B.; Zhu, J.J.; Wang, J.K.; Pan, X.C.; Gao, T. Black soil protection and utilization based on harmonization of mountain-rive-forest-farmland-lake-grassland-sandy land ecosystems and strategic construction of ecological barrier. *Bull. Chin. Acad. Sci.* **2021**, *36*, 1155–1164.
33. Zhou, G.; Liu, Q.J.; Xu, Z.Z.; Du, W.X.; Yu, J.; Meng, S.W.; Zhou, H.; Qi, L.H.; Shah, S. How can the shade intolerant Korean pine survive under dense deciduous canopy? *For. Ecol. Manag.* **2020**, *457*, 117735. [CrossRef]
34. Wu, H.H.; Xu, X.K.; Duan, C.T.; Li, T.S.; Cheng, W.G. Synergistic effects of dissolved organic carbon and inorganic nitrogen on methane uptake in forest soils without and with freezing treatment. *Sci. Rep.* **2016**, *6*, 32555. [CrossRef]
35. Wu, H.H.; Xu, X.K.; Cheng, W.G.; Fu, P.Q.; Li, F.Y. Antecedent soil moisture prior to freezing can affect quantity, composition and stability of soil dissolved organic matter during thaw. *Sci. Rep.* **2017**, *7*, 6380. [CrossRef] [PubMed]
36. Hao, Z.Q.; Zhang, J.; Song, B.; Ye, J.; Li, B.H. Vertical structure and spatial associations of dominant tree species in an old-growth temperate forest. *For. Ecol. Manag.* **2007**, *252*, 1–11. [CrossRef]
37. Xu, X.K.; Han, L.; Wang, Y.S.; Inubushi, K. Influence of vegetation types and soil properties on microbial biomass carbon and metabolic quotients in temperate volcanic and tropical forest soils. *Soil Sci. Plant Nutr.* **2007**, *53*, 430–440. [CrossRef]
38. Xu, X.K.; Yin, L.; Duan, C.T.; Jing, Y.S. Effect of N addition, moisture, and temperate on soil microbial respiration and microbial biomass in forest soil at different stages of litter decomposition. *J. Soils Sediments* **2016**, *16*, 1421–1439. [CrossRef]
39. Wang, N.; Wang, C.K.; Quan, X.K. Variations in fine root dynamics and turnover rates in five forest types in northeastern China. *J. For. Res.* **2020**, *31*, 871–884. [CrossRef]
40. Mariko, S.; Nishimura, N.; Mo, W.H.; Matsui, Y.; Kibe, T.; Koizumi, H. Winter CO_2 flux from soil and snow surfaces in a cool-temperate deciduous forest, Japan. *Ecol. Res.* **2000**, *15*, 363–372. [CrossRef]
41. Mo, W.H.; Lee, M.S.; Uchida, M.; Inatomi, M.; Saigusa, N.; Mariko, S.; Koizumi, H. Seasonal and annual variations in soil respiration in a cool-temperate deciduous broad-leaved forest in Japan. *Agri. For. Meteorol.* **2005**, *134*, 81–94. [CrossRef]
42. Wang, C.K.; Han, Y.; Chen, J.Q.; Wang, X.C.; Zhang, Q.Z.; Bond-Lamberty, B. Seasonality of soil CO_2 efflux in a temperate forest: Biophysical effects of snowpack and spring freeze–thaw cycles. *Agri. For. Meteorol.* **2013**, *77*, 83–92. [CrossRef]
43. Hicks, L.C.; Meir, P.; Nottingham, A.T.; Reay, D.S.; Stott, A.W.; Salinas, N.; Whitaker, J. Carbon and nitrogen inputs differently affect priming of soil organic matter in tropical lowland and montane soils. *Soil Biol. Biochem.* **2019**, *129*, 212–222. [CrossRef]
44. van Straaten, O.; Veldkamp, E.; Kohler, M.; Anas, I. Spatial and temporal effects of drought on soil CO_2 efflux in a cacao agroforestry system in Sulawesi, Indonesia. *Biogeosciences* **2010**, *7*, 1223–1235. [CrossRef]
45. Suseela, V.; Dukes, J.S. The responses of soil and rhizosphere respiration to simulated climatic changes vary by season. *Ecology* **2013**, *94*, 403–413. [CrossRef] [PubMed]
46. Zhang, X.; Zhang, Y.P.; Sha, L.Q.; Wu, C.S.; Tan, Z.H.; Song, Q.H.; Liu, Y.T.; Dong, L.Y. Effects of continuous drought stress on soil respiration in a tropical rainforest in southwest China. *Plant Soil* **2015**, *394*, 343–353. [CrossRef]
47. Bréchet, L.M.; Lopez-Sangil, L.; George, C.; Birkett, A.J.; Baxendale, C.; Trujillo, B.C.; Sayer, E.J. Distinct responses of soil respiration to experimental litter manipulation in temperate woodland and tropical forest. *Ecol. Evol.* **2018**, *8*, 3787–3796. [CrossRef]
48. Liu, L.; Estiarte, M.; Penuelas, J. Soil moisture as the key factor of atmospheric CH_4 uptake in forest soils under environmental change. *Geoderma* **2019**, *355*, 113920. [CrossRef]

9. Dijkstra, F.A.; Morgan, J.A.; von Fischer, J.C.; Follett, R.F. Elevated CO_2 and warming effects on CH_4 uptake in a semiarid grassland below optimum soil moisture. *J. Geophy. Res.* **2011**, *116*, G01007.
0. Kupka, D.; Pan, K.W.; Gruba, P. Initial responses of soil chemical properties to simulated warming in Norway spruce (Picea abies (L.) H.Karst.) stands in the Western Carpathians. *Geoderma* **2023**, *432*, 116400.
1. Li, Y.Q.; Qing, Y.X.; Lyu, M.K.; Chen, S.D.; Yang, Z.J.; Lin, C.F.; Yang, Y.S. Effects of artificial warming on different soil organic carbon and nitrogen pools in a subtropical plantation. *Soil Biol. Biochem.* **2018**, *124*, 161–167. [CrossRef]
2. Wu, H.H.; Xu, X.K.; Fu, P.Q.; Cheng, W.G.; Fu, C.S. Responses of soil WEOM quantity and quality to freeze-thaw and litter manipulation with contrasting soil water content: A laboratory experiment. *Catena* **2021**, *198*, 105058. [CrossRef]
3. Tang, S.R.; Cheng, W.G.; Hu, R.G.; Guigue, J.; Hattori, S.; Tawaraya, K.; Tokida, T.; Fukuoka, M.; Yoshimoto, M.; Sakai, H.; et al. Five-year soil warming changes soil C and N dynamics in a single rice paddy field in Japan. *Sci. Total Environ.* **2021**, *756*, 143845. [CrossRef] [PubMed]
4. Li, C.N.; Tu, B.; Kou, Y.P.; Wang, Y.S.; Li, X.Z.; Wang, J.M.; Li, J.B. The assembly of methanotrophic communities regulated by soil pH in a mountain ecosystem. *Catena* **2020**, *196*, 104883. [CrossRef]
5. Xu, X.; Inubushi, K. Responses of ethylene and methane consumption to temperature and pH in temperate volcanic forest soils. *Eur. J. Soil Sci.* **2009**, *60*, 489–498. [CrossRef]
6. Del Grosso, S.J.; Parton, W.J.; Mosier, A.R.; Ijima, D.S.; Potter, C.S.; Borken, W.; Brunmme, R.; Butterbach-Bahl, K.; Crill, P.M.; Dobbie, K.; et al. General CH_4 oxidation model and comparisons of CH_4 oxidation in natural and managed systems. *Glob. Biogeochem. Cy.* **2000**, *14*, 999–1019. [CrossRef]
7. Zhou, Y.M.; Hagedorn, F.; Zhou, C.L.; Jiang, X.J.; Wang, X.X.; Li, M.H. Experimental warming of a mountain tundra increases soil CO_2 effluxes and enhances CH_4 and N_2O uptake at Changbai Mountain, China. *Sci. Rep.* **2016**, *6*, 21108. [CrossRef]

forests

MDPI

Article

Effects of Snow Cover on Carbon Dioxide Emissions and Their $\delta^{13}C$ Values of Temperate Forest Soils with and without Litter

Xingkai Xu [1,2,*], Haohao Wu [1,3], Jin Yue [1], Shuirong Tang [4] and Weiguo Cheng [5]

1 State Key Laboratory of Atmospheric Boundary Layer Physics and Atmospheric Chemistry, Institute of Atmospheric Physics, Chinese Academy of Sciences, Beijing 100029, China; hhwu@niglas.ac.cn (H.W.); yuejin@mail.iap.ac.cn (J.Y.)

2 Department of Atmospheric Chemistry and Environmental Science, College of Earth and Planetary Sciences, University of Chinese Academy of Sciences, Beijing 100049, China

3 Key Laboratory of Watershed Geographic Sciences, Nanjing Institute of Geography and Limnology, Chinese Academy of Sciences, Nanjing 210008, China

4 College of Tropical Crops, Hainan University, Haikou 570228, China; srtang@hainanu.edu.cn

5 Faculty of Agriculture, Yamagata University, Tsuruoka 997-8555, Japan; cheng@tds1.tr.yamagata-u.ac.jp

* Correspondence: xingkai_xu@mail.iap.ac.cn

Abstract: The presence of litter and winter snow cover can affect the decomposition of organic matter in forest soils and changes in $\delta^{13}C$ values of soil-respired carbon dioxide (CO_2). However, limited information is available on the responses of CO_2 emissions from forest soils and their $\delta^{13}C$ values to snow cover and litter addition over the year. We experimentally manipulated snow cover to study the impacts of light and heavy artificial snow cover on soil heterotrophic respiration and its $\delta^{13}C$ values, using undisturbed large soil columns collected from two typical temperate forests in Northeastern China. Based on the average temperatures of surface forest soils in four seasons of the year in this study region, the simulations of autumn freeze–thaw, winter freeze, spring freeze–thaw, and the growing season were sequentially carried out under laboratory conditions. A set of novel analysis systems, including automated chamber equipment and laser spectroscopy analysis with high-frequency measurements for CO_2 concentrations and the $^{13}C/^{12}C$ isotopic ratios in CO_2, was used to study the effects of artificial snow cover and the presence of litter on soil heterotrophic respiration and its $\delta^{13}C$ values. During the autumn freeze–thaw simulation, there were larger CO_2 emissions and less negative $\delta^{13}C$ values of soil-respired CO_2 upon heavy snow cover than upon light snow cover, indicating that the presence of increased snow cover prior to winter freeze can increase the decomposition of organic C in subsurface soils under temperate forests. The $\delta^{13}C$ values of soil-respired CO_2 in all treatments were, on average, less negative as the simulated spring freeze–thaw proceeded, which was contrary to the variations of the $\delta^{13}C$ during the autumn freeze–thaw simulation. Soil heterotrophic respiration and its $\delta^{13}C$ values during the spring freeze–thaw simulation were, on average smaller upon heavy snow cover than upon light snow cover, which differed from those during the autumn freeze–thaw and growing season simulations respectively. Taken together, the results highlight that the effects of snow cover on soil heterotrophic respiration and its $\delta^{13}C$ values under temperate forests may vary with different seasons of the year and the presence of litter.

Keywords: forest soil carbon cycling; litter manipulation; $\delta^{13}C$ of soil respiration; snow manipulation soil moisture; stable isotope

check for updates

Citation: Xu, X.; Wu, H.; Yue, J.; Tang, S.; Cheng, W. Effects of Snow Cover on Carbon Dioxide Emissions and Their $\delta^{13}C$ Values of Temperate Forest Soils with and without Litter. *Forests* **2023**, *14*, 1384. https://doi.org/10.3390/f14071384

Academic Editor: Choonsig Kim

Received: 6 June 2023
Revised: 4 July 2023
Accepted: 5 July 2023
Published: 6 July 2023

1. Introduction

Permanent and seasonal snow cover regions occupy about 60% of the global land surface [1], which accounts for more than 70% of global soil organic carbon (C) stocks [2,3] Snow cover directly affects the heat exchange between topsoils and the atmosphere, reducing the impacts of soil temperature fluctuations. Thicker surface snow cover (>30 cm) may even prevent soil frost under severe cold conditions, supplying relatively stable living environments for underground microbial activity [4]. However, in the context of global climate change, the area of land snow cover in the Northern Hemisphere in spring is currently decreasing year by year and is predicted to reduce by 25% until the end of this century [5]. It has been confirmed that snow cover in the Northern Hemisphere reduced by 7% in March and 11% in April 2010 compared to that in 1970 and that this reduction is accelerating [6]. Seasonal snow cover patterns can have complex and diverse responses to global climate change, probably affecting above- and belowground processes in forest ecosystems via the redistribution of resources such as light, heat, water, and nutrients [7,8]. As a vital process of soil C cycling, soil respiration can account for more than 60% of forest ecosystem respiration [9,10], and a small change in soil respiration due to climate changes can even give rise to important changes in forest ecosystem respiration. The interaction between snowpack accumulation and increased air temperature affects soil freeze–thaw cycles, which can lead to changes in soil properties and soil CO_2 emissions [11].

In deciduous forests in the cold temperate zone, plant litterfall occurs as a pulse in late autumn and is considered one of the largest annual inputs of dissolved organic matter (DOM) into the soil [12,13], which can, in turn, affect soil CO_2 emissions and $\delta^{13}C$ values of soil-respired CO_2 [14–17]. Furthermore, the variations of winter snowpack and following freeze–thaw variations in non-growing seasons in the cold climate zone can substantially influence the decomposition of litter and the dynamics of DOM released into the soil [11,18–24]. More importantly, there has been a trend of decreasing winter snowpack in cold temperate and boreal zones, especially in East Asia, under the background of global warming [25–27]. These phenomena can, in turn, affect the changes in soil respiration and its $\delta^{13}C$ values in cold temperate forests over the year [15,16,24,28,29]. In recent decades, snow manipulation experiments have been widely carried out under different terrestrial ecosystems around the world, and there are some contrary results regarding the effects of snowpack on soil CO_2 emissions in forest ecosystems [11,28,30,31] due to different soil moisture levels as well as temperature and the duration of soil freeze–thaw events after snow manipulation [28,32]. For example, natural variations in snow cover did not affect annual soil CO_2 emission under a temperate forest in Austria [30], whereas soil CO_2 emission was reduced by snow removal under a temperate forest in northeastern China [28]. Based on a one-year winter snow manipulation experiment under a mid-temperate forest in northeastern China, 50% removal of snowpack did not affect soil respiration and microbial biomass C, whereas a 50% increase of snowpack temporarily enhanced the ratio of fungi to bacteria [31]. A recent meta-analysis showed that snow addition significantly increased soil respiration by 15.5%, whereas snow removal had no effect [33]. The deep continuous snow cover was reported to be closely associated with relatively high soil heterotrophic respiration in winter [34]. Unfortunately, to date, there have been few studies describing the impacts of snow cover and litter on soil heterotrophic respiration and its $\delta^{13}C$ values during non-growing and growing seasons over the year [15,16].

Based on the $\delta^{13}C$ measurements of soil respiration in trenched and non-trenched plots, Moyes et al. showed that seasonal $\delta^{13}C$ variability of approximately 4‰ enrichment in summer versus spring and fall soil respiration was ascribed to the heterotrophic CO_2 source [35]. Under a mature temperate forest in northeastern China, the $\delta^{13}C$ values of soil-respired CO_2 in the trenched plots were, on average, less negative (by 0.6‰) than those in the non-trenched plots in the growing season [36]. Probably, heterotrophic respiration from decomposing soil organic matter (SOM) and litter would affect the changes in $\delta^{13}C$ values of soil-respired CO_2, with relatively large contributions of litter decomposition and surface soil layers to the $\delta^{13}C$ in non-growing seasons [15,16]. Furthermore, aboveground litter inputs have an important impact on soil respiration and soil C cycling in various forest ecosystems over the year, mainly via their physical, chemical, and biological functions on soil properties [17]. However, to date, few studies have specifically documented how snow cover and the presence of litter affect soil CO_2 emissions and their $\delta^{13}C$ values in temperate forests over the year [14–16,36,37], which is indeed associated with evaluating more precisely the responses of forest soil C cycling to winter climate changes.

The Changbai mountain region of northeastern China belongs to a typical seasonal snow-covered area, with continuous surface snow cover lasting from late November to early April each year. Due to its unique geographical location and climatic conditions, the region is currently becoming sensitive to global climate change, which leads to changes in the aboveground snowpack and the quantity and quality of aboveground litter in forest ecosystems [38–40]. This can ultimately affect C cycling in forest soils and C fluxes between soils and the atmosphere in the region. The mature temperate mixed broadleaf and Korean pine forest (BKPF) belongs to the main zonal forest ecosystem in northeastern China, and its aboveground litter composition, root distribution, SOM, and microbial characteristics are quite different from those of nearby secondary white birch forest (WBF) soils [19,38,41,42]. Compared with the BKPF, the stand structure of WBF is simple, and the soil environment under the WBF stand is becoming more sensitive to solar radiation and air temperature due to the relatively low canopy [43], which can result in different snow cover and changes in soil temperature and moisture conditions under the two forest stands in the non-growing season. Probably, the above-mentioned factors can remarkably lead to different responses of soil respiration and its $\delta^{13}C$ values to changes in snow cover and litter decomposition between the two forests over the year.

We propose that winter snow cover can affect soil microbial respiration and its $\delta^{13}C$ during different seasons of the year, which would vary with forest stands and the presence of litter. To reduce the effects of other environmental factors as much as possible, laboratory-controlled incubation has been considered an effective method to study the interaction effects of snow cover and the presence of litter on soil CO_2 emissions and their $\delta^{13}C$ values. Based on the long-term laboratory-controlled incubation of undisturbed large soil columns collected from BKPF and WBF stands, the objectives of this study were to study the effects of snow cover and the presence of litter on soil heterotrophic respiration and its $\delta^{13}C$ values of typical temperate forests in the Changbai mountain region in northeastern China. The results would provide a nice basis for understanding the responses of soil C cycling in temperate forests to global winter climate change.

2. Materials and Methods

2.1. Site Description and Collection of Soil Columns

The study area is in the temperate forest stands near the National Research Station of Changbai Mountain Forest Ecosystems (42°24′ N and128°6′ E) (Figure S1), with flat terrain and an average altitude of 738 m. The Changbai mountain region is located on the northeastern edge of the global monsoon climate and is influenced by the continental climate, which is characterized by long cold winters and short summers. A basic message regarding the thickness of ground snowpack, daily average air temperatures in winter, and the status of soil freezing was given by Xu et al. [29]. Within the climate region, the mature temperate BKPF belongs to the main zonal forest ecosystem in northeastern China, and its aboveground litter composition, SOM, and soil microbial characteristics are quite different from those of nearby secondary WBF [19,29]. Dominant trees of the BKPF stand include Tuan linden (*Tilia amurensis* Rupr.), Mongolian oak (*Quercus mongolica* Fisch. ex Turcz), Manchurian ash (*Fraxinus mandshurica* Rupr.) and Korean pine (*Pinus koraiensis* Sieb. et Zucc.), mostly >200 years old, and their stand densities are 117.1, 37.1, 27.2 and 98.9 trees ha^{-1}, respectively [44]. The WBF stand is dominated by white birch (*Betula platyphylla* Suk.) and mountain poplar (*Populus davidiana* Dode), mostly >70 years old, and its stand density is 1402 stem ha^{-1}; both white birch and mountain polar account for approximately 51.5% and 23.1% basal area of tree species, respectively [43]. Due to the differences in vegetation cover and tree species phototaxis, both air temperature and understory PAR transmittances under WBF stand in the non-growing season are higher than those under BKPF stand [43], resulting in the relative rapid melting of surface snowpack each year. The temperate forest soil belongs to an Andosol, containing a 3–5 cm depth of organic layer and approximately 10 cm depth of A horizon. Main soil properties, such as pH, texture, SOM, and bulk density, as well as the properties of aboveground litter, were described by Xu et al. [45,46] and Wu et al. [19,42].

We established 12 2 m × 2 m small plots with the well-preserved aboveground litter layers under BKPF, and WBF stands each in October 2017, prior to the occurrence of the autumn freeze–thaw process, for collecting large soil columns at approximately 15 cm depth under the two forests. At the center of each small plot, a stainless-steel cylindrical grinding tool (20 cm in diameter) was firstly smashed in the soil at approximately 15 cm depth and then replaced by thickened UPVC tubes (20 cm in diameter and 20 cm in height), and the soil column was finally taken out as one undisturbed soil column with aboveground litter. All forest soil column samples were sealed by covers and transported to the laboratory immediately, and they were stored in the dark at 4 °C prior to laboratory incubation.

2.2. Setup of Incubation Experiments

Half of all undisturbed soil columns (n = 6) under BKPF and WBF stands each removed the litter layer, and the other half retained the litter layer (n = 6). Referring to the average monthly precipitation in winter in this study area from 2005 to 2015, two levels of snow cover were considered as light and heavy snow cover treatments, respectively (Table 1). Throughout the incubation, all soil columns with and without litter layers were covered four times with artificial snow made by the snowflake ice machine (FM50, Beijing Yijialin Technology Ltd., Beijing, China), and the scheme of artificial snow addition was shown in Table 1. The total amount of snow addition in the heavy snow cover treatment accounted for approximately 40% of the average annual snowfall to avoid water and nutrient leaching inside soil columns, which was twice the total amount of snow addition in the light snow cover treatment. Based on the average distribution of annual snowfall in autumn, winter, and spring in this study area, the total amount of snow cover in autumn and spring was assumed to be the same as in winter, and snow was added evenly twice in autumn. Based on the variations of soil temperature and air temperature in the field over the year (Figure S2), long-term incubation experiments were conducted by using cryogenic incubators (LRH-250CA, Shanghai Yiheng, Shanghai, China) to reasonably set up the changes in simulated temperature levels at four different seasons of the year (Table 2). During the whole incubation, the outer surface of all UPVC pipes containing soil columns was wrapped with insulation materials to simulate freeze–thaw fluctuations in the field as closely as possible.

Table 1. Amount and time of snow addition at the different stages of the incubation.

Levels of Snow Cover	Simulation Stages		
	Autumn Freeze–Thaw	Winter Freeze	Spring Freeze–Thaw
Light snow cover	35 g and 35 g on the first and 8th days of the incubation, respectively	160 g on the 15th day of the incubation	70 g on the 205th day of the incubation
Heavy snow cover	150 g and 150 g on the first and 8th days of the incubation, respectively	600 g on the 15th day of the incubation	300 g on the 205th day of the incubation

Table 2. Temperature setting at the four stages of the incubation.

Simulation Stage	Autumn Freeze–Thaw		Winter Freeze	Spring Freeze–Thaw		Growing Season
Duration (day)	7	7	180	7	7	61
Daily freeze–thaw alternation time (hour)	12/12		24 [(a)]	12/12		24 [(a)]
Temperature setting (°C)	−5/10	−10/5	−10	−10/5	−5/10	15

[(a)] 24 h each day at a constant temperature.

2.3. Measurements of CO_2 Emissions from Soil Columns and Their $\delta^{13}C$ Values

During the whole incubation, a CO_2 isotope spectrum analyzer (CCIA-38d-EP, Los Research Inc., Fremont, CA, USA) along with its supporting fully automated soil respiration analysis system (SF-3000, Beijing LICA United Technology Limited, Beijing, China) was nicely combined with cryogenic incubators (LRH-250CA, Shanghai Yiheng, Shanghai China), to determine CO_2 emissions from large soil columns and the $^{13}C/^{12}C$ isotope ratios of soil-respired CO_2 in the chamber headspace (Figure S3). The sampling frequency was 1 Hz for the measurement of CO_2 concentration and its $^{13}C/^{12}C$ ratio in the headspace, and the output results of the stable isotopic ratios were based on Pee Dee Belemnite (PDB) as the standard, expressed by $\delta^{13}C$:

$$\delta^{13}C\ (‰) = (R_{sample}/R_{standard} - 1) \times 100 \quad (1)$$

where R_{sample} and $R_{standard}$ represent the molar ratios of sample and standard $^{13}C/^{12}C$, respectively.

Both CO_2 concentrations and the $^{13}C/^{12}C$ isotope ratios in CO_2 were determined using the $\delta^{13}C$-CO_2 automatic analysis system (Figure S3), with a precision of 0.05 μmol mol^{-1} and 0.1‰ for a 5-min integration time, respectively. The $\delta^{13}C$ values of CO_2 produced via microbial respiration in soil columns were calculated based on the Keeling plot approach [47,48], namely using linear regressions of the $\delta^{13}C$ values of soil-respired CO_2 against the reciprocal of headspace CO_2 concentrations at second scales during the closure ranging from 90 to 200 s as mentioned afterward. The closure time as short as possible (e.g., within several minutes) can minimize the disturbance of the soil-atmosphere system and concentration dependence of $\delta^{13}C$ measurement of headspace CO_2 with high precision [49–51]. The effect of water vapor on the $\delta^{13}C$ measurement was nicely eliminated using a dehumidification device attached to the CO_2 isotope spectrum analyzer (CCIA-38d-EP, Los Research Inc., Fremont, CA, USA). Therefore, the $\delta^{13}C$-CO_2 automatic analysis system with high-frequency measurements can be nicely combined with laboratory incubation experiments to investigate the changes in soil CO_2 emissions and their $\delta^{13}C$ values using large soil columns (diameter = 20 cm) (Figure S3). Prior to each assay, the CO_2 isotope analyzer was calibrated using two standard CO_2 mixture gases containing two different CO_2 concentrations (CO_2 concentrations and its $\delta^{13}C$ values: 398.60 μL L^{-1} and −22.31‰; 794.30 μL L^{-1} and −22.34‰), 1% Argon (w/w), and balance gas of air. The standard CO_2 mixture gases were supplied by Air Products (Beijing Helium Pu North Branch Gas Industry Ltd., Beijing, China), and their $\delta^{13}C$ and $\delta^{18}O$ values were, respectively measured with a mass spectrometer combined with a Trace Gas System (Isoprime-100, Elementar Ltd., Langenselbold, Germany), using reference materials NBS 19 and IAEA-CO-8 by the Environmental Stable Isotope Laboratory, Chinese Academy of Agricultural Sciences. During the whole incubation, CO_2 emissions from soil columns in all treatments were also determined using a portable greenhouse gas analyzer (915-0011, Los Research Inc. Fremont, CA, USA) nicely coupled by a smart respiratory chamber (SC-11, Beijing LICA United Technology Limited, Beijing, China) (Figure S3). Due to sharp fluctuations of soil CO_2 emissions upon the simulated freeze–thaw processes, the frequency of CO_2 emission determination was six times each day during the periods of simulated autumn and spring freeze–thaw, respectively (Table 2). Within the period of winter freeze simulation (Table 2) soil CO_2 emissions in all treatments were measured only once every 60 days. During the period of growing season simulation (Table 2), soil CO_2 emissions in all treatments were measured initially five times each day, and then measured once every 5–7 days as it became gradually stable. Upon the determination of CO_2 emissions, the duration of the closure varied from 90 to 200 s to ensure the measurement accuracy as much as possible which was dependent on the size of soil microbial respiration at the different stages of the incubation. Throughout the incubation, the method of weighing and adding water was used at intervals to compensate for the evaporation loss of water in soil columns. During the non-freezing period of soil columns, soil volumetric water contents (v/v, %) at 5 cm depth inside soil columns in all treatments were determined at regular intervals using a portable MPkit (ICT, Sydney, NSW, Australia).

2.4. Statistical Analysis

We calculated CO_2 emission per unit area (m^{-2}) merely concerning the surface area of soil columns. The $\delta^{13}C$ values of soil-respired CO_2 in all treatments during the whole incubation were calculated according to the method of the Keeling plot [47,48]. Based on the two methods mentioned previously for measuring soil CO_2 emission (Figure S3), the results were almost the same and averaged at each sampling. Means and standard errors were calculated for selected variables at each sampling of the incubation based on three replicates. All variables were examined for homogeneity of variance using Levene's test and normality using Shapiro–Wilk test and log-transformed as needed. The soil CO_2 emissions and $\delta^{13}C$ values of soil-respired CO_2 in all treatments at each stage of the incubation were, on average, described with box plots using OriginPro 2021 (OriginLab Corporation, Northampton, MA, USA). Three-factor repeated analysis of variance (ANOVA) was used with forest type, snow cover, and the presence of litter to assess individual and interactive effects on cumulative soil CO_2 emissions at each stage of the incubation and throughout the incubation. Two-factor repeated ANOVA was used with snow cover and the presence of litter to assess individual and interactive effects on the average soil CO_2 emissions and average $\delta^{13}C$ values of soil-respired CO_2 at each stage of the incubation. When ANOVA was significant, pairwise comparisons of means were examined using Duncan's multiple-range test. The differences in soil volumetric water contents across all treatments, average soil CO_2 emissions, and average $\delta^{13}C$ values of soil-respired CO_2 between the first and second week of the simulated autumn and spring freeze–thaw simulations (Table 2) were, respectively, assessed at a significant level of $a = 0.05$ using Student's test. Least significant differences (LSDs) were calculated at a significant level of $a = 0.05$ to assess the differences between light and heavy snow cover-treated soil columns in soil CO_2 emissions and $\delta^{13}C$ values in CO_2 at each sampling of the incubation. The relationships between soil CO_2 emissions, $\delta^{13}C$ values of soil-respired CO_2, and soil moisture contents at 5 cm depth across soil columns were fitted with nonlinear and linear regressions. All statistical analyses were conducted using SPSS for Windows software (version 19.0) (IBM Corp., New York, NY, USA).

3. Results

3.1. Changes in Soil Moisture at the Different Stages of the Incubation

Prior to the incubation, volumetric water contents at 5 cm depth in all BKPF soil columns were, on average, 26.0 ± 1.0%, which was much larger than in all WBF soil columns (16.0 ± 1.0%). However, the spatial heterogeneity in initial volumetric water contents across all soil columns under each forest stand was negligible. Following the first addition of artificial snow, volumetric water contents of BKPF and WBF soil columns at 5 cm depth were, on average, increased by 29.0 ± 1.0% and 17.0 ± 1.0%, respectively (Figure 1), which were slightly larger than the background values as mentioned previously. Furthermore, following the second addition of artificial snow during the autumn freeze–thaw simulation, volumetric water contents of BKPF and WBF soil columns at 5 cm depth would increase substantially in the heavy snow cover treatment, especially in the absence of litter, and were mostly significantly larger than in the light snow cover treatment ($p < 0.05$) (Figure 1). Afterward, all soil columns treated with two different snow cover levels remained to be frozen until the second week of the spring freeze–thaw simulation. During the early spring freeze–thaw simulation, the task of the fourth snow addition was completed (Table 1 and Figure 2), and the following soil moisture did not change much. Therefore, soil volumetric water contents measured during the simulated growing season were averaged to represent the final volumetric water contents of soil columns in all treatments. During the simulated growing season, volumetric water contents of BKPF and WBF soil columns at 5 cm depth were, on average, 65.0 ± 0.5% and 56.0 ± 1.0% in the heavy snow cover treatment, and 40.0 ± 1.0% and 29.0 ± 1.0% in the light snow cover treatment, respectively (Figure 1).

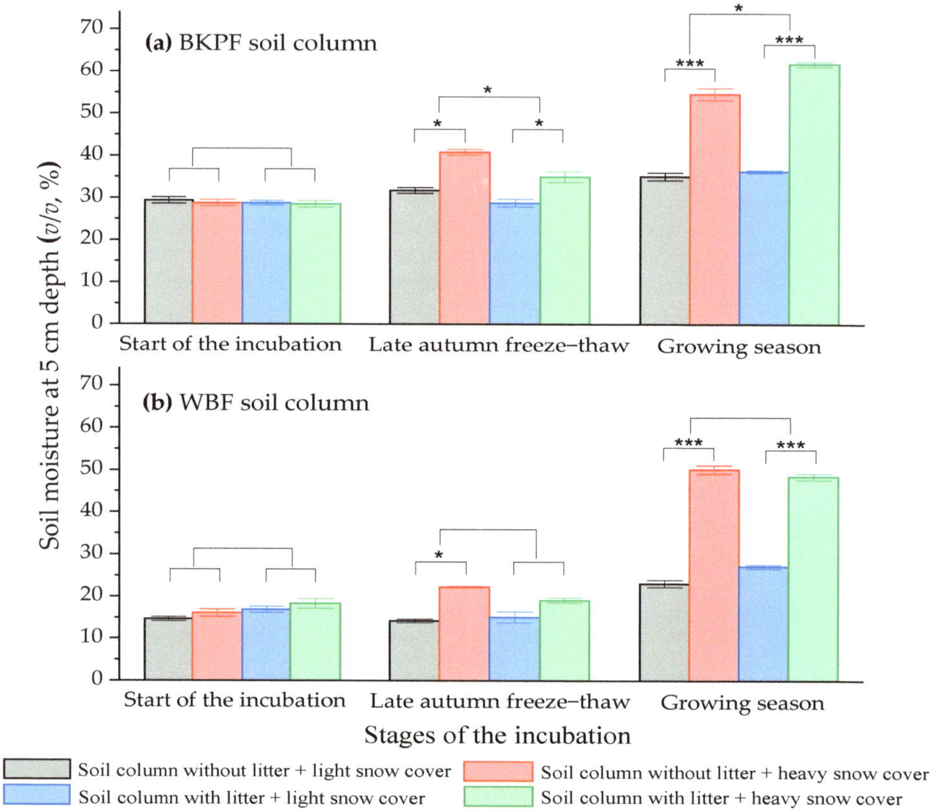

Figure 1. Changes in soil moisture at 5 cm depth in large forest soil columns during the different stages of the incubation. Data are shown as means ± 1 × standard errors (n = 3). *, $p < 0.05$; ***, $p < 0.001$.

3.2. Effect of Snow Cover and the Presence of Litter on CO_2 Emissions from Soil Columns

Prior to the incubation, CO_2 emissions from BKPF and WBF soil columns (n = 12, each forest type) at 15 °C were, on average, 1.48 ± 0.06 and 1.52 ± 0.09 µmol CO_2 m^{-2} s^{-1}, respectively. However, the soil CO_2 emissions across all soil columns varied little between the two forest stands and within a single stand. The results suggested that the spatial variability of background conditions was negligible in the subsequent analysis of soil CO_2 emissions in all treatments. During the autumn freeze–thaw simulation, CO_2 emissions from BKPF and WBF soil columns in all treatments showed a zigzag fluctuation with the freeze–thaw alterations (Figure 2), and with the decrease of freeze and thaw temperatures, the emissions had a decreasing trend ($p < 0.001$) (Figures 2 and 3). The increased snow cover could increase the soil CO_2 emissions, especially at the early stage of the autumn freeze–thaw simulation (Figures 2 and 3). This phenomenon resulted in relatively larger cumulative CO_2 emissions from soil columns with and without litter during the simulated autumn freeze–thaw, compared with those from the light snow cover treatment ($p < 0.05$) (Table 3). The results of ANOVA showed that the presence of litter and increased snow cover could significantly increase soil CO_2 emissions, but there were no significant differences between the two forest stands and no interaction effects of snow cover and the presence of litter on the soil CO_2 emissions (Table 3).

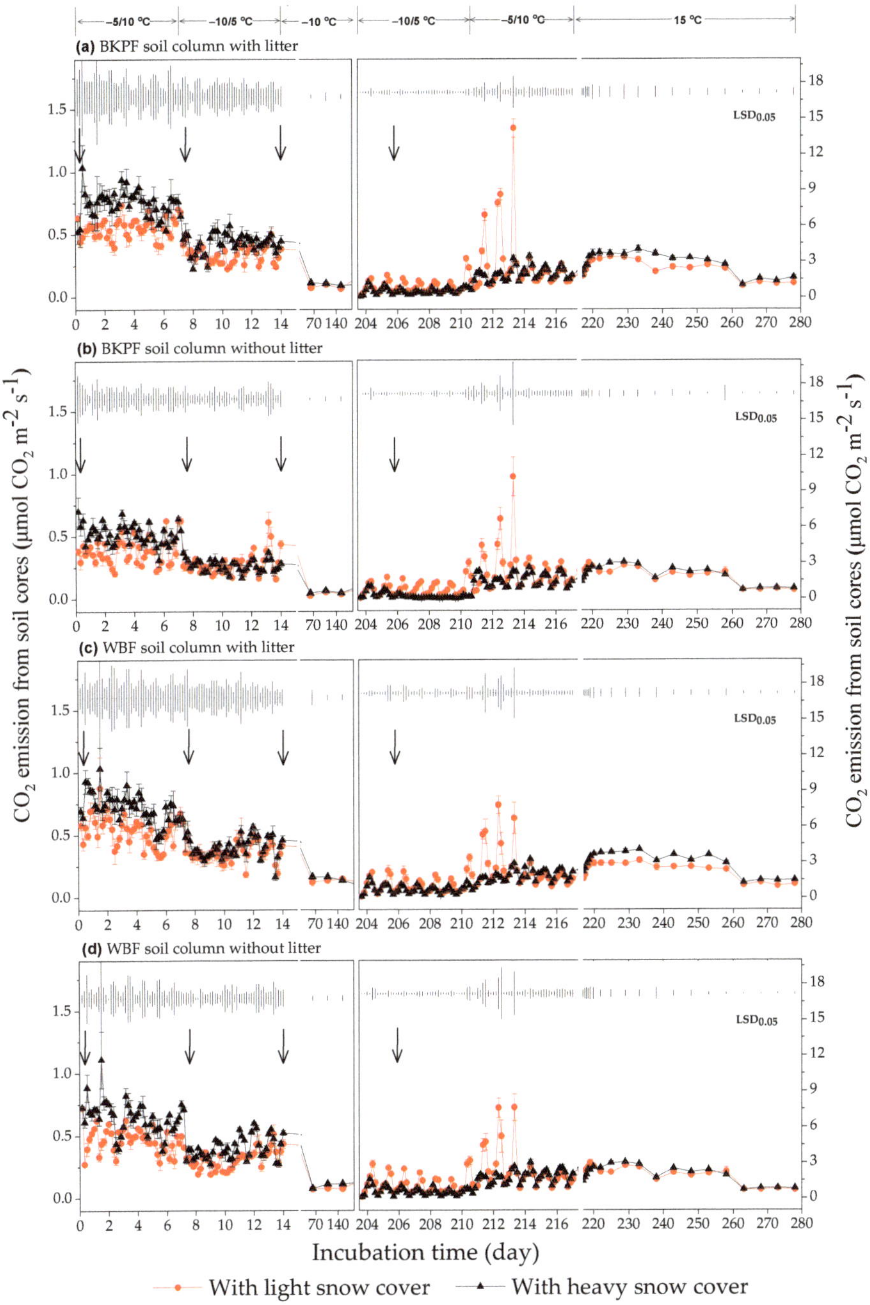

Figure 2. Dynamics of CO_2 emissions from snow-covered large forest soil columns with and without aboveground litter during the different stages of the incubation. Data are shown as means ± 1 × standard errors (n = 3). Vertical bars show the least significant difference at the a = 0.05 level ($LSD_{0.05}$) between light and heavy snow-covered forest soil columns at each sampling time. The arrows represent the time of snow addition during the simulated autumn freeze–thaw, winter freeze, and spring freeze–thaw periods, respectively.

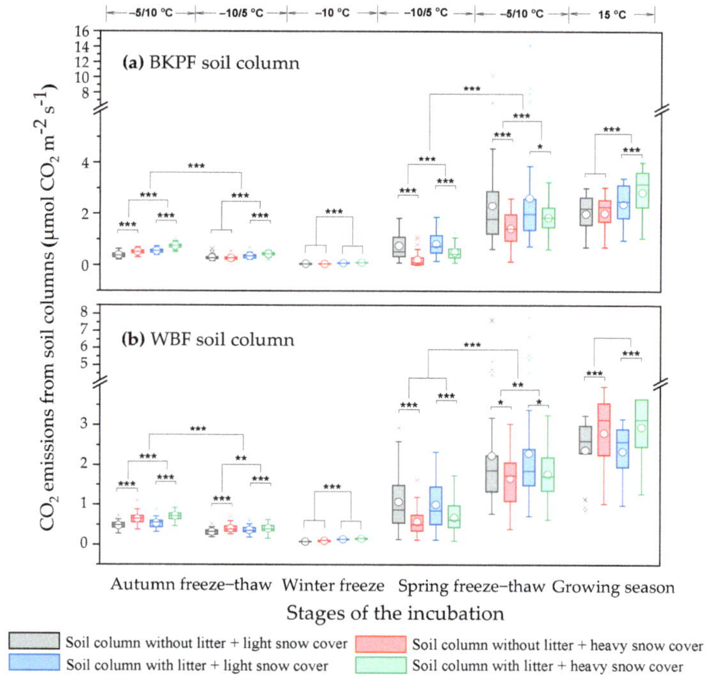

Figure 3. Box plots of the average CO_2 emissions from snow-covered large forest soil columns with and without aboveground litter during the different stages of the incubation. Boxes represent interquartile ranges (IQRs), and horizontal lines and circles within boxes indicate median and mean values, respectively. Lower and upper whiskers (x) represent 25 percentiles minus 1.5 IQR and 75 percentiles plus 1.5 IQR, respectively. *, $p < 0.05$; **, $p < 0.01$; ***, $p < 0.001$.

Table 3. Cumulative CO_2 emissions from forest soil columns during the different stages of the incubation and the results of ANOVA.

Forest Type (FT)	Litter Cover (L)	Levels of Snow Cover (S)	Simulation Stage				
			Autumn Freeze–Thaw	Winter Freeze	Spring Freeze–Thaw	Growing Season	Total
			($g\ CO_2$-$C\ m^{-2}$)				
BKPF	Without	Light	4.84 ± 0.11	8.43 ± 0.24	22.15 ± 1.55	128.67 ± 3.77	164.08 ± 5.21
		Heavy	5.74 ± 0.26	10.44 ± 0.77	11.98 ± 0.49	139.46 ± 2.41	167.62 ± 2.30
	With	Light	6.51 ± 0.31	16.02 ± 0.47	25.08 ± 0.30	161.92 ± 2.08	209.53 ± 2.34
		Heavy	8.58 ± 0.79	20.42 ± 1.68	17.13 ± 1.19	197.17 ± 13.47	243.30 ± 16.82
WBF	Without	Light	5.71 ± 0.11	15.03 ± 0.73	24.01 ± 1.51	154.06 ± 3.69	198.81 ± 5.59
		Heavy	7.67 ± 0.14	19.73 ± 0.81	16.30 ± 0.47	182.48 ± 2.67	226.17 ± 3.69
	With	Light	6.55 ± 0.65	27.04 ± 1.65	23.99 ± 2.05	157.92 ± 9.07	215.51 ± 12.40
		Heavy	8.20 ± 0.54	30.57 ± 1.77	17.83 ± 0.39	203.71 ± 3.14	260.31 ± 5.76
Three−way factorial ANOVA with FT, L, and S as fixed factors (*p*-value)							
	FT		0.338	<0.001	0.396	0.064	0.025
	L		0.032	<0.001	0.167	0.005	0.002
	S		0.018	0.040	<0.001	0.004	0.033
	FT × L		0.227	0.434	0.337	0.084	0.154
	FT × S		0.801	0.786	0.531	0.442	0.469
	L × S		0.733	0.856	0.576	0.259	0.326
	FT × L × S		0.564	0.594	0.922	0.845	0.789

During the simulated winter freeze period, CO_2 emissions from BKPF and WBF soil columns in all treatments remained at a low level, with a range from 0.03 to 0.17 μmol $CO_2\ m^{-2}\ s^{-1}$ (Figures 2 and 3). Based on the results of ANOVA, cumulative soil CO_2 emissions during the winter freeze simulation were significantly affected by forest stands, the presence of litter, and snow cover (Table 3). During the spring freeze–thaw simulation, CO_2 emissions from BKPF and WBF soil columns in all treatments showed a zigzag fluctuation with the soil freeze–thaw alterations (Figure 2), and with the increase of freeze and thaw temperatures, the emissions increased substantially, especially under light snow cover conditions ($p < 0.001$) (Figures 2 and 3). In contrast to that during the autumn freeze–thaw simulation, the increased snow cover significantly reduced CO_2 emissions from BKPF and WBF soil columns during the spring freeze–thaw simulation (Figures 2 and 3). The decrease in the increased snow cover-induced cumulative CO_2 emissions appeared larger in BKPF soil columns than in WBF soil columns, but the difference was not significant (Table 3). During the late period of the spring freeze–thaw simulation, there were several flush emissions of CO_2 from soil columns treated with the light snow cover, regardless of forest stands and the presence of litter (Figure 2).

During the simulated growing season, soil columns treated with heavy snow cover completely melted, and cumulative soil CO_2 emissions were significantly larger than those upon light snow cover, especially in the presence of litter ($p < 0.01$) (Table 3). Throughout the incubation, there were relatively large increases in the increased snow cover-induced cumulative CO_2 emissions from WBF soil columns, and the emissions were significantly affected by forest stands, snow cover, and the presence of litter (Table 3). In the absence of litter, cumulative CO_2 emissions from WBF soil columns treated with two different snow cover levels became larger than those from BKPF soil columns at the different stages of the incubation and throughout the incubation, especially during the winter freeze and growing season simulations ($p < 0.05$). However, regardless of snow cover levels, there were no significant differences in the cumulative soil CO_2 emissions in the presence of litter between BKPF and WBF stands, except during the winter freeze simulation (Table 3). Taken together, the results indicated that the impacts of snow cover and the presence of litter on soil heterotrophic respiration in temperate forests would vary with forest types and different seasons of the year.

3.3. Effect of Snow Cover and the Presence of Litter on the $\delta^{13}C$ Values of Soil-Respired CO_2

During the autumn freeze–thaw simulation, the $\delta^{13}C$ values of soil-respired CO_2 fluctuated with the soil freeze–thaw alterations (Figure 4), and with the decrease of freeze and thaw temperatures, $\delta^{13}C$ values of soil-respired CO_2 in all treatments decreased on average substantially ($p < 0.001$) (Figure 5). Furthermore, regardless of the presence of litter, the $\delta^{13}C$ values of soil-respired CO_2 in the heavy snow cover treatment became, on average, less negative than those in the light snow cover treatment ($p < 0.05$), especially at the early stage of the autumn freeze–thaw simulation ($p < 0.001$) (Figure 5). However, the effect of the presence of litter on the average $\delta^{13}C$ values of soil-respired CO_2 did not occur during the autumn freeze–thaw simulation (Figure 5). Together with the finding that the heavy snow cover treatment can significantly increase soil heterotrophic respiration (Figure 3), the results indicate that the increased snow cover may remarkably promote the decomposition of organic C in subsurface soils under temperate forest stands during the autumn freeze–thaw period under field conditions, regardless of the presence of litter.

In contrast to those during the autumn freeze–thaw simulation, $\delta^{13}C$ values of soil-respired CO_2 in all treatments during the spring freeze–thaw simulation appeared to have a V-shaped fluctuating trend with the freeze–thaw alterations (Figure 4), and with the increase of freeze and thaw temperatures, there were on average less negative $\delta^{13}C$ values of soil-respired CO_2 in all treatments ($p < 0.001$) (Figure 5). Furthermore, under heavy snow cover conditions, the $\delta^{13}C$ values of the CO_2 released from BKPF and WBF soil columns were, on average more negative, especially at the early stage of the simulated spring freeze–thaw ($p < 0.01$) (Figure 5). This indicates that the presence of increased snow cover in winter may accelerate the decomposition of organic C in surface mineral soils and/or litter layers under temperate forests in the field during the spring freeze–thaw period. During the simulated growing season, $\delta^{13}C$ values of soil-respired CO_2 in all treatments increased gradually, and then remained relatively less negative until the end of the incubation, compared with those at the other stages of the incubation (Figure 4). On average, the increased snow cover could result in less negative $\delta^{13}C$ of CO_2 released from BKPF and WBF soil columns during the simulated growing season ($p < 0.05$), which was the same as shown during the simulated autumn freeze–thaw (Figure 5). Throughout the incubation, there was a shift of about 4.2‰ ^{13}C-enrichment on average in the simulated growing season versus non-growing season soil respiration measured on all soil columns (Figure 5).

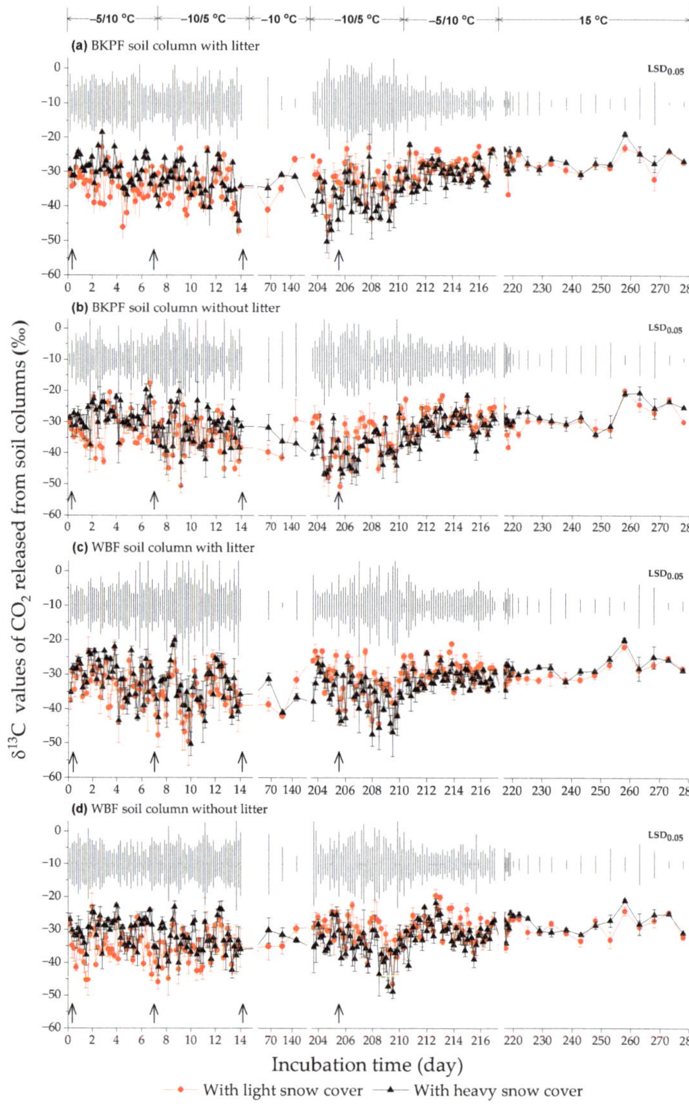

Figure 4. Dynamics of the $\delta^{13}C$ values of CO_2 released from snow-covered large forest soil columns with and without aboveground litter during the different stages of the incubation. Data are shown as means $\pm$ 1 $\times$ standard errors (n = 3). Vertical bars show the least significant difference at the a = 0.05 level ($LSD_{0.05}$) between light and heavy snow-covered forest soil columns at each sampling time. The arrows represent the time of snow addition during the simulated autumn freeze–thaw, winter freeze, and spring freeze–thaw periods, respectively.

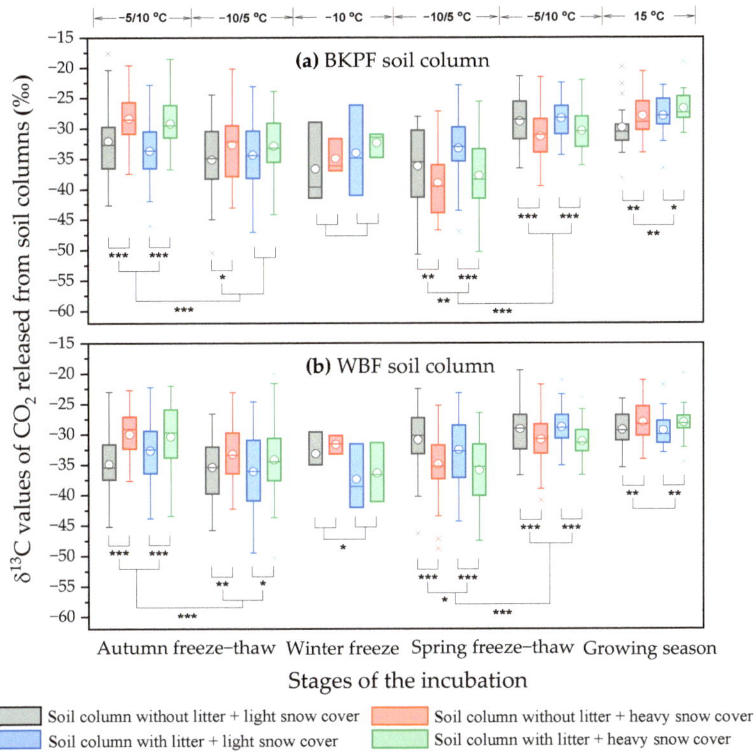

Figure 5. Box plots of the average $\delta^{13}C$ values of CO_2 released from snow-covered large forest soil columns with and without aboveground litter during the different stages of the incubation. (**a**) BKPF soil column; (**b**) WBF soil column. *, $p < 0.05$; **, $p < 0.01$; ***, $p < 0.001$. A description of box plots is provided in the caption of Figure 3.

With the development of litter decomposition, the effect of litter on the $\delta^{13}C$ values of soil-respired CO_2 would vary with forest stands and different stages of the incubation. The average $\delta^{13}C$ values of soil-respired CO_2 in WBF soil columns were significantly affected by the presence of litter during the winter freeze simulation and at the early stage of the spring freeze–thaw simulation, respectively (Figure 5b). However, the $\delta^{13}C$ values of the CO_2 emitted from BKPF soil columns were, on average, significantly affected by the presence of litter at the early stage of the spring freeze–thaw simulation and during the simulated growing season, respectively (Figure 5a).

3.4. Relationships among Soil CO_2 Emissions, $\delta^{13}C$ Values of Soil-Respired CO_2, and Soil Moisture

Regardless of snow cover and the presence of litter, there was a nice nonlinear relationship between soil CO_2 emissions and $\delta^{13}C$ values of soil-respired CO_2 throughout the incubation (Figure 6). Under experimental conditions, the $\delta^{13}C$ values of soil-respired CO_2 had a relatively large variability when the soil CO_2 emissions were less than 1.0 mmol CO_2 m^{-2} s^{-1} (Figure 6). The addition of different snow covers would result in a large range of volumetric water contents in BKPF and WBF soil columns at 5 cm depth during the different stages of the incubation (Figure 1). During the whole incubation, CO_2 emissions from soil columns and their corresponding $\delta^{13}C$ values increased with the soil moisture within the range from 14.2% to 65.2% (v/v), regardless of forest types and the presence of litter (Figure 7). Under experimental conditions, soil moisture at 5 cm depth could explain the 12% and 33% variability of the $\delta^{13}C$ values of soil-respired CO_2 in the absence and presence of litter, respectively (Figure 7b).

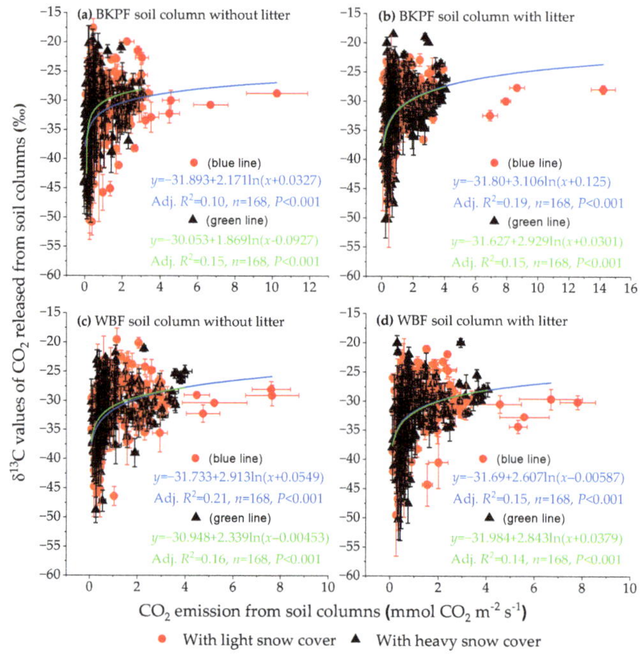

Figure 6. Relationships between CO_2 emissions from snow-covered large forest soil columns and their corresponding $\delta^{13}C$ values in the presence and absence of litter during the whole incubation. Data are shown as means ± 1 × standard errors (n = 3). The relationships between soil CO_2 emissions and their corresponding $\delta^{13}C$ values are fitted with nonlinear regressions. The blue and green lines in (**a–d**) are nonlinear regression fitting lines with light and heavy snow cover, respectively.

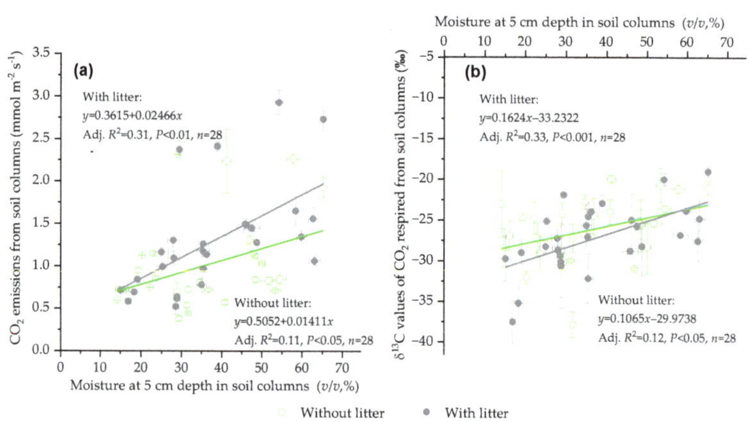

Figure 7. Relationships between CO_2 emissions from snow-covered large forest soil columns with and without aboveground litter (**a**) and their corresponding $\delta^{13}C$ values (**b**) during the whole incubation versus soil moisture at 5 cm depth in soil columns. Data are shown as means ± 1 × standard errors (n = 3). The relationships between soil CO_2 emissions and their corresponding $\delta^{13}C$ values against the soil moisture in the presence and absence of litter were fitted with linear regressions, respectively.

4. Discussion

4.1. Effects of Snow Cover on Soil CO_2 Emissions Vary with the Presence of Litter and Forest Stands

As the simulated autumn freeze–thaw proceeded, CO_2 emission rates of BKPF and WBF soil columns decreased on average substantially due to a decrease in freeze-and-thaw temperatures (Figure 3), and cumulative soil CO_2 emissions significantly varied with levels of snow cover and the presence of litter (Table 3). During the autumn freeze–thaw simulation, heavy snow cover led to an increase in volumetric soil water contents due to the initial melting of artificial snow (Figure 1) and the incomplete freezing status of soil columns prior to the winter freeze. This would increase CO_2 emissions from soil columns upon the addition of heavy snow cover, especially during the early autumn freeze–thaw simulation (Figures 2 and 3), hence enhancing cumulative soil CO_2 emissions (Table 3). Additionally, changes in freeze and thaw temperatures and following microbial activity, as well as the supply of labile dissolved organic C (DOC) (e.g., carbohydrates) initially released into the soil upon melting of artificial snow, could promote soil CO_2 emissions during the early autumn freeze–thaw simulation (Figure 3). The soil labile C pools released initially upon artificial snow cover can be partly supported by the results reported by Zhu et al. [52], who documented that the release of only O-alkyl C (e.g., carbohydrates) from fir and birch litter was increased by the presence of snow cover.

Based on the differences between CO_2 emissions from soil columns without and with litter, litter-derived CO_2 emissions following two levels of snow cover were coarsely calculated. During the autumn freeze–thaw simulation, the litter-derived CO_2 emissions from BKPF soil columns became, on average greater than those from WBF soil columns, especially upon heavy snow cover (Table 3). In this study area, the BKPF stand lies in a climax community of forest succession, with a greater SOM content and lower bulk density than the adjacent secondary WBF stand [19]. Furthermore, the thickness of litter layers under the BKPF stand is normally greater than that under the WBF stand due to the relatively slow degradable needles of Pinus koraiensis trees and different litter leaf accumulations of tree species [38,40]. The different properties of litter layers under the two forest stands and their impacts on the soils upon initial melting of artificial snow would explain the relatively great litter-induced CO_2 emissions from BKPF soil columns during the autumn freeze–thaw simulation (Table 3). The relatively high soil C availability (e.g., the microbial biomass C divided by soil organic C) under WBF stand [19] would explain larger cumulative CO_2 emissions from WBF soil columns without litter and with two levels of snow cover during the different stages of the incubation and throughout the incubation, compared with those from BKPF soil columns (Table 3). However, except during the winter freeze simulation, no significant differences in the cumulative CO_2 emissions were observed in the two forest soil columns with litter. In the presence of litter, the relatively great cumulative CO_2 emissions from WBF soil columns during the winter freeze simulation (Table 3) were partially attributed to the relatively great DOC concentrations and microbial degradable C pools (e.g., glucose C) in the water extracts of organic layers under WBF stand in autumn than under BKPF stand [42].

On the contrary to that, during the autumn freeze–thaw simulation, the reduced snow cover during the spring freeze–thaw simulation could increase soil CO_2 emission (Figure 2). This can result from more freeze–thaw cycles and greater soil temperature fluctuations under relatively low soil moisture conditions [53]. Following long-term winter freeze simulation, frequent soil freeze–thaw cycles could increase the contents of DOC released into the soil at thaw [22,24,29], especially at the late stage of the spring freeze–thaw simulation (Figure 2). The melting of artificial spring snow and following increased soil moisture would promote the activity of soil microorganisms and an associated flush of nutrients [22,54–57], which can, in turn, increase the soil CO_2 emissions during the spring freeze–thaw simulation, especially when the soil water content is below optimum [24,58]. This can explain several substantial flush CO_2 emissions from soil columns treated with light snow cover during the late spring freeze–thaw simulation, regardless of the presence of litter and forest stands (Figure 2). As the artificial snow cover increased, the complete melting of soil columns was detarded with the simulated spring freeze–thaw fluctuations These phenomena resulted in relatively small cumulative CO_2 emissions from heavy snow cover-treated soil columns during the spring freeze–thaw simulation, compared with those from light snow cover-treated soil columns (Table 3). Since the incubation entered the simulated growing season, soil columns treated with heavy snow cover showed relatively larger CO_2 emissions compared with soil columns treated with light snow cover, especially in the presence of litter (Figures 2 and 3, Table 3). Furthermore, the increased snow cover-induced cumulative CO_2 emissions from WBF soil columns with and without litter were larger than those from BKPF soil columns during the simulated growing season (Table 3) Taken together, the results indicate that the impacts of snow cover on CO_2 emissions from forest soils can vary with forest stands and the presence of aboveground litter as well as different seasons of the year.

4.2. Changes in the $\delta^{13}C$ Values of Soil-Respired CO_2 and Influencing Mechanisms

The $\delta^{13}C$ values of soil-respired CO_2 in all treatments became, on average, less negative during the first week than during the second week of the autumn freeze–thaw simulation (Figure 5). This was probably ascribed to an initial supply of DOC (e.g., carbohydrates) containing less depleted ^{13}C in the soils upon the melting of artificial snow relative to the $\delta^{13}C$ of soil organic C [59]. At the start of the simulated autumn freeze–thaw, there was an increase in labile C pools (e.g., carbohydrates) released into the soil [19,21], especially under high soil moisture conditions as caused by the initial melting of heavy artificial snow. The selective degradation of carbohydrates by microorganisms in our microcosm incubation was thus most likely to explain relatively less negative $\delta^{13}C$ of the CO_2 respired initially during the autumn freeze–thaw simulation [60] (Figure 5). While the autumn freeze–thaw simulation proceeded, labile, isotopically heavier C compounds (e.g., carbohydrates) in soils could be preferentially and rapidly decomposed by microorganisms, and then the relatively refractory organic matter (e.g., lignin) with relatively low $\delta^{13}C$ was likely to be selectively decomposed by microorganisms surviving in soils under freeze–thaw conditions [61]. Based on the measurements of $\delta^{13}C$ values and specific ultraviolet absorbance at 254 nm ($SUVA_{254}$) of DOC collected from the intact organic horizons across the latitudinal transect over the year, the $\delta^{13}C$ values of DOC would decrease on average with the increase in the $SUVA_{254}$ values [59]. Kalbitz et al. [62] also reported that greater $SUVA_{254}$ values of soil-dissolved organic matter in forest floors were associated with more negative $\delta^{13}C$ values. The index of $SUVA_{254}$ values normally reflects the aromaticity and bioavailability of DOC in soil extracts and leachates [63], and the $SUVA_{254}$ values can increase following the simulated freeze–thaw fluctuations mainly due to microbial decomposition of labile DOC pools [19,21]. These could be responsible for the relatively lower $\delta^{13}C$ values of soil-respired CO_2 in all treatments at the late stage of the autumn freeze–thaw simulation. Based on relatively larger CO_2 emissions and less negative $\delta^{13}C$ of soil-respired CO_2 upon heavy snow cover than upon light snow cover (Figures 4 and 5) and the fractionation variations of $\delta^{13}C$ in SOM occur at different soil layers [37], it can be thus reasonably concluded that the presence of increased snow cover prior to winter freeze can increase the decomposition of organic C in subsurface soils under temperate forests in late autumn annually, which deserves further study in the future under field conditions to explore the magnitude of the soil C flush emission and the critical intrinsic influencing mechanisms involved.

Our results indicate a relatively great range of $\delta^{13}C$ values of soil-respired CO_2 across all forest soil columns when soil CO_2 emissions are less than 1.0 mmol CO_2 m^{-2} s^{-1} (Figure 6). Furthermore, during the simulated autumn and spring freeze–thaw as well as winter freeze stages, $\delta^{13}C$ values of soil-respired CO_2 at each sampling were mostly characterized by relatively large standard errors (Figure 4). The large uncertainty of $\delta^{13}C$ values of soil-respired CO_2 during the non-growing season simulation corresponded with the results reported by Bowling et al. [64], who reported that the $\delta^{13}C$ values of nocturnal whole-forest respiration in non-growing season ranged from −15‰ to −40‰ and varied more greatly than in growing season in a subalpine forest of Niwot Ridge, USA. This large uncertainty in the $\delta^{13}C$-CO_2 values would result from a small range in CO_2 concentrations and $\delta^{13}C$ values in non-growing seasons upon using the Keeling plot approach [65] and the effects of diffusive transport-induced fractionation under non-steady state conditions caused by the freeze–thaw simulations [16,51].

The $\delta^{13}C$ values of soil-respired CO_2 in all treatments became, on average, less negative as the simulated spring freeze–thaw proceeded ($p < 0.001$) (Figure 5), which was contrary to the variations of the $\delta^{13}C$ values during the autumn freeze–thaw simulation. Furthermore, there were, on average, relatively more negative $\delta^{13}C$ of soil-respired CO_2 upon heavy snow cover than upon light snow cover during the spring freeze–thaw simulation, which differed from the relatively less negative $\delta^{13}C$ of soil-respired CO_2 upon heavy snow cover during the autumn freeze–thaw and growing season simulations, respectively (Figure 5). These phenomena could result in a shift of about 4.2‰ ^{13}C-enrichment on average in the simulated growing season versus non-growing season soil respiration measured on all soil columns during the whole incubation (Figure 5). This shift is well supported by the results reported by Moyes et al. [35], who documented that a seasonal pattern of about 4‰ ^{13}C-enrichment in summer versus spring and fall soil respiration under deciduous trees can be attributed to the seasonal variability of soil heterotrophic respiration.

As the autumn and spring freeze–thaw simulation proceeded, the $\delta^{13}C$ values of soil-respired CO_2 in all treatments became, on average, more negative at low temperatures (−10/5 °C) than at high temperatures (−5/10 °C), regardless of snow cover and the presence of litter (Figure 5), which indicates temperature-dependent changes in the $\delta^{13}C$ values of soil-respired CO_2 under experimental conditions. This would result from the microbial-mediated decomposition of different C compounds available in the soils at the different stages of the incubation, which is likely related to the structure of the soil microbial community, as proposed by Andrews et al. [66]. Under frequent freeze–thaw disturbances caused by changes in snow cover, soil fungi, which are normally considered K-strategic microorganisms, can have a better performance than bacteria in substrate utilization in mid-latitude systems [4,67,68], and fungal biomass typically has lower $\delta^{13}C$ values than bacterial biomass [69,70]. Normally, K-strategic microorganisms (mainly fungi) can degrade recalcitrant organic matter, whereas r-strategic microorganisms (mainly bacteria) depend on easily degradable organic matter and sufficient nutrients [71]. Furthermore, organic carbon compounds with different biodegradability contain different C isotopic compositions, with a decrease in the $\delta^{13}C$ values from labile to recalcitrant organic matters [61]. At the early stage of soil freeze–thaw in autumn, r-strategic microorganisms can preferentially utilize labile C pools (e.g., carbohydrates) with more enrichment in ^{13}C, resulting in relatively less negative $\delta^{13}C$ of soil-respired CO_2 during the first week than during the second week of the autumn freeze–thaw simulation (Figure 5). As the autumn freeze–thaw proceeded or after a long-term severe winter freeze, soil microbial biomass decreased, and the structure of the microbial community would have a shift towards fungi-dominated microorganisms [29,67,68], due to less sensitivity of fungi than bacteria following soil freeze–thaw or freeze processes, especially under cold winter conditions [68,72,73]. Therefore, microorganisms surviving after a severe winter freeze can release extracellular enzymes to degrade recalcitrant organic C compounds [21,74]. Probably, a shift from more labile to more recalcitrant respiratory substances, or changes in soil microbial biomass and microbial community after long-term severe winter freeze can lead to more negative $\delta^{13}C$ of the CO_2 respired initially during the spring freeze–thaw simulation (Figure 5). The temperature-dependent changes in the $\delta^{13}C$ values of soil-respired CO_2 in this study differed obviously from the results reported in previous studies without freeze–thaw disturbances [66,75]. During the 98-day laboratory incubation using forest soils, Gauthier et al. [75] observed a linear increase in $\delta^{13}C$ values of soil-respired CO_2 with decreasing temperature from 28 °C to 8 °C. Andrews et al. [66] also observed a larger increase in the $\delta^{13}C$ values of soil-respired CO_2 at 4 °C than at 22 °C, which would be related to changes in the soil microbial community composition. Taken together, the results indicate that changes in the amount and quality of organic C compounds released into the soil upon freeze–thaw disturbances [19,21,76], soil aggregate distribution [76], and microbial properties such as the structure of the microbial community and microbial activity as well as soil enzyme activity [21,23,24,57,68,74,77] can most likely influence the temperature-dependent changes in the $\delta^{13}C$ values of soil-respired CO_2.

The $\delta^{13}C$ values of soil microbial respiration can reflect the decomposition of SOM and plant substrates and have been reported to vary greatly with environmental conditions [78,79]. Our results show that the $\delta^{13}C$ values of soil-respired CO_2 under experimental conditions can increase with the increase in volumetric soil water contents at 5 cm depth within a range from 14.2% to 65.2% (v/v) (Figure 7b). This is different from the results reported by Barbour et al. [79], who observed a negative relationship between the $\delta^{13}C$ values of soil-respired CO_2 and volumetric soil water content at a pasture site in New Zealand during nearly 10-day measurement. In a temperate desert steppe, the $\delta^{13}C$ values of soil-respired CO_2 significantly decreased with increasing soil moisture within the range from 2.0% to 28.0% (v/v) (determination coefficient of regression, R^2 = 0.24) [80]. However Phillips et al. reported that there were no changes in the $\delta^{13}C$ values of heterotrophic soil-respired CO_2 with a large soil moisture range [81]. The differences would result from different site-specific experimental conditions and other soil properties. Furthermore, 12% and 33% variations in the $\delta^{13}C$ values of soil-respired CO_2 in this study were, respectively explained by volumetric soil water contents, in the absence and presence of litter (Figure 7b). This result indicates that in addition to soil moisture, other soil properties such as different organic C sources, microbial biomass, and microbial community may affect the $\delta^{13}C$ of soil heterotrophic respiration under experimental conditions, which deserves further study in the future under field conditions.

5. Conclusions and Future Perspectives

The set of novel analysis systems was well used to study the changes in heterotrophic respiration of undisturbed large soil columns and their corresponding $\delta^{13}C$ values under laboratory conditions. Based on laboratory simulation incubation experiments of four different seasons over the year, the effects of artificial snow cover on forest soil heterotrophic respiration and its $\delta^{13}C$ values could vary with different seasons of the year, the presence of litter, and forest types. During the autumn freeze–thaw simulation, there were relatively larger CO_2 emissions and less negative $\delta^{13}C$ values of soil-respired CO_2 upon heavy snow cover than upon light snow cover, indicating that the presence of increased snow cover prior to winter freeze may increase the decomposition of organic C in subsurface soils under temperate forests. The $\delta^{13}C$ values of soil-respired CO_2 in all treatments became, on average less negative as the simulated spring freeze–thaw proceeded, which was contrary to the variations of the $\delta^{13}C$ values observed during the autumn freeze–thaw simulation. The $\delta^{13}C$ values of soil-respired CO_2 during the simulated spring freeze–thaw were, on average more negative upon heavy snow cover than upon light snow cover, which was contrary to the variations of the $\delta^{13}C$ values observed during the simulated autumn freeze–thaw and growing season periods, respectively. Regardless of snow cover and the presence of litter, a shift of about 4.2‰ ^{13}C-enrichment on average in the simulated growing season versus non-growing season soil heterotrophic respiration was observed on all large soil columns and it would be attributed to changes in soil moisture, amount and quality of organic C compounds released into the soil, and soil microbial properties under experimental conditions. Normally, seasonal changes in $\delta^{13}C$ values of soil heterotrophic and autotrophic respiration in terrestrial ecosystems can, to some extent, reflect the responses of SOM decomposition to environmental conditions [16,35,78,79]. The results of this study highlight the impacts of winter snow cover and the presence of litter on soil respiration and its $\delta^{13}C$ values under temperate forests over the year, which deserves further study in the future under field conditions to explore the critical intrinsic influencing mechanisms, by moderately considering the functions of soil physicochemical and microbial properties as well as priming effects caused by fine root biomass in regulating the $\delta^{13}C$ of soil respiration and soil C dynamics [11,16,36,61,76,77,82].

Supplementary Materials: The following supporting information can be downloaded at: https://www.mdpi.com/article/10.3390/f14071384/s1, Figure S1: Map of the study area located near the National Research Station of Changbai Mountain Forestry Ecosystem in Jilin Province, northeastern China. Figure S2: Dynamics of maximum and minimum daily air temperature as well as daily average soil temperature at 15 cm depth during the period from September 2011 to November 2012 in the study area and four subdivisions for the late autumn freeze–thaw period, winter freeze period, spring freeze–thaw period, and the growing season, respectively. Figure S3: Measurements of soil CO_2 emissions and $\delta^{13}C$ values of CO_2 released from large soil columns collected from two temperate forest stands, using a set of novel analysis systems.

Author Contributions: Conceptualization, methodology, formal analysis, resources, writing—original draft preparation, writing—review and editing, supervision, funding acquisition, X.X.; methodology, formal analysis, investigation, writing—review and editing, H.W.; formal analysis, writing—review and editing, J.Y.; formal analysis, writing—review and editing, S.T.; formal analysis, writing—review and editing, W.C. All authors have read and agreed to the published version of the manuscript.

Funding: This research was funded by the National Natural Science Foundation of China (grant numbers: 41175133, 41775163, 41975121, and 42275130) and by the State Key Laboratory of Atmospheric Boundary Layer Physics and Atmospheric Chemistry (No. LAPC-KF-2023-02).

Data Availability Statement: Not applicable.

Conflicts of Interest: The authors declare no conflict of interest.

References

1. Zhang, T.; Barry, R.G.; Armstrong, R.L. Application of satellite remote sensing techniques to frozen ground studies. *Polar Geogr.* **2004**, *28*, 163–196. [CrossRef]
2. Tarnocai, C.; Canadell, J.G.; Schuur, E.A.G.; Kuhry, P.; Mazhitova, G.; Zimov, S. Soil organic carbon pools in the northern circumpolar permafrost region. *Glob. Biogeochem. Cycles* **2009**, *23*, GB2023. [CrossRef]
3. Brooks, P.D.; Grogan, P.; Templer, P.H.; Groffman, P.; Öquist, M.G.; Schimel, J. Carbon and nitrogen cycling in snow-covered environments. *Geogr. Compass* **2011**, *5*, 682–699. [CrossRef]
4. Uchida, M.; Mo, W.; Nakatsubo, T.; Tsuchiya, Y.; Horikoshi, T.; Koizumi, H. Microbial activity and litter decomposition under snow cover in a cool-temperate broad-leaved deciduous forest. *Agric. For. Meteorol.* **2005**, *134*, 102–109. [CrossRef]
5. IPCC. Summary for policymakers. In *Climate Change 2020: The Physical Science Basis*; Cambridge University Press: New York, NY, USA, 2021.
6. Brown, R.D.; Robinson, D.A. Northern Hemisphere spring snow cover variability and change over 1922–2010 including an assessment of uncertainty. *Cryosphere* **2011**, *5*, 219–229. [CrossRef]
7. Kreyling, J.; Haei, M.; Laudon, H. Absence of snow cover reduces understory plant cover and alters plant community composition in boreal forests. *Oecologia* **2012**, *168*, 577–587. [CrossRef]
8. Robroek, B.J.M.; Heijboer, A.; Jassey, V.E.J.; Hefting, M.M.; Rouwenhorst, T.G.; Buttler, A.; Bragazza, L. Snow cover manipulation effects on microbial community structure and soil chemistry in a mountain bog. *Plant Soil* **2013**, *369*, 151–164. [CrossRef]
9. Wu, J.B.; Guan, D.X.; Wang, M.; Pei, T.F.; Han, S.J.; Jin, C.J. Year-round soil and ecosystem respiration in a temperate broad-leaved Korean pine forest. *For. Ecol. Manag.* **2006**, *223*, 35–44. [CrossRef]
10. Liu, X.P.; Dong, W.X.; Wood, J.D.; Wang, Y.Y.; Li, X.X.; Zhang, Y.M.; Hu, C.S.; Gu, L.H. Aboveground and belowground contributions to ecosystem respiration in a temperate deciduous forest. *Agric. For. Meteorol.* **2022**, *314*, 108807. [CrossRef]
11. Xu, X.K. Effect of freeze-thaw disturbance on soil C and N dynamics and GHG fluxes of East Asia forests: Review and future perspectives. *Soil Sci. Plant Nutr.* **2022**, *68*, 15–26. [CrossRef]
12. Kalbitz, K.; Solinger, S.; Park, J.H.; Michalzik, B.; Matzner, E. Controls on the dynamics of dissolved organic matter in soils: A review. *Soil Sci.* **2000**, *165*, 277–304. [CrossRef]
13. Wardle, D.A.; Walker, L.R.; Bardgett, R.D. Ecosystem properties and forest decline in contrasting long-term chronosequences. *Science* **2004**, *305*, 509–513. [CrossRef] [PubMed]
14. Ngao, J.; Epron, D.; Brechet, C.; Granier, A. Estimating the contribution of leaf litter decomposition to soil CO_2 efflux in a beech forest using ^{13}C-depleted litter. *Glob. Chang. Biol.* **2005**, *11*, 1768–1776. [CrossRef]
15. Sakata, T.; Ishizuka, S.; Takahashi, M. Separation of soil respiration into CO_2 emission sources using ^{13}C natural abundance in a deciduous broad-leaved forest in Japan. *Soil Sci. Plant Nutr.* **2007**, *53*, 328–336. [CrossRef]
16. Mauritz, M.; Celis, G.; Ebert, C.; Hutchings, J.; Ledman, J.; Natali, S.M.; Pegoraro, E.; Salmon, V.G.; Schadel, C.; Taylor, M. Using stable carbon isotopes of seasonal ecosystem respiration to determine permafrost carbon loss. *J. Geophys. Res.-Biogeosci.* **2019**, *124*, 46–60. [CrossRef]

17. Zhang, Y.J.; Zou, J.L.; Meng, D.L.; Dang, S.N.; Zhou, J.H.; Osborne, B.; Ren, Y.Y.; Liang, T.; Yu, K.K. Effect of soil microorganisms and labile C availability on soil respiration in response to litter inputs in forest ecosystems: A meta-analysis. *Ecol. Evol.* **2020**, *10*, 13602–13612. [CrossRef]
18. Kiikkila, O.; Kitunen, V.; Spetz, P.; Smolander, A. Characterization of dissolved organic matter in decomposing Norway spruce and silver birch litter. *Eur. J. Soil Sci.* **2012**, *63*, 476–486. [CrossRef]
19. Wu, H.H.; Xu, X.K.; Cheng, W.G.; Fu, P.Q.; Li, F.Y. Antecedent soil moisture prior to freezing can affect quantity, composition and stability of soil dissolved organic matter during thaw. *Sci. Rep.* **2017**, *7*, 6380. [CrossRef] [PubMed]
20. Hobara, S.; Osono, T.; Hirose, D.; Noro, K.; Hirota, M.; Benner, R. The roles of microorganisms in litter decomposition and soil formation. *Biogeochemistry* **2014**, *118*, 471–486. [CrossRef]
21. Wu, H.H.; Xu, X.K.; Fu, P.Q.; Cheng, W.G.; Fu, C.S. Responses of soil WEOM quantity and quality to freeze-thaw and litter manipulation with contrasting soil water content: A laboratory experiment. *Catena* **2021**, *198*, 105058. [CrossRef]
22. Tan, B.; Wu, F.Z.; Yang, W.Q.; He, X.H. Snow removal alters soil microbial biomass and enzyme activity in a Tibetan alpine forest. *Appl. Soil Ecol.* **2014**, *76*, 34–41. [CrossRef]
23. Nakajima, M.; Cheng, W.G.; Tang, S.R.; Hori, Y.; Yaginuma, E.; Hattori, S.; Hanayama, S.; Tawaraya, K.; Xu, X.K. Modeling aerobic decomposition of rice straw during off-rice season in an Andisol paddy soil in a cold temperate region, Japan: Effects of soil temperature and moisture. *Soil Sci. Plant Nutr.* **2016**, *62*, 90–98. [CrossRef]
24. Wu, H.H.; Xu, X.K.; Cheng, W.G.; Han, L. Dissolved organic matter and inorganic N jointly regulate greenhouse gases fluxes from forest soils with different moistures during a freeze-thaw period. *Soil Sci. Plant Nutr.* **2020**, *66*, 163–176. [CrossRef]
25. Yu, Z.; Liu, S.R.; Wang, J.X.; Sun, P.S.; Liu, W.G.; Hartley, D.S. Effects of seasonal snow on the growing season of temperate vegetation in China. *Glob. Chang. Biol.* **2013**, *19*, 2182–2195. [CrossRef] [PubMed]
26. Shibata, H. Impact of winter climate change on nitrogen biogeochemistry in forest ecosystems: A synthesis from Japanese case studies. *Ecol. Indic.* **2016**, *65*, 4–9. [CrossRef]
27. Zhou, B.T.; Wang, Z.Y.; Shi, Y.; Xu, Y.; Han, Z.Y. Historical and future changes of snowfall events in China under warming background. *J. Clim.* **2018**, *31*, 5873–5889. [CrossRef]
28. Wang, C.K.; Han, Y.; Chen, J.Q.; Wang, X.C.; Zhang, Q.Z.; Bond-Lamberty, B. Seasonality of Soil CO_2 Efflux in a Temperate Forest: Biophysical Effects of Snowpack and Spring Freeze–Thaw Cycles. *Agric. For. Meteorol.* **2013**, *177*, 83–92. [CrossRef]
29. Xu, X.K.; Duan, C.T.; Wu, H.H.; Li, T.S.; Cheng, W.G. Effect of intensity and duration of freezing on soil microbial biomass, extractable C and N pools, and N_2O and CO_2 emissions from forest soils in cold temperate region. *Sci. China Earth Sci.* **2016**, *59*, 156–169. [CrossRef]
30. Schindlbacher, A.; Jandl, R.; Schindlbacher, S. Natural variations in snow cover do not affect the annual soil CO_2 efflux from a mid-elevation temperate forest. *Glob. Chang. Biol.* **2014**, *20*, 622–632. [CrossRef]
31. Gao, D.C.; Hagedorn, F.; Zhang, L.; Liu, J.; Qu, G.F.; Sun, J.F.; Peng, B.; Fan, Z.Z.; Zheng, J.Q.; Jiang, P.; et al. Small and transient response of winter soil respiration and microbial communities to altered snow depth in a mid-temperate forest. *Appl. Soil Ecol.* **2018**, *130*, 40–49. [CrossRef]
32. Tang, S.R.; Cheng, W.G.; Hu, R.G.; Nakajima, M.; Guigue, J.; Kimani, S.M.; Sato, S.; Tawaraya, K.; Xu, X.K. Decomposition of soil organic carbon influenced by soil temperature and moisture in Andisol and Inceptisol paddy soils in a cold temperate region of Japan. *J. Soils Sediments* **2017**, *17*, 1843–1851. [CrossRef]
33. Li, W.; Wu, J.; Bai, E.; Jin, C.; Wang, A.; Yuan, F.; Guan, D. Response of terrestrial carbon dynamics to snow cover change: A meta-analysis of experimental manipulation (II). *Soil Biol. Biochem.* **2016**, *103*, 388–393. [CrossRef]
34. Blankinship, J.C.; Hart, S.C. Consequences of manipulated snow cover on soil gaseous emission and N retention in the growing season: A meta-analysis. *Ecosphere* **2012**, *3*, 1–20. [CrossRef]
35. Moyes, A.B.; Gaines, S.J.; Siegwolf, R.T.W.; Bowling, D.R. Diffusive fractionation complicates isotopic partitioning of autotrophic and heterotrophic sources of soil respiration. *Plant Cell Environ.* **2010**, *33*, 1804–1819. [CrossRef]
36. Diao, H.Y.; Wang, A.Z.; Yuan, F.H.; Guan, D.X.; Wu, J.B. Autotrophic respiration modulates the carbon isotope composition of soil respiration in a mixed forest. *Sci. Total Environ.* **2022**, *807*, 150834. [CrossRef]
37. Diao, H.Y.; Wang, A.Z.; Yuan, F.H.; Guan, D.X.; Dai, G.H.; Wu, J.B. Environmental effects on carbon isotope discrimination from assimilation to respiration in a coniferous and broad-leaved mixed forest of northeast China. *Forests* **2020**, *11*, 1156. [CrossRef]
38. Zheng, J.P.; Guo, Z.L.; Xu, C.Y.; Fan, C.N. Seasonal dynamics of litter accumulation in major forest communities on the northern slope of Changbai Mountain, Northeast China. *Acta Ecol. Sin.* **2011**, *31*, 4299–4307.
39. Ma, L.J.; Qin, D.H. Spatial-temporal characteristics of observed key parameters for snow cover in China during 1957–2009. *J. Glaciol. Geocryol.* **2012**, *34*, 1–11.
40. Li, D.S.; Zheng, J.Q.; Wang, X.X.; Zheng, X.B.; Han, S.J. Effects of nitrogen addition and water manipulation on leaf litter decomposition. *J. Beijing For. Univ.* **2016**, *38*, 44–52.
41. Yang, L.Y.; Li, W.H.; Wu, S.T. Fine root decomposition and nitrogen mineralization of the primitive Korean pine and broadleaved forests as well as its secondary forests in the Changbaishan Mountains, northeastern China. *J. Beijing For. Univ.* **2007**, *29*, 10–15.
42. Wu, H.H.; Xu, X.K.; Duan, C.T.; Li, T.S.; Cheng, W.G. Synergistic effects of dissolved organic carbon and inorganic nitrogen on methane uptake in forest soils without and with freezing treatment. *Sci. Rep.* **2016**, *6*, 32555. [CrossRef] [PubMed]
43. Zhou, G.; Liu, Q.J.; Xu, Z.Z.; Du, W.X.; Yu, J.; Meng, S.W.; Zhou, H.; Qin, L.H.; Shah, S. How can the shade intolerant Korean pine survive under dense deciduous canopy? *For. Ecol. Manag.* **2020**, *457*, 117735. [CrossRef]

44. Hao, Z.Q.; Zhang, J.; Song, B.; Ye, J.; Li, B.H. Vertical structure and spatial associations of dominant tree species in an old-growth temperate forest. *For. Ecol. Manag.* **2007**, *252*, 1–11. [CrossRef]
45. Xu, X.K.; Han, L.; Wang, Y.S.; Inubushi, K. Influence of vegetation types and soil properties on microbial biomass carbon and metabolic quotients in temperate volcanic and tropical forest soils. *Soil Sci. Plant Nutr.* **2007**, *53*, 430–440. [CrossRef]
46. Xu, X.K.; Yin, L.; Duan, C.T.; Jing, Y.S. Effect of N addition, moisture, and temperate on soil microbial respiration and microbial biomass in forest soil at different stages of litter decomposition. *J. Soils Sediments* **2016**, *16*, 1421–1439. [CrossRef]
47. Keeling, C.D. The concentration and isotopic abundances of atmospheric carbon dioxide in rural areas. *Geochim. Cosmochim. Acta* **1958**, *13*, 322–334. [CrossRef]
48. Keeling, C.D. The concentration and isotopic abundances of carbon dioxide in rural and marine air. *Geochim. Cosmochim. Acta* **1961**, *24*, 277–298. [CrossRef]
49. Kammer, A.; Tuzson, B.; Emmenegger, L.; Knohl, A.; Mohn, J.; Hagedorn, F. Application of a quantum cascade laser-based spectrometer in a closed chamber system for real-time $\delta^{13}C$ and $\delta^{18}O$ measurements of soil-respired CO_2. *Agric. For. Meteorol.* **2011**, *151*, 39–48. [CrossRef]
50. Wen, X.F.; Meng, Y.; Zhang, X.Y.; Sun, X.M.; Lee, X. Evaluating calibration strategies for isotope ratio infrared spectroscopy for atmospheric $^{13}CO_2/^{12}CO_2$ measurement. *Atmos. Meas. Technol.* **2013**, *6*, 1491–1501. [CrossRef]
51. Chen, C.H.; Wei, J.; Wang, J.Y.; Guo, Q.J.; Wen, X.F. Disentangling diel variations in the carbon isotopic composition of soil CO_2 efflux: Insights from a multi-chamber system and a non-steady model. *J. Soil Sci. Plant Nutr.* **2022**, *22*, 3096–3109. [CrossRef]
52. Zhu, J.J.; Wu, Q.X.; Wu, F.Z.; Yue, K.; Ni, X.Y. Decline in carbon decomposition from litter after snow removal is driven by a delayed release of carbohydrates. *Plant Soil* **2022**, *481*, 83–95. [CrossRef]
53. Groffman, P.M.; Driscoll, C.T.; Fahey, T.J.; Hardy, J.P.; Fitzhugh, R.D.; Tierney, G.L. Colder soils in a warmer world: A snow manipulation study in a northern hardwood forest ecosystem. *Biogeochemistry* **2001**, *56*, 135–150. [CrossRef]
54. Brooks, P.; Williams, M.; Schmidt, S. Inorganic nitrogen and microbial biomass dynamics before and during spring snowmelt. *Biogeochemistry* **1998**, *43*, 1–15. [CrossRef]
55. Lee, M.S.; Lee, J.S.; Koizumi, H. Temporal variation in CO_2 efflux from soil and snow surfaces in a Japanese cedar (Cryptomeria japonica) Plantation, central Japan. *Ecol. Res.* **2008**, *23*, 777–785. [CrossRef]
56. Gao, D.C.; Peng, B.; Fan, Z.Z.; Pei, G.T.; Bai, E. Different winter soil respiration between two mid-temperate plantation forests. *For. Ecol. Manag.* **2018**, *409*, 390–398. [CrossRef]
57. Wu, Q.Q. Season-dependent effect of snow depth on soil microbial biomass and enzyme activity in a temperate forest in Northeast China. *Catena* **2020**, *195*, 104760. [CrossRef]
58. Yohannes, Y.; Shibistova, O.; Abate, A.; Fetene, M.; Guggenberger, G. Soil CO_2 efflux in an Afromontane forest of Ethiopia as driven seasonality and tree species. *For. Ecol. Manag.* **2011**, *261*, 1090–1098. [CrossRef]
59. Philben, M.; Bowering, K.; Podrebarac, F.A.; Laganière, J.; Edwards, K.; Ziegler, S.E. Enrichment of ^{13}C with depth in soil organic horizons is not explained by CO_2 or DOC losses during decomposition. *Geoderma* **2022**, *424*, 116004. [CrossRef]
60. Benner, R.; Fogel, M.L.; Sprague, E.K.; Hodson, R.E. Depletion of ^{13}C in lignin and its implications for stable carbon isotope studies. *Nature* **1987**, *329*, 708–710. [CrossRef]
61. Bowling, D.R.; Pataki, E.E.; Randerson, J.T. Carbon isotopes in terrestrial ecosystem pools and CO_2 fluxes. *New Phytol.* **2008**, *178*, 24–40. [CrossRef]
62. Kalbitz, K.; Glaser, B.; Bol, R. Clear-cutting of a Norway spruce stand: Implications for controls on the dynamics of dissolved organic matter in the forest floor. *Eur. J. Soil Sci.* **2004**, *55*, 401–413. [CrossRef]
63. Xu, X.K.; Luo, X.B.; Jiang, S.H.; Xu, Z.J. Biodegradation of dissolved organic carbon in soil extracts and leachates from a temperate forest stand and its relationship to ultraviolet absorbance. *Chin. Sci. Bull.* **2012**, *57*, 912–920. [CrossRef]
64. Bowling, D.R.; Massman, W.J.; Schaeffer, S.M.; Burns, S.P.; Monson, R.K.; Williams, M.W. Biological and physical influences on the carbon isotope content of CO_2 in a subalpine forest snowpack, Niwot Ridge, Colorado. *Biogeochemistry* **2009**, *95*, 37–59. [CrossRef]
65. Zobitz, J.M.; Keener, J.P.; Schnyder, H.; Bowling, D.R. Sensitivity analysis and quantification of uncertainty for isotopic mixing relationships in carbon cycle research. *Agric. For. Meteorol.* **2006**, *136*, 56–75. [CrossRef]
66. Andrews, J.A.; Matamala, R.; Westover, K.M.; Schlesinger, W.H. Temperature effects on the diversity of soil heterotrophs and the $\delta^{13}C$ of soil-respired CO_2. *Soil Biol. Biochem.* **2000**, *32*, 699–706. [CrossRef]
67. Monson, R.K.; Lipson, D.L.; Burns, S.P.; Turnipseed, A.A.; Delany, A.C.; Williams, M.W.; Schmidt, S.K. Winter forest soil respiration controlled by climate and microbial community composition. *Nature* **2006**, *439*, 711–714. [CrossRef]
68. Aanderud, Z.T.; Jones, S.E.; Schoolmaster, D.R., Jr.; Fierer, N.; Lennon, J.T. Sensitivity of soil respiration and microbial communities to altered snowfall. *Soil Biol. Biochem.* **2013**, *57*, 217–227. [CrossRef]
69. Dijkstra, P.; Ishizu, A.; Doucett, R.; Hart, S.C.; Schwartz, E.; Menyailo, O.V.; Hungate, B.A. ^{13}C and ^{15}N natural abundance of the soil microbial biomass. *Soil Biol. Biochem.* **2006**, *38*, 3257–3266. [CrossRef]
70. Werth, M.; Kuzyakov, Y. ^{13}C fractionation at the root–microorganisms–soil interface: A review and outlook for partitioning studies. *Soil Biol. Biochem.* **2010**, *42*, 1372–1384. [CrossRef]
71. Fontaine, S.; Mariotti, A.; Abbadie, L. The priming effect of organic matter: A question of microbial competition? *Soil Biol. Biochem.* **2003**, *35*, 837–843. [CrossRef]
72. Schadt, C.W.; Martin, A.P.; Lipson, D.A.; Schmidt, S.K. Seasonal dynamics of previously unknown fungal lineages in tundra soils. *Science* **2003**, *301*, 1359–1361. [CrossRef]

73. Buckeridge, K.M.; Grogan, P. Deepened snow alters soil microbial nutrient limitations in arctic birch hummock tundra. *Appl. Soil Ecol.* **2008**, *39*, 210–222. [CrossRef]
74. Weintraub, M.N.; Scott-Denton, L.E.; Schmidt, S.K.; Monson, R.K. The effects of tree rhizodeposition on soil exoenzyme activity, dissolved organic carbon, and nutrient availability in a subalpine forest ecosystem. *Oecologia* **2007**, *154*, 327–338. [CrossRef]
75. Gauthier, A.; Amiotte-Suchet, P.; Nelson, P.N.; Léeêque, J.; Zeller, B.; Hénault, C. Dynamics of the water extractable organic carbon pool during mineralisation in soils from a Douglas fir plantation and an oak-beech forest—An incubation experiment. *Plant Soil* **2010**, *330*, 465–479. [CrossRef]
76. Tang, S.R.; Yuan, P.; Tawaraya, K.; Tokida, T.; Fukuoka, M.; Yoshimoto, M.; Sakai, H.; Hasegawa, T.; Xu, X.K.; Cheng, W.G. Winter nocturnal warming affects the freeze-thaw frequency, soil aggregate distribution, the contents and decomposability of C and N in paddy fields. *Sci. Total Environ.* **2022**, *802*, 149870. [CrossRef]
77. Böstrom, B.; Comstedt, D.; Ekblad, A. Can isotopic fractionation during respiration explain the ^{13}C-enriched sporocarps of ectomycorrhizal and saprotrophic fungi? *New Phytol.* **2007**, *177*, 1012–1019. [CrossRef]
78. Amundson, R.; Stern, L.; Baisden, T.; Wang, Y. The isotopic composition of soil and soil-respired CO_2. *Geoderma* **1998**, *82*, 83–114. [CrossRef]
79. Barbour, M.M.; Hunt, J.E.; Kodama, N.; Laubach, J.; McSeveny, T.M.; Rogers, G.N.D.; Tcherkez, G.; Wingate, L. Rapid changes in δ^{13}C of ecosystem-respired CO_2 after sunset are consistent with transient ^{13}C enrichment of leaf respired CO_2. *New Phytol.* **2011**, *190*, 990–1002. [CrossRef] [PubMed]
80. Guo, N.; Lv, S.J.; Lv, G.Y.; Xu, X.B.; Yao, H.Y.; Yu, Z.H.; Qiu, X.; Wang, Z.Y.; Wang, C.J. Effects of warming and precipitation on soil CO_2 flux and its stable carbon isotope composition in the temperate desert steppe. *Sustainability* **2022**, *14*, 3351. [CrossRef]
81. Phillips, C.L.; Nickerson, N.; Risk, D.; Kayler, Z.E.; Anderson, C.; Mix, A.; Bond, B.J. Soil moisture effects on the carbon isotope composition of soil respiration. *Rapid Commun. Mass Spectrom.* **2010**, *24*, 1271–1280. [CrossRef]
82. Babur, E.; Dindaroglu, T.; Danish, S.; Haggblom, M.M.; Ozlu, E.; Gozukara, G.; Uslu, O.S. Spatial responses of soil carbon stocks, total nitrogen, and microbial indices to post-wildfire in the Mediterranean red pine forest. *J. Environ. Manag.* **2022**, *320*, 115939. [CrossRef] [PubMed]

forests

MDPI

Article

Effects of Nitrogen Deposition on Leaf Litter Decomposition and Soil Organic Carbon Density in Arid and Barren Rocky Mountainous Regions: A Case Study of Yimeng Mountain

Baishu Kong [1,†], Jilei Zhou [2,†], Liguo Qi [3], Shuying Jiao [4], Lujie Ma [1], Wenwen Geng [1], Yuhao Zhao [1], Ting Gao [1], Jie Gong [1], Kun Li [1,*] and Chuanrong Li [1,*]

1 Mountain Tai Forest Ecosystem Research Station of State Forestry Administration, Key Laboratory of State Forestry Administration for Silviculture of the Lower Yellow River, College of Forestry, Shandong Agricultural University, Tai'an 271018, China; 2021010047@sdau.edu.cn (B.K.); 2021110208@sdau.edu.cn (L.M.); 2021110207@sdau.edu.cn (W.G.); 2022110215@sdau.edu.cn (Y.Z.); 2022110213@sdau.edu.cn (T.G.); 2022120319@sdau.edu.cn (J.G.)

2 Shandong Forestry Protection and Development Service Centre, Jinan 250014, China; zhoujl666@shandong.cn

3 Mount Tai Scenic Spot Management Committee, Tai'an 271000, China; 2021120332@sdau.edu.cn

4 College of Resources and Environment, Shandong Agricultural University, Tai'an 271018, China; jiaoshuying@sdau.edu.cn

* Correspondence: kunli@sdau.edu.cn (K.L.); chrli@sdau.edu.cn (C.L.)

† These authors contributed equally to this work.

Abstract: The ecological impact of nitrogen (N) deposition has gained significance since the advent of the industrial revolution. Although numerous studies have examined the impact of N deposition on soil organic carbon (SOC), certain arid and barren rocky mountainous regions, which experience more pronounced N limitations, have been overlooked. This study was conducted in the Yimeng Mountains, examining eight treatments created by four N addition levels (0 kg N ha^{-1} yr^{-1}, 50 kg N ha^{-1} yr^{-1}, 100 kg N ha^{-1} yr^{-1} and 200 kg N ha^{-1} yr^{-1}) and two tree species (*Quercus acutissima* Carruth. and *Pinus thunbergii* Parl.). The research revealed variations in the effect of N addition on leaf litter decomposition and SOC density (SOCD) between different tree species. Notably, N addition stimulated the decomposition of leaf litter from *Quercus acutissima* Carruth. However, the decomposition of *Pinus thunbergii* Parl. leaf litter was enhanced at N addition levels below 100 kg N ha^{-1} yr^{-1}, while it was hindered at levels exceeding 100 kg N ha^{-1} yr^{-1}. In the *Quercus acutissima* Carruth. forest, the N addition levels of 50 kg N ha^{-1} yr^{-1}, 100 kg N ha^{-1} yr^{-1} and 200 kg N ha^{-1} yr^{-1} resulted in decreases in SOCD by 10.57%, 22.22% and 13.66%, respectively, compared to 0 kg N ha^{-1} yr^{-1}. In the *Pinus thunbergii* Parl. forest, the N addition levels of 50 kg N ha^{-1} yr^{-1}, 100 kg N hm^{-2} ha^{-1} and 200 kg N ha^{-1} yr^{-1} led to increases in SOCD by 49.53%, 43.36% and 60.87%, respectively, compared to 0 kg N ha^{-1} yr^{-1}. Overall, N addition decreases the SOCD of *Quercus acutissima* Carruth., but it increases the SOCD of *Pinus thunbergii* Parl., attributed to the alteration in soil enzyme stoichiometry and nutrient cycling by N addition. This study fills a theoretical gap concerning leaf litter decomposition and SOC sequestration in arid and barren rocky mountainous regions under global climate change.

Keywords: nitrogen deposition; leaf litter decomposition; enzyme stoichiometry; soil organic carbon density

Citation: Kong, B.; Zhou, J.; Qi, L.; Jiao, S.; Ma, L.; Geng, W.; Zhao, Y.; Gao, T.; Gong, J.; Li, K.; et al. Effects of Nitrogen Deposition on Leaf Litter Decomposition and Soil Organic Carbon Density in Arid and Barren Rocky Mountainous Regions: A Case Study of Yimeng Mountain. *Forests* **2023**, *14*, 1351. https://doi.org/10.3390/f14071351

Academic Editor: Frank S. Gilliam

Received: 21 May 2023
Revised: 26 June 2023
Accepted: 28 June 2023
Published: 30 June 2023

1. Introduction

The emission of reactive nitrogen (Nr) into the atmosphere has experienced a significant increase due to the industrial revolution and economic development [1,2]. Nr emissions can enter ecosystems through dry and wet deposition, leading to issues, such as soil acidification and lake eutrophication [3]. Increased nitrogen (N) deposition promotes plant growth and enhances ecosystem productivity because N is a limiting nutrient for vegetation and soil microorganisms [4]. However, excessive Nr can reduce ecosystem biodiversity and disrupt geo-biochemical cycles [5]. Atmospheric N deposition has undergone significant changes in the last four decades, primarily attributable to the regulation of NOy emissions through the Clean Air Act [6]. While total atmospheric N deposition has decreased in Europe and North America, it remains 3–4 times higher in China compared to these regions. This disparity is attributed to the absence of effective NHx emission management policies in China, leading to an increase in NHx that offsets the reduction in NOy [7]. Studies have indicated that certain regions in China are still experiencing increasing or excessive N deposition, surpassing critical thresholds [8]. This emphasizes the ongoing ecological significance of artificial N additions in high N deposition regions of China, aiding in the prediction of future environmental impacts associated with N deposition.

Forest soils play a crucial role as a carbon (C) pool, containing more than 40% of the total soil organic carbon (SOC) in terrestrial ecosystems [9]. Leaf litter decomposition is a crucial process for enhancing SOC levels within forest soils. A meta-analysis of SOC inputs to temperate forest soils revealed that more than 41% of the annual SOC increase was contributed by leaf litter [10]. The main pathways of SOC formation from leaf litter are (1) the conversion of leaf litter into complex compounds on the forest floor and (2) the transfer of leaf litter and humus to mineral soils. They can become part of aggregates and clay minerals, thereby contributing to their stabilization [11]. These pathways are mainly influenced by the natural environment, biological factors and the inherent characteristics of leaf litter [12]. Among them, N deposition plays a significant role in leaf litter decomposition. Studies have indicated that modest or moderate levels of N deposition can enhance leaf litter decomposition, but beyond a certain threshold, N deposition can impede this process [13]. The extent of N limitation within the local ecosystem determines this threshold. Song et al. observed that low levels of N addition significantly accelerated leaf litter decomposition [14]. He et al. demonstrated that N addition alleviated N limitation in soil microorganisms and enhanced enzyme activity, thereby promoting leaf litter decomposition [15]. However, prolonged or excessive N addition generally exhibited negative effects on litter decomposition [16]. N addition induces soil acidification and elevates the osmotic pressure of the soil solution, which has detrimental effects on microbial growth [17]. Ramirez et al. discovered that prolonged N addition caused a 35% reduction in total soil microbial biomass and an 11% decline in microbial respiration rate [18]. Consequently, the response of leaf litter decomposition to N deposition may lead to different results depending on the extent of N limitation within the ecosystem.

Most research on the impacts of N deposition on litter decomposition and SOC pool has concentrated on forest and grassland ecosystems within temperate and subtropical zones [19–21]. However, this response has been neglected in some arid and barren rocky mountainous regions with higher N limitation. The Yimeng Mountain region, located in the central-eastern region of China, experiences high levels of N deposition and is characterized by its typical arid and barren rocky mountain terrain. The soils in this region have thin layers and poor physical structure, leading to a low nutrient retention capacity [22,23]. Meanwhile, the rainfall in the region has caused severe soil erosion, resulting in reduced local nutrient effectiveness [24]. N is the most affected nutrient, being lost through leaching and other pathways, resulting in severe limitations for local above-ground vegetation and soil microorganisms [25,26]. This exacerbates the influence of N deposition on local leaf litter decomposition. It is known that nutrient effectiveness modifies the nutrient utilization strategy of microorganisms. Soil microorganisms optimize resource allocation to obtain the most limited resource and modify soil nutrient cycling through the production of various extracellular enzymes [27,28]. This means that soil microorganisms change their C use strategies due to differences in soil N and phosphorus (P) effectiveness, which affects soil C sequestration. Enzyme stoichiometry is a tool for determining microbial nutrient limitation and nutrient utilization strategies. Sinsabaugh et al. found that log-transformed β-1,4-glucosidase (βG), β-1,4-N-acetyl-glucosaminidase (NAG) and acid phosphatase (ACP) activities converge to 1:1:1 [29]. Therefore, the deviation direction of enzyme stoichiometry indicates the microbial demand for a specific nutrient

Based on the above, we propose the following hypotheses: (1) N addition may enhance leaf litter decomposition in the Yimeng Mountains due to a more substantial local N limitation. (2) N addition may augment SOC sequestration in the Yimeng Mountains by modifying soil nutrient cycling. (3) The effects of N addition on leaf litter decomposition and the SOC pool in the Yimeng Mountains may vary depending on the tree species. The research objectives include investigating leaf litter decomposition and nutrient release, soil nutrient effectiveness, enzyme activity and SOC density (SOCD) in *Quercus acutissima* Carruth. and *Pinus thunbergii* Parl. forests through N addition experiments.

2. Materials and Methods

2.1. Study Area

The experimental site was selected in Dawa Forestry, Yimeng Mountains (35°32′46″ N, 117°54′47″ E). The local climate corresponds to a monsoon climate of medium latitudes, characterized by an average annual temperature of 13.2 °C, an average annual rainfall of 750 mm and a frost-free period of about 196 days. The mother rocks predominantly consist of granite gneiss and limestone, while the soil is mainly brown and cinnamon soil with a pH of 5.0. The dominant tree species in the study area are *Quercus acutissima* Carruth. and *Pinus thunbergii* Parl. Understory shrubs are mainly represented by *Vitex negundo* L. *var. heterophylla* (Franch.) Rehd. The herbaceous species in the forest understory mainly consist of *Achnatherum pekinense* (Hance) Ohwi and *Ophiopogon bodinieri* Levl. *Quercus acutissima* Carruth. and *Pinus thunbergii* Parl. forests with similar altitudes (470 and 434 m), similar slopes (18 and 22°) and similar forest density (689 and 765 trees ha^{-1}) were selected as experimental sites in Dawa Forestry. The average tree height and diameter at the breast height of *Quercus acutissima* Carruth. were 9.41 m and 18.43 cm, respectively, and those of *Pinus thunbergii* Parl. were 7.71 m and 14.45 cm, respectively.

2.2. Experimental Design

The study employed the multiplicative method to simulate N deposition, incorporating four N addition treatments: no nitrogen addition (0 kg N ha^{-1} yr^{-1}, CK), low N addition (50 kg N ha^{-1} yr^{-1}, LN), medium N addition (100 kg N ha^{-1} yr^{-1}, MN) and high N addition (200 kg N ha^{-1} yr^{-1}, HN). These treatments were set based on local atmospheric N deposition monitoring values [30]. The orthogonal experiment involved two tree species, namely *Quercus acutissima* Carruth. and *Pinus thunbergii* Parl., and included the four N addition treatments, resulting in a total of eight treatments: PCK, PLN, PMN, PHN, QCK, QLN, QMN and QHN. A total of 3 replications were set up for each treatment, with a total of 24 plots (10×10 m^2, at least 10 m apart). Ammonium nitrate solution (NH_4NO_3) was applied starting in October 2020 and sprayed every two months. Specifically, NH_4NO_3 was uniformly mixed with 10 L of water and sprayed within the corresponding plots using a backpack sprayer. The same volume of deionized water was applied for the CK treatment to maintain consistency and avoid confounding effects from additional precipitation.

2.3. Leaf Litter Collection and Litter Bag Experiment

Fresh leaf litter (Table 1) was collected from *Quercus acutissima* Carruth. and *Pinus thunbergii* Parl. using a litter trap method at the experimental site in October 2020. The fresh leaf litter was transported to the laboratory and dried to constant weight at 65 °C. Samples weighing 10.00 g were put into litter bags of 20×15 cm^2 (aperture 1 mm). The bags were attached to the soil surface of each plot with a wire so that the bags were in close contact with the soil. Leaf litter bags were spaced at least 5 cm apart to avoid interaction. Litter bags from each treatment were collected at bimonthly intervals. After removing soil adhering to the surface of the leaf litter, the bags were dried at 65 °C for 48 h. The residual mass of leaf litter after decomposition was measured. The carbon (C), nitrogen (N) and phosphorus (P) content of the leaf litter were determined using the potassium dichromate external heating method, Kjeldahl method and Mo-Sb colorimetric method, respectively.

The lignin and cellulose content of the leaf litter was determined using Klason's method and the ice bath digestion-anthrone colorimetric method, respectively [13].

Table 1. Initial leaf litter characteristics.

Tree Species	C (%)	N (%)	P (%)	Lignin (%)	Cellulose (%)	C:N	Lignin:N
Quercus acutissima Carruth.	47.60 ± 2.79 a	0.81 ± 0.02 a	0.12 ± 0.01 b	6.39 ± 0.44 a	14.29 ± 1.59 b	59.17 ± 4.08 b	7.93 ± 0.52 a
Pinus thunbergii Parl.	48.91 ± 2.99 a	0.77 ± 0.01 b	0.13 ± 0.01 a	5.92 ± 0.22 b	18.14 ± 2.08 a	63.75 ± 3.81 a	7.72 ± 0.28 a

Note: The same alphabets of the same list show no significance ($p > 0.05$) and, on the contrary, having significance ($p < 0.05$).

2.4. Soil Sample Collection and Laboratory Analysis

Soil samples were collected in December 2021 using the "S"-shaped sampling method. A total of 9 points were selected within each plot for soil sample collection using a soil drill (8 cm diameter) in the 0–20 cm topsoil layer. All soil samples within each plot were combined and thoroughly mixed to create a representative soil sample. The mixed soil samples were then transported to the laboratory. We divided the samples into two parts after removing roots and stones. One part was stored at −80 °C to determine enzyme activity and inorganic nitrogen (IN). The other part was dried, crushed and sieved through a 0.15 mm sieve to determine SOC and available phosphorus (AP).

Soil enzyme activity was determined according to Deforest's method [31]. Acetic acid buffer was obtained by mixing sodium acetate with deionized water and adjusting the solution pH to 5 with acetic acid to match the soil pH at the study site. Enzyme activity was determined using three 4-methylumbelliferone (MUB)-based fluorogenic enzyme substrates: 4-MUB-β-D-glucopyranoside for β-1,4-glucosidase (βG), 4-MUB-N-acetyl-β-D-glucosaminide for β-1,4-N-acetyl-glucosaminidase (NAG) and 4-MUB-phosphate for acid phosphatase (ACP). Soil homogenates and substrates were added to 96-well plates, incubated in the dark (25 °C, 4 h) and soil enzyme activities were determined using a microplate reader (Synergy HTX, Hudson, MA, USA). Soil IN content was determined by continuous flow injection analyzer (AA3-A001-02E, Bran-Luebbe, Germany) after mixing 25 mL of 1 mol L^{-1} KCl solution and 3 g of fresh soil samples. SOC was determined by the potassium dichromate external heating method. Soil AP was extracted by $NaHCO_3$ solution and determined by Mo-Sb colorimetric method.

2.5. Statistical Analysis

The mass remaining and the C, N and P remaining are expressed in terms of the relative ratio of the initial litter [32]. The mass remaining (M_r) and the nutrient remaining (N_r) of the litter were calculated using Formulas (1) and (2), respectively.

$$M_r(\%) = M_t/M_0 \times 100 \quad (1)$$

$$N_r(\%) = \frac{M_t \times N_t}{M_0 \times N_0} \times 100\% \quad (2)$$

where M_t (g) is the residual mass at t (d) days; M_0 (g) is the initial mass of litter; N_t (%) is the nutrient element concentration of litter at t (d) days; N_0 (%) is the initial nutrient element concentration of litter.

The leaf litter decomposition coefficient was calculated using the Olson [33] exponential decay model (3).

$$M_t/M_0 = e^{-Kt} \quad (3)$$

where k is the decomposition coefficient (K value); t (d) is the time.

Enzymes C:N, C:P and N:P were calculated by the Formulas (4)–(6), respectively.

$$\text{Enzyme C : N} = \ln \beta\text{G} \div \ln \text{NAG} \quad (4)$$

$$\text{Enzyme C : P} = \ln \beta\text{G} \div \ln \text{ACP} \quad (5)$$

$$\text{Enzyme N : P} = \ln \text{NAG} \div \ln \text{ACP} \quad (6)$$

SOC density (SOCD) (kg m^{-2}) was calculated using Equation (7):

$$\text{SOCD}\left(\text{kg m}^{-2}\right) = \text{soil bulk density}\left(\text{g cm}^{-3}\right) \times \text{soil depth (cm)} \times \text{SOC (g kg}^{-1}\text{)} \quad (7)$$

The collected data were assessed for the normal distribution and homogeneity of variances. Independent samples *t*-tests were used to compare the differences in the initial leaf litter chemistry between the two tree species. One-way analysis of variance (ANOVA) and Duncan's multiple analysis ($\alpha = 0.05$) were conducted to determine the impact of N addition on leaf litter decomposition, soil physicochemical properties, soil enzyme activity and enzyme stoichiometry. Pearson correlation analysis was performed to examine the relationships between leaf litter nutrient remaining, soil enzyme activity and soil nutrients. The above operations were performed by the software SPSS 23.0. Figures were made using Origin 2018.

3. Results

3.1. Leaf Litter Decomposition

N addition accelerated the decomposition process of the *Quercus acutissima* Carruth. leaf litter with increasing N addition levels (Figure 1). The Mr of the *Quercus acutissima* Carruth. leaf litter for each treatment after one year showed QCK (68.13%) > QLN (61.63%) > QHN (58.57%) > QMN (55.43%). The Mr of *Pinus thunbergii* Parl. leaf litter decomposition showed PHN (69.5%) > PMN (68.47%) > PCK (63.57%) > PLN (61.37%). Notably, the PLN promoted the decomposition of *Pinus thunbergii* Parl. leaf litter, whereas the PMN and PHN hindered this process.

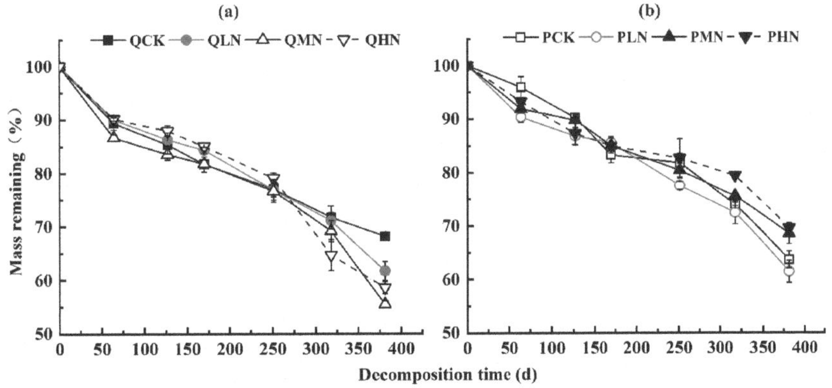

Figure 1. Effect of different treatments on the leaf litter Mr of *Quercus acutissima* Carruth. (**a**) and *Pinus thunbergii* Parl. (**b**) The points of different shapes are averaged. The length of the error bars is the standard deviation. QCK, QLN, QMN and QHN: *Quercus acutissima* Carruth. under no N addition, low N addition, medium N addition and high N addition, respectively. PCK, PLN, PMN and PHN: *Pinus thunbergii* Parl. under no N addition, low N addition, medium N addition and high N addition, respectively. Mr: mass remaining.

The Olson exponential decay model was determined to be a more appropriate description for the leaf litter decomposition process of the two tree species (Table 2). The larger the

K value, the faster the leaf litter mass loss. Comparing the two tree species, the *Quercus acutissima* Carruth. leaf litter exhibited higher K values than the *Pinus thunbergii* Parl. leaf litter, indicating a more rapid decomposition rate for *Quercus acutissima* Carruth.

Table 2. Decomposition constants (K value) and decomposition age of *Quercus acutissima* Carruth and *Pinus thunbergii* Parl.

Treatment	Decomposition Equation	Decomposition Constant (K Value)	50% Decomposition Age (a)	95% Decomposition Age (a)	Correlation Coefficient
QCK	$y = e^{-0.386x}$	0.386	1.796	7.761	0.967
QLN	$y = e^{-0.421x}$	0.421	1.646	7.116	0.968
QMN	$y = e^{-0.484x}$	0.484	1.432	6.190	0.913
QHN	$y = e^{-0.46x}$	0.460	1.507	6.512	0.931
PCK	$y = e^{-0.367x}$	0.367	1.889	8.163	0.937
PLN	$y = e^{-0.416x}$	0.416	1.666	7.201	0.958
PMN	$y = e^{-0.34x}$	0.340	2.039	8.811	0.982
PHN	$y = e^{-0.315x}$	0.315	2.200	9.510	0.940

Note: QCK, QLN, QMN and QHN: *Quercus acutissima* Carruth. under no N addition, low N addition, medium N addition and high N addition, respectively. PCK, PLN, PMN and PHN: *Pinus thunbergii* Parl. under no N addition, low N addition, medium N addition and high N addition, respectively.

3.2. Nutrients Remaining in Leaf Litter

N addition had a facilitative effect on the C release from the *Quercus acutissima* Carruth leaf litter (Figure 2a). The C remaining after one year showed QCK (50.09%) > QLN (42.15%) > QMN (38.57%) > QHN (38.04%). Conversely, high N addition inhibited C release from *Pinus thunbergii* Parl. leaf litter, while low N addition promoted C release from it. The C remaining of *Pinus thunbergii* Parl. leaf litter after one year exhibited PMN (69.79%) > PHN (63.49%) > PCK (58.25%) > PLN (55.28%) (Figure 2d). N addition had an inhibitory effect on the N release from both species of leaf litter (Figure 2b,e). For *Quercus acutissima* Carruth., high N treatments began to inhibit the N release from the leaf litter on day 127, while the other N addition treatments showed inhibitory effects after day 170. For *Pinus thunbergii* Parl. leaf litter, the inhibition started between days 170 and 251. On average, the leaf litter N remaining of *Pinus thunbergii* Parl. after one year was 11.02% lower than that of *Quercus acutissima* Carruth. The trends of P remaining in the *Quercus acutissima* Carruth and *Pinus thunbergii* Parl. leaf litter were similar to those of the N remaining (Figure 2c,f) On average, N addition reduced the P remaining from the *Quercus acutissima* Carruth. leaf litter by 16.42% and increased that from the *Pinus thunbergii* Parl. leaf litter by 4.22%. This indicates that N addition showed a facilitative effect on P release from *Quercus acutissima* Carruth. leaf litter and an inhibitory effect on *Pinus thunbergii* Parl. leaf litter.

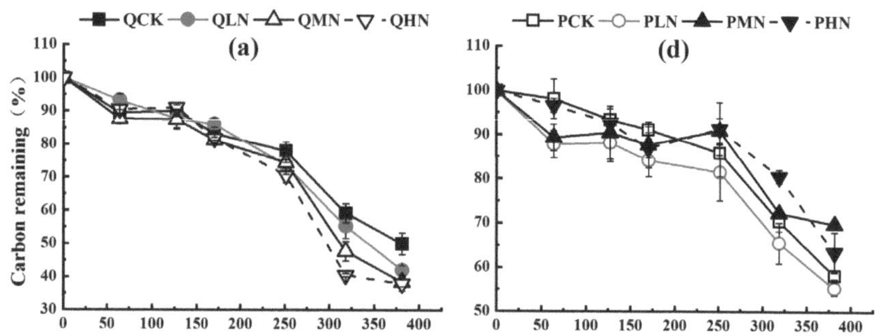

Figure 2. *Cont.*

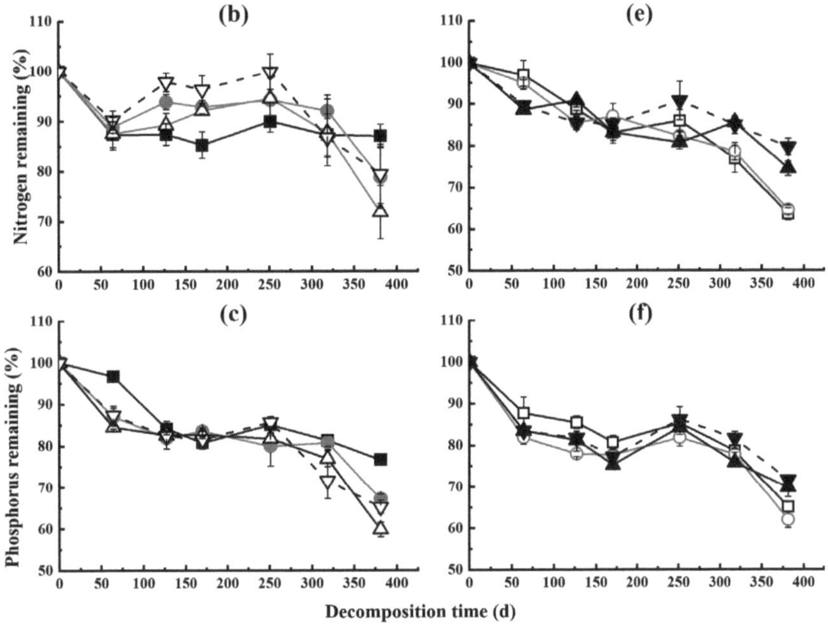

Figure 2. Effect of different treatments on C, N and P remaining in the leaf litter of *Quercus acutissima* Carruth. (**a**–**c**) and *Pinus thunbergii* Parl. (**d**–**f**). The points of different shapes are averaged. The length of the error bars is the standard deviation. QCK, QLN, QMN and QHN: *Quercus acutissima* Carruth. under no N addition, low N addition, medium N addition and high N addition, respectively. PCK, PLN, PMN and PHN: *Pinus thunbergii* Parl. under no N addition, low N addition, medium N addition and high N addition, respectively. C: carbon. N: nitrogen. P: phosphorus.

3.3. Soil Available Nutrient and Carbon Density

N addition led to a decrease in the SOC content of *Quercus acutissima* Carruth. forest (Figure 3). The SOC content of QLN, QMN and QHN decreased by 10.57%, 22.22% and 13.66%, respectively, compared to QCK. In contrast, N addition increased the SOC content in the *Pinus thunbergii* Parl. forest. PLN, PMN and PHN increased the SOC content by 49.53%, 43.36% and 60.86%, respectively, compared to PCK. The effect of N addition on soil IN content was not significant in both the *Quercus acutissima* Carruth. and *Pinus thunbergii* Parl. forests. Among all N addition treatments, PMN and PHN increased the soil IN content more, by 61.88% and 38.00%, respectively, compared to PCK. The effect of N addition on soil AP content was not significantly different. However, the soil AP content was higher in the *Quercus acutissima* Carruth. forest than that of the *Pinus thunbergii* Parl. forest. For SOCD (Figure 4), the *Quercus acutissima* Carruth. and *Pinus thunbergii* Parl. forests showed different results. N addition decreased the SOCD in the *Quercus acutissima* Carruth. forest but increased the SOCD in the black pine forest, although the differences among the treatments were not significant.

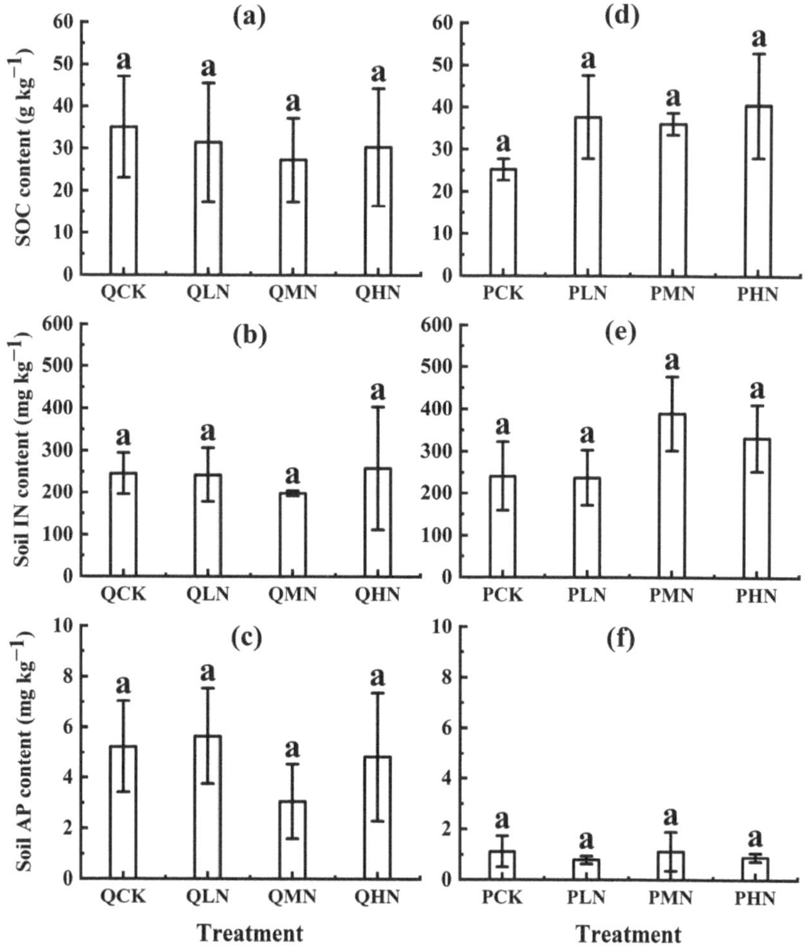

Figure 3. Effect of different treatments on SOC, IN and AP content of *Quercus acutissima* Carruth. (**a–c**) and *Pinus thunbergii* Parl. forests (**d–f**) soil. The length of the error bars is the standard deviation. Different letters indicate significant differences between different treatments ($p < 0.05$) for the same tree species. QCK, QLN, QMN and QHN: *Quercus acutissima* Carruth. forest under no N addition, low N addition, medium N addition and high N addition, respectively. PCK, PLN, PMN and PHN: *Pinus thunbergii* Parl. forest under no N addition, low N addition, medium N addition and high N addition, respectively. SOC: soil organic carbon. IN: inorganic nitrogen. AP: available phosphorus.

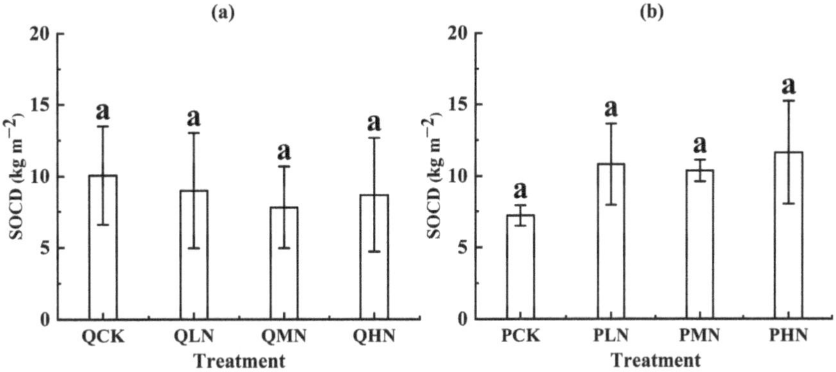

Figure 4. Effect of different treatments on SOCD of *Quercus acutissima* Carruth. (**a**) and *Pinus thunbergii* Parl. forests (**b**). The length of the error bars is the standard deviation. Different letters indicate significant differences between different treatments ($p < 0.05$) for the same tree species. QCK, QLN, QMN and QHN: *Quercus acutissima* Carruth. forest under no N addition, low N addition, medium N addition and high N addition, respectively. PCK, PLN, PMN and PHN: *Pinus thunbergii* Parl. forest under no N addition, low N addition, medium N addition and high N addition, respectively. SOCD: soil organic carbon density.

3.4. Enzyme Activity and Enzyme Stoichiometry

Both low and high N additions significantly increased the soil βG enzyme activity in the *Quercus acutissima* Carruth. forest (Figure 5). However, our study revealed no significant variations in the soil NAG enzyme activity among all N addition levels. The effect of N addition on soil ACP enzyme activity in the *Quercus acutissima* Carruth. forest showed inhibition at low N addition, promotion at medium N addition and no significant difference at high N addition. In the *Pinus thunbergii* Parl. forest, all N addition levels significantly increased the soil βG enzyme activity. The soil NAG enzyme activity increased with an increasing N addition level. PLN, PMN and PHN increased the NAG enzyme activity by 16.98%, 39.54% and 86.41%, respectively, compared to PCK. All N addition levels inhibited the soil ACP enzyme activity in the *Pinus thunbergii* Parl. forest, with significant differences for PLN and PHN compared to PCK.

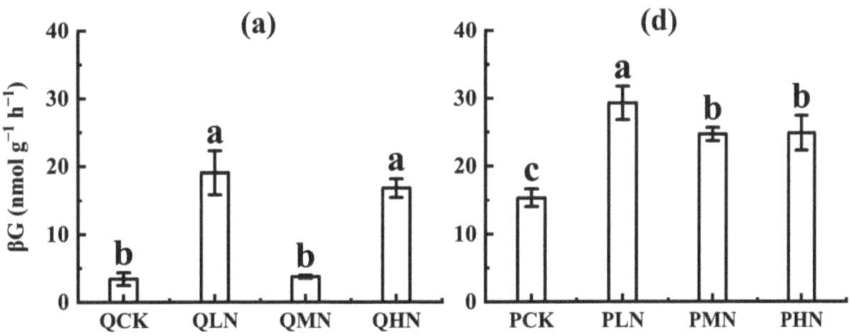

Figure 5. *Cont.*

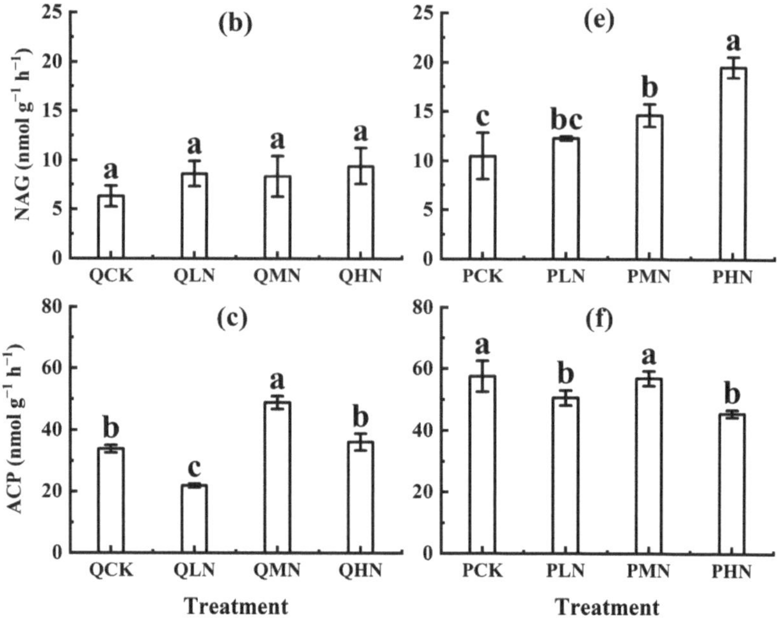

Figure 5. Effect of different treatments on soil enzyme activities of *Quercus acutissima* Carruth. (**a–c**) and *Pinus thunbergii* Parl. forests (**b–f**). The length of the error bars is the standard deviation. Different letters indicate significant differences between different treatments ($p < 0.05$) for the same tree species. QCK, QLN, QMN and QHN: *Quercus acutissima* Carruth. forest under no N addition, low N addition, medium N addition and high N addition, respectively. PCK, PLN, PMN and PHN: *Pinus thunbergii* Parl. forest under no N addition, low N addition, medium N addition and high N addition, respectively βG: β-1,4-glucosidase. NAG: β-1,4-N-acetyl-glucosaminidase. ACP: acid phosphatase.

The enzyme stoichiometry closer to one indicates that soil microorganisms are less nutrient limited (Table 3). In the *Quercus acutissima* Carruth. forest, QHN changed the microbial limiting nutrient from N to C compared to QCK. However, QHN brought the soil enzyme C:N ratio closer to one. Soil microorganisms exhibited P limitation in all treatments of the *Quercus acutissima* Carruth. forest because the enzyme C:P and N:P ratios were less than one. In the *Pinus thunbergii* Parl. forest, the soil enzyme C:N ratio was greater than one, while the enzyme C:P and N:P ratios were less than one in all treatments, indicating that the soil microbial nutrients limitations were C and P. Among them, high N addition brought the soil enzyme C:N, C:P and N:P ratios closer to one, which alleviated the C and P limitation of soil microorganisms.

3.5. The Correlation between Leaf Litter, Soil Nutrients and Enzyme Activity

In the *Quercus acutissima* Carruth. forest, a significantly positive correlation was observed between SOC content and AP content. Conversely, βG enzyme activity and ACP enzyme activity exhibited a significant negative correlation (Figure 6). The C and N remaining from the *Quercus acutissima* Carruth. leaf litter were highly significantly and positively correlated with the P remaining. Meanwhile, this leaf litter N remaining was also highly significantly positively correlated with soil AP content. In the *Pinus thunbergii* Parl leaf litter, there were significant or highly significant positive correlations among the C, N and P remaining. Meanwhile, they all showed significant positive correlations with the soil IN content. The SOC content of the *Pinus thunbergii* Parl. forest was only significantly positively correlated with βG activity.

Table 3. Soil enzyme stoichiometry of *Quercus acutissima* Carruth. and *Pinus thunbergii* Parl. forests.

Treatment	Enzyme C:N Ratio	Enzyme C:P Ratio	Enzyme N:P Ratio
QCK	0.67 ± 0.22 b	0.34 ± 0.09 c	0.52 ± 0.05 b
QLN	1.37 ± 0.03 a	0.95 ± 0.06 a	0.70 ± 0.05 a
QMN	0.63 ± 0.09 b	0.34 ± 0.01 c	0.54 ± 0.06 b
QHN	1.27 ± 0.10 a	0.79 ± 0.02 b	0.62 ± 0.06 ab
PCK	1.18 ± 0.11 b	0.67 ± 0.01 c	0.58 ± 0.05 c
PLN	1.35 ± 0.03 a	0.86 ± 0.02 a	0.64 ± 0.01 b
PMN	1.20 ± 0.02 b	0.79 ± 0.02 b	0.66 ± 0.02 b
PHN	1.08 ± 0.04 b	0.84 ± 0.03 a	0.78 ± 0.02 a

Note: Data are expressed as mean ± standard deviation. Different letters indicate significant differences between different treatments ($p < 0.05$) for the same tree species. QCK, QLN, QMN and QHN: *Quercus acutissima* Carruth. forest under no N addition, low N addition, medium N addition and high N addition, respectively. PCK, PLN, PMN and PHN: *Pinus thunbergii* Parl. forest under no N addition, low N addition, medium N addition and high N addition, respectively. C: carbon. N: nitrogen. P: phosphorus.

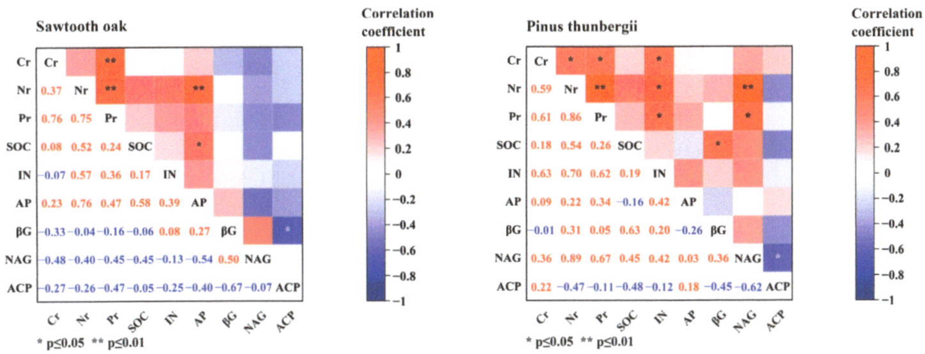

Figure 6. The heatmap of Pearson's correlation coefficients between leaf litter, soil nutrients and enzymatic activity in *Quercus acutissima* Carruth. and *Pinus thunbergii* Parl. Cr, Nr and Pr indicate the nutrient remaining of carbon, nitrogen and phosphorus in leaf litter, respectively. SOC, IN and AP indicate the organic carbon, inorganic nitrogen and available phosphorus content of the soil, respectively. βG: β-1,4-glucosidase. NAG: β-1,4-N-acetyl-glucosaminidase. ACP: acid phosphatase.

4. Discussion

4.1. Response of Leaf Litter Decomposition to N Addition

According to references [34–37], the impact of N addition on leaf litter decomposition can be categorized into three main conclusions: promotion, inhibition or no significant effect. In our study, N addition promoted the decomposition of *Quercus acutissima* Carruth. leaf litter. This can be attributed to the relatively low lignin content of *Quercus acutissima* Carruth. litter, which made it less susceptible to the inhibitory effects of N addition. When the exogenous N was added to the system, N could combine with lignin and its degradation intermediates, inhibiting lignin decomposition [38]. N addition altered the C:N ratio of leaf litter, thus promoting the decomposition of *Quercus acutissima* Carruth. leaf litter [39]. For *Pinus thunbergii* Parl. leaf litter, low N addition promoted decomposition, while medium and high N inhibited decomposition. High N addition can limit the growth of white rot fungi, which play a crucial role in lignin degradation [40]. Additionally, high N levels can inhibit the secretion of ligninolytic enzymes, further hindering the decomposition of *Pinus thunbergii* Parl. leaf litter [41]. The results of our study align with previous research, specifically a worldwide meta-analysis by Knorr et al. This indicated that leaf litter decomposition patterns exhibit consistency in areas with high N limitation and globally, albeit with increased levels of extremity and significance. Knorr et al. found that high N addition inhibited the decomposition of leaf litter with high lignin content but promoted the decomposition of leaf litter with low lignin content [42]. In summary, the impact of N addition on leaf litter decomposition in the Yimeng Mountains was influenced by both N levels and the inherent lignin content of the leaf litter.

4.2. Response of Soil Nutrient Cycling to N Addition

Research has shown that the impacts of N addition on SOC content are consisten with the trend of leaf litter decomposition and C release. This is because the subsurfac C input is mainly from the mass loss of leaf litter [43]. However, the opposite result wa observed in this study, i.e., N addition caused varying decreases in SOC content and SOCI in the *Quercus acutissima* Carruth. forest. In contrast, SOC content and SOCD in the *Pinu thunbergii* Parl. forest increased with increasing N addition. We considered that SOC accumulation was mainly influenced by both or a combination of increased SOC suppl and decreased SOC mineralization. *Quercus acutissima* Carruth. leaf litter is a high-qualit C source with high cellulose content, which can be easily decomposed and utilized b microorganisms [44]. Meanwhile, N addition was observed to enhance soil microbia activity and respiration, leading to increased SOC mineralization. The effect of N additio on SOC mineralization became more pronounced with higher N addition levels [45]. Ou findings also supported the view that the soil microbial nutrient utilization strategy of th *Quercus acutissima* Carruth. forest changed from N to C with the increased N addition leve This indicated that SOC mineralization was enhanced in the *Quercus acutissima* Carruth forest with increasing N effectiveness, resulting in lower SOC sequestration. Conversel *Pinus thunbergii* Parl. leaf litter has lower cellulose content and higher lignin content, whic is difficult for soil microorganisms to utilize [46]. This caused the *Pinus thunbergii* Parl. lea litter to be stored in the soil for a long time. The SOC input exceeded the SOC output in th *Pinus thunbergii* Parl. forest, which increased subsurface SOC sequestration [47].

N addition reduced the microbial demand for N from leaf litter, resulting in higher N remaining in the leaf litter compared to the treatment without N addition [48]. Chomel et al observed that N addition changed the microbial community structure to fungal dominate and reduced the microbial demand for N in leaf litter [49]. This is consistent with thi study that the increased N addition enhanced the inhibition of N release from leaf litte Furthermore, N addition significantly alleviated N limitation by soil microorganisms in th *Quercus acutissima* Carruth. and *Pinus thunbergii* Parl. forests. The microbial nutrient us strategy shifted from N to P due to N addition, alleviating the soil microbial N limitatio This change in nutrient limitation strategy led to a greater emphasis on P utilization b microorganisms, resulting in enhanced P release from leaf litter [50,51]. The promotion of I release from leaf litter became more significant with higher levels of N addition. The soi enzyme C: P and N: P ratios, which were less than one, further support the idea that P wa a limiting nutrient for soil microorganisms in these ecosystems [52].

4.3. Response of SOCD to N Addition

SOCD is an important indicator to measure the SOC pool. This study indicatec that the effect of N addition on SOCD was influenced by the tree species and leaf litte characteristics. This is because the tree species and leaf litter characteristics significantl influence soil C, N and P cycling, which alters the SOC content and stability [53]. In th *Quercus acutissima* Carruth. forest, SOC content was significantly positively correlate with soil AP content, suggesting that P availability plays an important role in influencin SOC content. However, increasing soil AP content may not be an effective strategy fo increasing the local SOC pool in the context of N addition. This is because increasin soil AP content can lead microorganisms to invest more resources in acquiring C [54 Li et al. showed that P increased soil microbial activity, accelerated SOC turnover an reduced SOC stability in the N addition context [55]. In the *Pinus thunbergii* Parl. fores the SOC content was significantly positively correlated with soil βG enzyme activit N addition alleviated soil microbial N limitation, leading to increased βG enzyme activit and accelerated cellulose decomposition in leaf litter. The lignin in the *Pinus thunbergi* Parl. leaf litter was transferred to the soil with leaf litter decomposition, leading to ar increase in SOCD and SOC stability [56]. Therefore, the response of SOCD to N addition in the Yimeng Mountains can depend on the tree species and the characteristics of their leaf litter. N addition decreases the SOCD of tree species with high cellulose content and low lignin content in leaf litter. Meanwhile, it increases the SOCD of tree species with low cellulose content and high lignin content in leaf litter. Overall, these findings highlight the complex interactions between N deposition, tree species, leaf litter characteristics and soil nutrient cycling in influencing SOCD and SOC stability. Further research is needed to fully understand the underlying mechanisms behind these responses to maintain or enhance the SOC pool in forest ecosystems.

5. Conclusions

The results highlight the importance of considering tree species and leaf litter characteristics when studying the effects of N addition on leaf litter decomposition and SOCD in arid and barren rocky mountainous regions, such as the Yimeng Mountains. The different responses of *Quercus acutissima* Carruth. and *Pinus thunbergii* Parl. leaf litter to N addition reflect the influence of leaf litter lignin content. The N addition promoted the decomposition of *Quercus acutissima* Carruth. leaf litter with high cellulose content and low lignin content. It also accelerated SOC turnover, leading to significant C loss and SOCD reduction. In addition, N addition showed a different effect on *Pinus thunbergii* Parl. leaf litter with low cellulose content and high lignin content. It promoted cellulose decomposition and facilitated the transfer of lignin to the soil, contributing to increased SOCD. Furthermore, N addition had contrasting effects on N and P release from leaf litter. The N addition alleviated the N limitation by soil microorganisms, which inhibited the N release from the leaf litter of both tree species. The N addition promoted P release from the leaf litter due to the microorganisms exhibiting P limitation. Overall, the findings demonstrate that N addition alters leaf litter decomposition, nutrient cycling and SOCD in the Yimeng Mountains, with the response depending on the tree species and leaf litter characteristics. This emphasizes the importance of considering these factors in understanding and managing nutrient dynamics and SOC sequestration in forest ecosystems.

Supplementary Materials: The following supporting information can be downloaded at: https://www.mdpi.com/article/10.3390/f14071351/s1.

Author Contributions: B.K. and J.Z. contributed equally to this work. B.K.: Conceptualization, Data Curation, Formal Analysis, Investigation, Writing—Original Draft, Writing—Review and Editing. J.Z.: Investigation, Writing—Review and Editing. L.Q.: Funding Acquisition, Project Administration, Resources. S.J.: Funding Acquisition, Resources. L.M.: Investigation. W.G.: Investigation. Y.Z.: Investigation. T.G.: Investigation. J.G.: Investigation. K.L.: Project Administration, Resources, Supervision, Writing—Review and Editing. C.L.: Funding Acquisition, Project Administration, Resources, Supervision, Writing—Review and Editing. All authors have read and agreed to the published version of the manuscript.

Funding: This study was supported by the National Natural Science Foundation of China (31570705), the Carbon stock assessment project of major dominant tree species in Shandong Province, the Open Research Fund Program of Shandong Provincial Key Laboratory of Eco-Environmental Science for the Yellow River Delta (Binzhou University) (2022KFJJ03) and the Natural Science Foundation of Shandong Province, China (ZR2020MC173 and ZR2021QD124).

Data Availability Statement: The data presented in this study are available in Supplementary Material.

Conflicts of Interest: The authors declare that they have no known competing financial interest or personal relationships that could have appeared to influence the work reported in this paper.

References

1. Sun, K.; Gao, Y.; Guo, X.; Zhang, J.; Zeng, X.; Ma, M.; Chen, Y.; Luo, K.; Yao, X.; Gao, H. The enhanced role of atmospheric reduced nitrogen deposition in future over East Asia–Northwest Pacific. *Sci. Total Environ.* **2022**, *833*, 155146. [CrossRef]
2. Galloway, J.N.; Townsend, A.R.; Erisman, J.W.; Bekunda, M.; Cai, Z.; Freney, J.R.; Martinelli, L.A.; Seitzinger, S.P.; Sutton, M.A.J.S. Transformation of the Nitrogen Cycle: Recent Trends, Questions, and Potential Solutions. *Science* **2008**, *320*, 889–892. [CrossRef]

3. Li, Y.; Schichtel, B.A.; Walker, J.T.; Schwede, D.B.; Chen, X.; Lehmann, C.M.B.; Puchalski, M.A.; Gay, D.A.; Collett, J.L. Increasing importance of deposition of reduced nitrogen in the United States. *Proc. Natl. Acad. Sci. USA* **2016**, *113*, 5874–5879. [CrossRef] [PubMed]
4. Fleischer, K.; Rebel, K.T.; Van Der Molen, M.K.; Erisman, J.W.; Wassen, M.J.; Van Loon, E.E.; Montagnani, L.; Gough, C.M.; Herbst, M.; Janssens, I.A.; et al. The contribution of nitrogen deposition to the photosynthetic capacity of forests. *Glob. Biogeochem. Cycles* **2013**, *27*, 187–199. [CrossRef]
5. Walter, C.A.; Adams, M.B.; Gilliam, F.S.; Peterjohn, W.T. Non-random species loss in a forest herbaceous layer following nitrogen addition. *Ecology* **2017**, *98*, 2322–2332. [CrossRef] [PubMed]
6. Gilliam, F.S.; Burns, D.A.; Driscoll, C.T.; Frey, S.D.; Lovett, G.M.; Watmough, S.A. Decreased atmospheric nitrogen deposition in eastern North America: Predicted responses of forest ecosystems. *Environ. Pollut.* **2019**, *244*, 560–574. [CrossRef] [PubMed]
7. Liu, L.; Zhang, X.; Xu, W.; Liu, X.; Zhang, Y.; Li, Y.; Wei, J.; Lu, X.; Wang, S.; Zhang, W.; et al. Fall of oxidized while rise of reduced reactive nitrogen deposition in China. *J. Clean. Prod.* **2020**, *272*, 122875. [CrossRef]
8. Zhao, Y.H.; Zhang, L.; Chen, Y.F.; Liu, X.J.; Xu, W.; Pan, Y.P.; Duan, L. Atmospheric nitrogen deposition to China: A model analysis on nitrogen budget and critical load exceedance. *Atmos. Environ.* **2017**, *153*, 32–40. [CrossRef]
9. IPCC. *Contribution of Working Group I to the Fourth Assessment Report of the Intergovernmental Panel on Climate Change*; Cambridge University Press: Cambridge, UK; New York, NY, USA, 2007.
10. Freschet, G.T.; Cornwell, W.K.; Wardle, D.A.; Elumeeva, T.G.; Liu, W.; Jackson, B.G.; Onipchenko, V.G.; Soudzilovskaia, N.A.; Tao, J.; Cornelissen, J.H. Linking litter decomposition of above- and below-ground organs to plant-soil feedbacks worldwide *J. Ecol.* **2013**, *101*, 943–952. [CrossRef]
11. Prescott, C.E. Litter decomposition: What controls it and how can we alter it to sequester more carbon in forest soils? *Biogeochemistry* **2010**, *101*, 133–149. [CrossRef]
12. Laiho, R.; Laine, J.; Trettin, C.C.; Finér, L. Scots pine litter decomposition along drainage succession and soil nutrient gradients in peatland forests, and the effects of inter-annual weather variation. *Soil Biol. Biochem.* **2004**, *36*, 1095–1109. [CrossRef]
13. Yin, R.; Liu, Q.; Tian, S.; Potapov, A.; Zhu, B.; Yang, K.; Li, Z.; Zhuang, L.; Tan, B.; Zhang, L.; et al. Nitrogen deposition stimulates decomposition via changes in the structure and function of litter food webs—ScienceDirect. *Soil Biol. Biochem.* **2022**, *166*, 108522. [CrossRef]
14. Song, X.Z.; Zhou, G.M.; Gu, H.H.; Qi, L.H. Management practices amplify the effects of N deposition on leaf litter decomposition of the Moso bamboo forest. *Plant Soil* **2015**, *395*, 391–400. [CrossRef]
15. He, Z.M.; Yu, Z.P.; Huang, Z.Q.; Davis, M.; Yang, Y.S. Litter decomposition, residue chemistry and microbial community structure under two subtropical forest plantations: A reciprocal litter transplant study. *Appl. Soil Ecol.* **2016**, *101*, 84–92. [CrossRef]
16. Frey, S.D.; Ollinger, S.; Nadelhoffer, K.E.; Bowden, R.; Brzostek, E.; Burton, A.; Caldwell, B.A.; Crow, S.; Goodale, C.L.; Grandy, A.S.; et al. Chronic nitrogen additions suppress decomposition and sequester soil carbon in temperate forests. *Biogeochemistry* **2014**, *121*, 305–316. [CrossRef]
17. Xia, M.X.; Talhelm, A.F.; Pregitzer, K.S. Chronic nitrogen deposition influences the chemical dynamics of leaf litter and fine roots during decomposition. *Soil Biol. Biochem.* **2017**, *112*, 24–34. [CrossRef]
18. Ramirez, K.S.; Craine, J.M.; Fierer, N. Consistent effects of nitrogen amendments on soil microbial communities and processes across biomes. *Glob. Chang. Biol.* **2012**, *18*, 1918–1927. [CrossRef]
19. Wang, Q.; Kwak, J.H.; Choi, W.J.; Chang, S.X. Decomposition of trembling aspen leaf litter under long-term nitrogen and sulfur deposition: Effects of litter chemistry and forest floor microbial properties. *For. Ecol. Manag.* **2018**, *412*, 53–61. [CrossRef]
20. Peng, Y.; Li, Y.J.; Song, S.Y.; Chen, Y.Q.; Chen, G.T.; Tu, L.H. Nitrogen addition slows litter decomposition accompanied by accelerated manganese release: A five-year experiment in a subtropical evergreen broadleaf forest. *Soil Biol. Biochem.* **2022**, *165*, 108511. [CrossRef]
21. Zhang, Y.J.; Jin, Y.H.; Xu, J.W.; He, H.S.; Tao, Y.; Yang, Z.P.; Bai, Y.Y. Effects of exogenous N and endogenous nutrients on alpine tundra litter decomposition in an area of high nitrogen deposition. *Sci. Total Environ.* **2022**, *805*, 150388. [CrossRef]
22. Liu, X.; Zhang, G.C.; Heathman, G.C.; Wang, Y.Q.; Huang, C.H. Fractal features of soil particle-size distribution as affected by plant communities in the forested region of Mountain Yimeng, China. *Geoderma* **2009**, *154*, 123–130. [CrossRef]
23. Yan, T.T.; Zhao, W.J.; Zhu, Q.K.; Xu, F.J.; Gao, Z.K. Spatial distribution characteristics of the soil thickness on different land use types in the Yimeng Mountain Area, China. *Alex. Eng. J.* **2021**, *60*, 511–520. [CrossRef]
24. Zhang, R.H.; Xia, L.; Heathman, G.C.; Yao, X.Y.; Hu, X.L.; Zhang, G.C. Assessment of soil erosion sensitivity and analysis of sensitivity factors in the Tongbai–Dabie mountainous area of China. *Catena* **2013**, *101*, 92–98. [CrossRef]
25. Lasota, J.; Małek, S.; Jasik, M.; Bońska, E. Effect of planting method on C:N:P stoichiometry in soils, young silver fir (*Abies alba* Mill.) and stone pine (*Pinus cembra* L.) in the upper mountain zone of karpaty mountains. *Ecol. Indic.* **2021**, *129*, 107905. [CrossRef]
26. Qin, Y.Y.; Feng, Q.; Adamowski, J.F.; Zhu, M.; Zhang, X.F. Community level response of leaf stoichiometry to slope aspect in a montane environment: A case study from the Central Qilian Mountains, China. *Glob. Ecol. Conserv.* **2021**, *28*, e01703. [CrossRef]
27. Rosinger, C.; Rousk, J.; Sandén, H. Can enzymatic stoichiometry be used to determine growth-limiting nutrients for microorganisms?—A critical assessment in two subtropical soils. *Soil Biol. Biochem.* **2019**, *128*, 115–126. [CrossRef]
28. Xu, M.P.; Jian, J.N.; Wang, J.Y.; Zhang, Z.J.; Yang, G.H.; Han, X.H.; Ren, C.J. Response of root nutrient resorption strategies to rhizosphere soil microbial nutrient utilization along *Robinia pseudoacacia* plantation chronosequence. *For. Ecol. Manag.* **2021**, *489*, 119053. [CrossRef]
29. Sinsabaugh, R.L.; Lauber, C.L.; Weintraub, M.N.; Ahmed, B.; Zeglin, L.H. Stoichiometry of soil enzyme activity at global scale *Ecol. Lett.* **2008**, *11*, 1252–1264. [CrossRef]
30. Xu, W.; Luo, X.S.; Pan, Y.P.; Zhang, L.; Tang, A.H.; Shen, J.L.; Zhang, Y.; Li, K.H.; Wu, Q.H.; Yang, D.W.; et al. Quantifying atmospheric nitrogen deposition through a nationwide monitoring network across China. *Atmos. Chem. Phys.* **2015**, *15*, 12345–12360. [CrossRef]

31. Deforest, J.L. The influence of time, storage temperature, and substrate age on potential soil enzyme activity in acidic forest soils using MUB-linked substrates and l -DOPA. *Soil Biol. Biochem.* **2009**, *41*, 1180–1186. [CrossRef]
32. Liu, Q.; Yin, R.; Tan, B.; You, C.M.; Zhang, L.; Zhang, J.; Xu, Z.F.; Schadler, M.; Scheu, S. Nitrogen addition and plant functional type independently modify soil mesofauna effects on litter decomposition. *Soil Biol. Biochem.* **2021**, *160*, 108340. [CrossRef]
33. Olson, J.S. Energy Storage and the Balance of Producers and Decomposers in Ecological Systems. *Ecology* **1963**, *44*, 322–331. [CrossRef]
34. Deforest, J.L. Chronic phosphorus enrichment and elevated pH suppresses *Quercus* spp. leaf litter decomposition in a temperate forest—ScienceDirect. *Soil Biol. Biochem.* **2019**, *135*, 206–212. [CrossRef]
35. Wu, J.P.; Liu, W.F.; Zhang, W.X.; Shao, Y.H.; Duan, H.L.; Chen, B.D.; Wei, X.H.; Fan, H.B. Long-term nitrogen addition changes soil microbial community and litter decomposition rate in a subtropical forest. *Appl. Soil Ecol.* **2019**, *142*, 43–51. [CrossRef]
36. Wu, C.S.; Shu, C.J.; Zhang, Z.J.; Zhang, Y.; Liu, Y.Q. Effect of N deposition on the home-field advantage of wood decomposition in a subtropical forest. *Ecol. Indic.* **2022**, *140*, 109043. [CrossRef]
37. Vivanco, L.; Austin, A.T. Nitrogen addition stimulates forest litter decomposition and disrupts species interactions in Patagonia, Argentina. *Clobal Chang. Biol.* **2011**, *17*, 1963–1974. [CrossRef]
38. Wu, A.; Yin, R.; Xu, Z.; Zhang, L.; You, C.; Liu, Y.; Li, H.; Wang, L.; Liu, S.; Zhang, Y.; et al. Forest gaps slow lignin and cellulose degradation of fir (*Abies faxoniana*) twig litter in an alpine forest. *Geoderma* **2019**, *424*, 116010. [CrossRef]
39. Pei, G.T.; Liu, J.; Peng, B.; Gao, D.C.; Wang, C.; Dai, W.W.; Jiang, P.; Bai, E. Nitrogen, lignin, C/N as important regulators of gross nitrogen release and immobilization during litter decomposition in a temperate forest ecosystem. *For. Ecol. Manag.* **2019**, *440*, 61–69. [CrossRef]
40. Frey, S.D.; Knorr, M.; Parrent, J.L.; Simpson, R.T. Chronic nitrogen enrichment affects the structure and function of the soil microbial community in temperate hardwood and pine forests. *For. Ecol. Manag.* **2004**, *196*, 159–171. [CrossRef]
41. Micks, P.; Downs, M.R.; Magill, A.H.; Nadelhoffer, K.J.; Aber, J.D. Decomposing litter as a sink for 15N-enriched additions to an oak forest and a red pine plantation. *For. Ecol. Manag.* **2004**, *196*, 71–87. [CrossRef]
42. Knorr, M.; Frey, S.D.; Curtis, P.S. Nitrogen additions and litter decomposition: A meta-analysis. *Ecology* **2005**, *86*, 3252–3257. [CrossRef]
43. Rubino, M.; Dungait, J.A.J.; Evershed, R.P.; Bertolini, T.; De Angelis, P.; D'Onofrio, A.; Lagomarsino, A.; Lubritto, C.; Merola, A.; Terrasi, F.; et al. Carbon input belowground is the major C flux contributing to leaf litter mass loss: Evidences from a 13C labelled-leaf litter experiment. *Soil Biol. Biochem.* **2010**, *42*, 1009–1016. [CrossRef]
44. Lin, H.; He, Z.H.; Hao, J.W.; Tian, K.; Jia, X.Q.; Kong, X.S.; Akbar, S.; Bei, Z.L.; Tian, X.J. Effect of N addition on home-field advantage of litter decomposition in subtropical forests—ScienceDirect. *For. Ecol. Manag.* **2017**, *398*, 216–225. [CrossRef]
45. Duan, P.P.; Xiao, K.C.; Wang, K.L.; Li, D.J. Responses of soil respiration to nitrogen addition are mediated by topography in a subtropical karst forest. *Catena* **2023**, *221*, 106759. [CrossRef]
46. Marshall, J.D.; Peichl, M.; Tarvainen, L.; Lim, H.; Lundmark, T.; Nasholm, T.; Oquist, M.; Linder, S. A carbon-budget approach shows that reduced decomposition causes the nitrogen-induced increase in soil carbon in a boreal forest. *For. Ecol. Manag.* **2021**, *502*, 119750. [CrossRef]
47. Ngaba, M.J.Y.; Uwiragiye, Y.; Bol, R.; Vries, W.D.; Zhou, J.B. Low-level nitrogen and short-term addition increase soil carbon sequestration in Chinese forest ecosystems. *Catena* **2014**, *215*, 106333. [CrossRef]
48. Dias, T.; Oakley, S.; Alarcón-Gutiérrez, E.; Ziarelli, F.; Trindade, H.; Martins, M.A.; Sheppard, L.; Ostle, N.; Cruz, C. N-driven changes in a plant community affect leaf-litter traits and may delay organic matter decomposition in a Mediterranean maquis. *Soil Biol. Biochem.* **2013**, *58*, 163–171. [CrossRef]
49. Chomel, M.; Baldy, V.; Guittonny-Larchevque, M. Effect of mixing herbaceous litter with tree litters on decomposition and N release in boreal plantations. *Plant Soil* **2016**, *398*, 229–241. [CrossRef]
50. Bai, X.J.; Dippold, M.A.; An, S.S.; Wang, B.R.; Zhang, H.X.; Loeppmann, S. Extracellular enzyme activity and stoichiometry: The effect of soil microbial element limitation during leaf litter decomposition. *Ecol. Indic.* **2021**, *121*, 107200. [CrossRef]
51. Zhang, Y.; Shengzhe, E.; Wang, Y.; Su, S.; Bai, L.; Wu, C.; Zeng, X. Long-term manure application enhances the stability of aggregates and aggregate-associated carbon by regulating soil physicochemical characteristics. *Catena* **2021**, *203*, 105342. [CrossRef]
52. Yang, Y.; Liang, C.; Wang, Y.Q.; Cheng, H.; An, S.S.; Chang, S.X. Soil extracellular enzyme stoichiometry reflects the shift from P- to N-limitation of microorganisms with grassland restoration. *Soil Biol. Biochem.* **2020**, *149*, 107928. [CrossRef]
53. Yang, Y.; Liu, B.R.; An, S.S. Ecological stoichiometry in leaves, roots, litters and soil among different plant communities in a desertified region of Northern China. *Catena* **2018**, *166*, 328–338. [CrossRef]
54. Li, J.W.; Liu, Y.L.; Hai, X.Y.; Shangguan, Z.P.; Deng, L. Dynamics of soil microbial C:N:P stoichiometry and its driving mechanisms following natural vegetation restoration after farmland abandonment. *Sci. Total Environ.* **2019**, *693*, 133613. [CrossRef] [PubMed]
55. Li, H.; Chen, Y.; Lu, Z.; Wang, F.; Lambers, H.; Zhang, J.; Qin, G.; Zhou, J.; Wu, J.; Zhang, L.; et al. Nitrogen deposition in low-phosphorus tropical forests benefits soil C sequestration but not stabilization. *Ecol. Indic.* **2023**, *146*, 109761. [CrossRef]
56. Wang, H.; Liu, S.; Wang, J.; You, Y.; Yang, Y.; Shi, Z.; Huang, X.; Zheng, L.; Li, Z.; Ming, A.; et al. Mixed-species plantation with *Pinus massoniana* and *Castanopsis hystrix* accelerates C loss in recalcitrant coniferous litter but slows C loss in labile broadleaf litter in southern China. *For. Ecol. Manag.* **2018**, *422*, 207–213. [CrossRef]

forests

MDPI

Review

Effect of Changes in Throughfall on Soil Respiration in Global Forest Ecosystems: A Meta-Analysis

Xingkai Xu [1,2]

[1] State Key Laboratory of Atmospheric Boundary Layer Physics and Atmospheric Chemistry, Institute of Atmospheric Physics, Chinese Academy of Sciences, Beijing 100029, China; xingkai_xu@mail.iap.ac.cn

[2] Department of Atmospheric Chemistry and Environmental Science, College of Earth and Planetary Sciences, University of Chinese Academy of Sciences, Beijing 100049, China

Abstract: To date, there has been limited knowledge about how soil carbon dioxide (CO_2) emission from forest ecosystems at a global scale respond to the altered precipitation, and the key influencing mechanisms involved. Thirty-seven studies conducted under throughfall manipulation conditions in forest ecosystems around the globe were selected in this meta-analysis, with a total of 103 paired observations. Experimental categories such as climate types, forest types, soil texture, and the area size of changes in throughfall manipulation were included to qualify the responses of annual soil CO_2 emissions to the altered throughfall. The responses of the annual soil CO_2 emissions to the altered throughfall would be more sensitive in temperate forests than those in tropical and subtropical forests, probably due to the relatively long residence time of soil carbon (C) and the seasonal freeze-thaw events in temperate forests, as well as the relatively high concentration of non-structural carbohydrates in the belowground part of temperate terrestrial plants. A relatively large positive response of the soil CO_2 emissions to the increased throughfall was observed in Mediterranean forest due to small precipitation during the growing season and mostly coarse-textured soils. Besides climate types, the sizes of the effect of the altered throughfall on the soil CO_2 emissions ($\ln R_{CO2}$) varied with forest types and soil texture categories. Based on the regression analysis of the $\ln R_{CO2}$ value against the changes in throughfall, the annual soil CO_2 emissions in forest ecosystems at a global scale would be increased by 6.9%, provided that the change in annual precipitation was increased by 10%. The results of structural equation modeling analysis indicate that fine root biomass and soil microbial biomass, along with the changes in annual precipitation, would substantially affect the altered throughfall-induced annual soil CO_2 emissions in global forest ecosystems. The findings of this meta-analysis highlight that the measurement of soil respiration components, the priming effects of soil organic C decomposition, and C allocation between the aboveground and belowground parts of different tree species under the altered precipitation conditions, deserve more attention in the future.

Keywords: forest soil; fine root biomass; microbial biomass carbon; soil carbon dioxide emission; soil respiration; throughfall manipulation

Citation: Xu, X. Effect of Changes in Throughfall on Soil Respiration in Global Forest Ecosystems: A Meta-Analysis. *Forests* **2023**, *14*, 1037. https://doi.org/10.3390/f14051037

Academic Editor: Cate Macinnis-Ng

Received: 19 April 2023
Revised: 12 May 2023
Accepted: 15 May 2023
Published: 17 May 2023

1. Introduction

Forests cover about 31% of the world's land area [1], and as the largest carbon (C) sink among terrestrial ecosystems, forest ecosystems account for more than 86% of the C pool in vegetation and more than 73% of the C pool in soil [2,3]. Because two-thirds of forest ecosystem respiration results from soil respiration [4–6], a small change in soil respiration due to climate changes, particularly in heterotrophic respiration, can even result in an important change in the C balance of forest ecosystems [5,7,8], probably leading to impacts on global C cycling, atmospheric carbon dioxide (CO_2) concentration, and the climate system [9–11].

Because of climate change (especially global warming), air circulation and the hydrological cycle have largely been intensified, leading to rapid shifts in precipitation regimes across the globe. For example, global precipitation was estimated to increase by 7.4 ± 2.6% with each 1 °C increment in temperature over the period 1987–2006 [12]. The altered precipitation may directly and indirectly affect terrestrial C dynamics, and then ecosystem structure and function [13], which may influence interactions with other global change drivers (e.g., elevated CO_2 and climate warming). Therefore, understanding the responses of soil C dynamics to precipitation changes is of key importance to accurately predict the rate and extent of climate change [14].

Soil respiration, a vital process of soil C dynamics, is strongly affected by soil water availability and its association with the activities of roots and microorganisms [5,8,11], which is closely related to changing precipitation patterns [15]. Normally, throughfall (TF) is much larger than stemflow as a component of total precipitation under the forest canopy, and the amount of TF input varies with the characteristics of precipitation and the structure of forest ecosystems, such as tree species compositions and stand density [16]. In recent decades, precipitation manipulation experiments via TF transferring have widely occurred in forest ecosystems in different climate zones around the globe. The TF reduction was reported to decrease soil respiration in some temperate and tropical forests [15–22], whereas elevated soil respiration was measured in some tropical forests [23–25] and in subtropical broadleaf forests [26–29]. In Mediterranean forests, soil CO_2 emissions upon TF reduction were reported to be variable, whereas the emissions increased upon TF addition [30–33]. A previous meta-analysis showed that decreased precipitation across all terrestrial ecosystems reduced soil respiration by 12%, whereas increased precipitation increased soil respiration by 45% [34]. In spite of this, a doubling precipitation treatment did not affect soil respiration in a mixed subtropical coniferous and broadleaf forest [35]. The different direction and magnitude of the soil CO_2 emissions' responses to precipitation changes is likely due to the considerable spatial heterogeneity in edaphic conditions (e.g., soil organic matter content, texture, and bulk density) and biotic communities within and across forest ecosystems, as well as climate types [19,22,36–40]. However, to date, there has been limited knowledge available about whether the responses of CO_2 emissions from forest soils to altered precipitation vary with climate types.

Soil CO_2 emissions usually have a peak under moderate soil moisture conditions, and decrease when soil moisture becomes either relatively wetter or drier, especially in tropical forests [25,41–43]. Besides soil moisture, the changes in root activity and root biomass and soil microbial biomass C under altered precipitation conditions would have important effects on the precipitation-induced soil CO_2 emissions from forest ecosystems [41,42,44,45]. Furthermore, soil texture, the aboveground litter layer of forest stands, and ambient temperature and humidity can substantially affect the changes in the soil water retention characteristics following a precipitation event, thus influencing CO_2 emissions from forest soils [19,46]. Probably, the responses of the soil CO_2 emissions from forest ecosystems to altered precipitation are dependent on experimental categories such as climate types, forest types, soil texture, and the area sizes of changes in TF manipulation. However, to date, limited knowledge has been available about how the annual soil CO_2 emissions in forest ecosystems at a global scale respond to the altered precipitation among the different experimental categories, and the key influencing mechanisms involved.

Based on 103 previously published pairs of observation data from 37 throughfall manipulation field experiments in forest ecosystems at a global scale in recent decades, a meta-analysis was performed to (1) quantify the effect sizes on the altered TF-induced annual soil CO_2 emissions ($\ln R_{CO2}$) among the different experimental categories, and (2) to find whether the $\ln R_{CO2}$ values vary with climate types, forest types, soil texture, and the area sizes of changes in TF manipulation, as well as the related mechanisms affecting the $\ln R_{CO2}$, by fully incorporating the results of nearly 60 references published previously. The third objective of this meta-analysis was to explore the dominant factors regulating the $\ln R_{CO2}$ values in global forest ecosystems under altered precipitation conditions.

2. Materials and Methods

2.1. Data Collection and Quality Control

Data on CO_2 emissions from forest soils under throughfall manipulation conditions were collected by searching the published literature in the databases of ISI-Web of Science and Google Scholar, as well as the China National Knowledge Infrastructure, using combinations of keywords such as "throughfall" or "rainfall" or "precipitation" and "carbon dioxide" or "CO_2" or "respiration" and "forest soil" in the topic field for the period between 1 January 1990 and 31 December 2021. Information was recorded relating to the nation, climate zone, forest type, annual precipitation, annual average air temperature, soil properties such as total organic C and total N, pH, bulk density, soil texture, the duration of measurement, and CO_2 emission rate. For a few study cases where soil bulk density values were not available, these were estimated according to an equation developed by Post and Kwon [47]. Experimental datasets that met the following criteria were finally included in this review: (1) only field-based measurements covering at least one complete growing season or over the year; (2) studies including paired control and TF reduction or TF addition treatments, with at least three independent repeated observations per treatment; (3) field measurements with a combination of TF manipulation and other factors (e.g., warming and nitrogen deposition) were excluded, considering the fact that these factors altered soil properties and thus the soil CO_2 emission; (4) the accessibility of cumulative CO_2 emission data with experimental unit area. Based on these criteria, 37 studies conducted under TF manipulation conditions in forest ecosystems were selected from around the global (Table S1). Data were collected from the tables and figures of these selected studies. The data shown in figures were extracted using Getdata Graph Digitizer (version 2.26) (available at https://getdata-graph-digitizer.software.informer.com/, accessed on 10 March 2020) Either the mean or the experimental unit-level data were collected, depending on data availability, resulting in a total of 103 paired observations. Among them, 80 paired observations reported changes in the soil CO_2 emissions upon TF reduction, and the rest reported those following TF addition. Besides the soil CO_2 emission data, soil properties (e.g., moisture and temperature, pH, microbial biomass C, biomass of fine roots, exchangeable NH_4-N and NO_3-N contents, and dissolved organic C content) under TF manipulation conditions were collected, considering data availability from the literature, in order to study the causal relationships among the effect sizes on the soil CO_2 emissions and soil properties under altered TF conditions.

Experimental conditions, such as climate type, forest type, soil texture, and area size of changes in TF manipulation, were also collected (Table S1). These data were then sorted into groups of experimental conditions and analyzed as categorical variables (Table 1). Climate types were classified into subtropical, tropical, temperate, Mediterranean, and boreal. Forest types were sorted into coniferous, broadleaf, and mixed forests. Soil texture was classified into three groups: fine (silt clay and clay), medium (silt clay loam, clay loam silt loam, and loam), and coarse (sandy loam, sandy clay loam, loamy sand, and sand) soils. The area sizes of changes during TF manipulation were sorted into three groups: size 1 ($\leq$5 m $\times$ 5 m), size 2 (>5 m $\times$ 5 m and <20 m $\times$ 20 m), and size 3 ($\geq$20 m $\times$ 20 m) The numbers of paired observations within these experimental categories are shown in Table 1. These experimental categories were used to qualify the effect sizes of the soil CO_2 emissions and soil properties, as mentioned afterwards.

Table 1. Amount of data analyzed when evaluating the lnR_{CO2} values under different experimental conditions and the statistical analysis methods.

Experimental Categories	Specific Conditions (Number of Paired Observation Data Points) [a]	Methods of Statistical Analysis
Climate conditions	Climate types (103): Boreal (6), Mediterranean (11), Temperate (36), Tropical (20), and Subtropical (30)	Analysis of variance (ANOVA)
	Annual rainfall (103): 493 to 3528 mm	Structural equation modeling (SEM)
	Annual mean air temperature (103): −3.5 to 28.0 °C	SEM
Forest types	Coniferous (13), Broadleaf (62), and Mixed forest (28)	ANOVA
Soil properties	Texture (103) [b]: Fine (19), Medium (36), and Coarse (48)	ANOVA
	pH (103): 3.2 to 7.5	SEM
	SOC (103): 1.4 to 83.0 g C kg^{-1}	SEM
	TN (103): 0.1 to 11.3 g N kg^{-1}	SEM
	Ratio of SOC to TN (103): 5.8 to 44.7	SEM
	Bulk density (103): 0.5 to 1.6 g cm^{-3}	SEM
TFsize [c]	Sizes (103): Size 1 ($\leq$5 m × 5 m) (60), Size 2 (>5 m × 5 m and <20 m × 20 m) (18), and Size 3 ($\geq$20 m × 20 m) (25)	ANOVA

[a] The amount of data differed across conditions according to data availability within the literature. [b] Fine (silt clay and clay), Medium (silt clay loam, clay loam, silt loam, and loam), and Coarse (sandy loam, sandy clay loam, loamy sand, and sand). [c] TFsize, the size of throughfall manipulation in area across the different experiments.

2.2. Data Analysis

Based on previous paired observation data from TF manipulation field experiments in forest ecosystems at a global scale (Table S1), a meta-analysis was performed using the software MetaWin 2.1 [48], in order to compare the effect sizes of the altered TF-induced annual soil CO_2 emissions across the experimental categories. Missing standard deviations of a few datasets were replaced by the average values of standard deviations of other groups of data using the software MetaWin 2.1. The natural log of the response ratio (lnR_{CO2}), defined as the "effect size", was used to assess the responses of the annual soil CO_2 emissions to decreased or increased TF treatments using the following equation [49,50]:

$$LnR = \ln (X_t/X_c) \quad (1)$$

where X_t and X_c are the means of annual soil CO_2 emissions for the decreased or increased TF treatment and control groups, respectively.

The confidence interval (CI) of the effect size was generated using bootstrapping (999 iterations). The effect was considered significant if the 95% CI of the effect size did not exceed zero. To facilitate our explanation, the mean effect size (lnR_{CO2}) was transformed back to the percentage change of the altered TF treatments using the following equation [50]:

$$(e^{LnR} - 1) \times 100\% \quad (2)$$

The whole dataset was checked for homogeneity of variance and normalization of distribution with Levene's test and the Kolmogorov–Smirnov test, respectively. The results show that the variance was homogeneous, and the distribution was normal (Figure 1). Differences in the LnR_{CO2} values among the experimental categories were assessed with analysis of variance (ANOVA) using the general linear model of the IBM SPSS Statistics 19 (IBM Corp., New York, NY, USA) (Table 2). When the ANOVA was significant, pairwise comparisons of means were undertaken using Duncan's multiple range test. The LnR_{CO2} values of all paired observations in this meta-analysis were plotted against the percentage changes in TF and fitted with linear regressions. The effect sizes of soil properties under the altered TF conditions were calculated using similar methods to those mentioned, such as the lnR_{CO2}. Here, the effect sizes of soil properties, such as soil moisture, fine root biomass, microbial biomass C, exchangeable NH_4-N and NO_3-N contents, dissolved organic C contents, and pH values (namely, $lnR_{moisture}$, lnR_{roots}, lnR_{mbc}, lnR_{NH4}, lnR_{NO3}, lnR_{DOC}, and lnR_{pH}) were assessed with ANOVA using the general linear model of the IBM SPS Statistics 19 (IBM Corp., New York, NY, USA) (Table S2). Soil temperature data among the experimental categories were not included because of the lack of significant differences. The sizes of effects on the soil properties selected among the experimental categories are described with box plots.

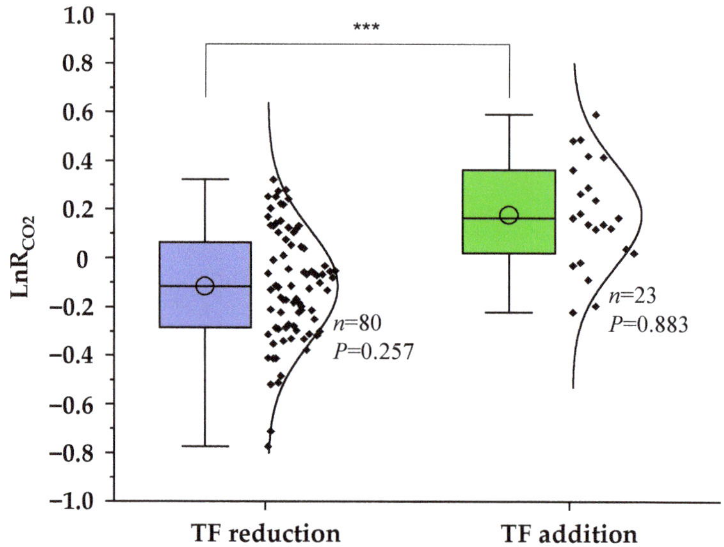

Figure 1. Box normal plots of the lnR_{CO2} values of TF reduction and TF addition treatments across all experimental categories. Boxes and horizontal lines and circles within boxes represent interquartile ranges, as well as median and mean values, respectively. Differences in the means of the lnR_{CO2} values between TF reduction and TF addition were assessed by *t*-test statistics. ***, $p < 0.001$. At the $a = 0.05$ level, the data on TF reduction and TF addition were significant when drawn from a normally distributed population, respectively.

Table 2. Results of ANOVA on the effects of the experimental categories on the lnR_{CO2} values for the whole dataset.

Experimental Categories	Effects	lnR_{CO2} Upon TF Reduction		lnR_{CO2} Upon TF Addition	
		F Values	*p* Values	*F* Values	*p* Values
Climate conditions	Climate types	1.439	0.230	10.261	0.000
	Annual rainfall	NA [a]		NA	
	Annual mean air temperature	NA		NA	
Forest types	Forest types	7.955	0.001	1.819	0.188
Soil properties	Texture	13.616	0.000	0.607	0.444
	pH	NA		NA	
	SOC	NA		NA	
	TN	NA		NA	
	Ratio of SOC to TN	NA		NA	
	Bulk density	NA		NA	
TFsize	Sizes	0.775	0.464	0.031	0.970

[a] Not applicable as structural equation modeling analyses were conducted.

The percentage changes in TF (ΔTP), $lnR_{moisture}$, lnR_{mbc}, lnR_{roots}, and lnR_{DOC}) were selected as predictors to establish a priori SEM analyses to evaluate the direct and indirect effects of ΔTP, and the effect sizes of selected soil properties on the lnR_{CO2} values. Another

SEM analysis was performed to investigate the direct and indirect effects of annual rainfall, annual mean air temperature, the ratio of soil total organic C to total N (SOCtoTN), and soil bulk density on the $\ln R_{CO2}$ values under the experimentally altered TF conditions (Table 1). The standardized total effects of selected variables on the $\ln R_{CO2}$ values have been explained. The overall goodness of fit for the SEM analysis was tested using the Chi-squared test and the root-mean-square error of approximation. The SEM analysis was conducted with the software AMOS 24 (SPSS Inc., Chicago, IL, USA).

3. Results

3.1. Overall Pattern of Soil CO_2 Emission by TF Manipulation

The sizes of effects on the annual soil CO_2 emissions ($\ln R_{CO2}$) in forest ecosystems under altered TF conditions at a global scale are shown in Figure 2. On average, decreased rainfall significantly depressed the annual soil CO_2 emission by 8.8% (95% CI: −13.0% to −4.4%) (Figure 2a), while increased rainfall significantly increased the emissions by 19.7% (95% CI: 9.7% to 30.8%) (Figure 2b). Under decreased TF conditions, annual soil CO_2 emissions in temperate and subtropical forests were significantly reduced by 16.0% and 9.1% (95% CI: −22.4% to −9.1% and −16.0% to −1.5%), respectively, whereas no significant changes in the soil CO_2 emissions were observed in tropical, Mediterranean, and boreal forests (Figure 2a). Following the TF addition, annual soil CO_2 emissions in temperate and Mediterranean forests were significantly increased by 43.9% and 43.4% (95% CI: 26.7% to 63.4% and 21.4% to 69.5%), respectively, whereas there were no significant changes in the soil CO_2 emissions in subtropical and tropical forests (Figure 2b). Furthermore, the averages of the $\ln R_{CO2}$ values under increased TF conditions became larger in temperate and Mediterranean forests than those in tropical and subtropical forests ($p < 0.05$) (Figure 2b and Table 2). Overall, the responses of the annual soil CO_2 emissions to the altered rainfall would be more sensitive in temperate forests than those in tropical and subtropical forests (Figure 2), and a relatively large positive response of the soil CO_2 emissions to increased rainfall was observed in Mediterranean forests (Figure 2b).

Regardless of the experimental categories, the $\ln R_{CO2}$ values upon TF addition were on average larger than those upon TF reduction ($p < 0.001$) (Figure 1). There were significant effects of forest types on the $\ln R_{CO2}$ values upon TF reduction (Table 2). Under decreased TF conditions, the annual soil CO_2 emissions in coniferous and mixed forests were, on average, significantly reduced by 17.4% and 20.6% (95% CI: −27.4% to −6.0% and −28.0% to −12.5%), respectively, whereas no significant changes in the soil CO_2 emissions occurred in broadleaf forests (Figure 2a). Compared with the control, the averages of $\ln R_{CO2}$ values in broadleaf and mixed forests under increased TF conditions were significantly increased ($p < 0.05$), whereas no changes in the $\ln R_{CO2}$ values occurred in coniferous forests (Figure 2b), which indicates that the responses of the annual soil CO_2 emissions to increased rainfall would be more sensitive in broadleaf and mixed forests than in coniferous forests.

Soil texture has a vital effect on the soil CO_2 emission by regulating its biophysical properties. On average, decreased rainfall significantly depressed the CO_2 emissions from medium-textured soils by 22.8% (95% CI: −29.7% to −15.2%), but it had no significant effects on either fine or coarse-textured soils (Figure 2a). Furthermore, the $\ln R_{CO2}$ values in medium-textured soils became on average much smaller than the values in coarse and fine-textured soils ($p < 0.05$) (Figure 2a and Table 2), suggesting that the responses of the CO_2 emissions to decreased rainfall would be more sensitive in medium-textured soils. Contrary to those under decreased TF conditions, the CO_2 emissions from coarse and medium-textured soils were on average increased by 29.9% and 12.8% (95% CI: 12.1% to 50.5% and –0.4% to 27.7%) under increased TF conditions, respectively, and a significant increase in the $\ln R_{CO2}$ values occurred in coarse-textured soils compared with the control ($p < 0.05$) (Figure 2b).

Regarding the TFsize category, there was, on average, generally a reduction in the $\ln R_{CO2}$ values upon TF reduction, and an increase in the values with increased TF (Figure 2).

However, the differences in lnR_{CO2} values within the TFsize category are not significant (Table 2).

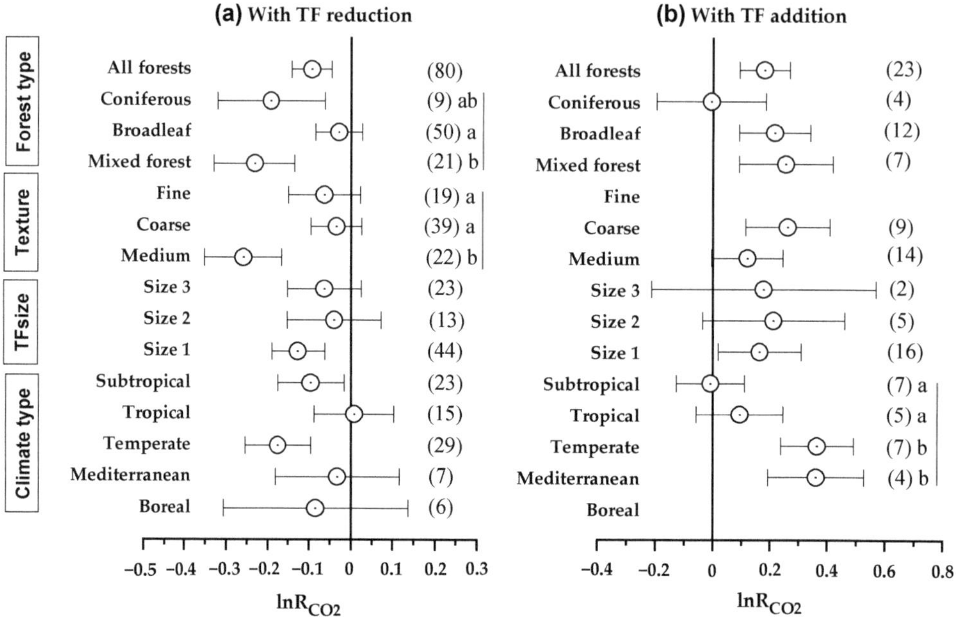

Figure 2. Effect sizes of TF reduction (**a**) and TF addition (**b**) treatments on the annual CO_2 emissions from forest soils (lnR_{CO2}) among the experimental categories. The number of paired observations is shown in parenthesis, and the results of ANOVA are provided in Table 2. Different letters next to red lines indicate statistical differences among the experimental categories at the $a = 0.05$ level Error bars represent 95% confidence intervals (CI) of the lnR_{CO2} values based on normal distribution and the effect was considered significant if the 95% CI of the effect size did not exceed zero. TFsize represents the size of throughfall manipulation in area, and sizes 1–3 are explained in Table 1. Soil texture categories are documented in Table 1.

3.2. *Relationship between lnR_{CO2} Values and Changes in Throughfall*

The sizes of effect on the annual soil CO_2 emissions and lnR_{CO2} values under altered TF conditions generally conformed to a normal distribution ($p = 0.20$), and the variance was homogeneous ($p = 0.31$ with Levene's test) (Figure 3a). A 28% variation in the lnR_{CO2} values for the whole dataset can be explained by changes in throughfall (Figure 3b). Based on the linear regression of the lnR_{CO2} values against the changes in throughfall (Figure 3b), the annual soil CO_2 emission in forest ecosystems at a global scale would be increased by 6.9%, provided that the change in annual rainfall was increased by 10%. Furthermore, the slope values of linear regressions within the soil texture category indicate that the largest response of CO_2 emissions to the changes in annual rainfall across global forest ecosystems was shown by medium-textured soils (Figure 3b).

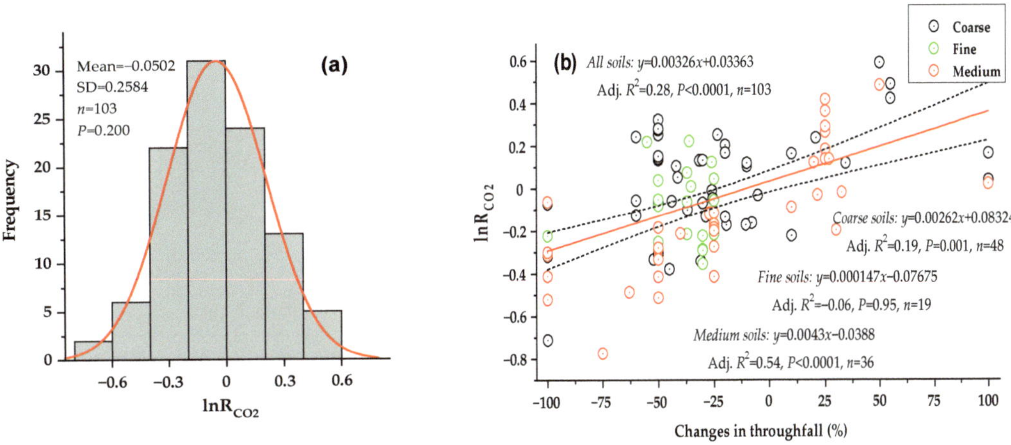

Figure 3. Frequency distribution of effect sizes on the annual soil CO_2 emission (lnR_{CO2}) (**a**) and the relationship between the lnR_{CO2} values and the changes in throughfall (**b**) for the whole dataset. The linear regression of the lnR_{CO2} values against the changes in throughfall for the whole dataset is shown in (**b**). Dashed lines represent upper and lower 95% confidence intervals, respectively. Linear regression equations of the lnR_{CO2} values against the changes in throughfall for coarse, fine, medium-textured soils, and all soils are respectively shown in (**b**).

3.3. Effect of TF Manipulation on Soil Properties

The results of ANOVA on the sizes of effects on soil properties among the experimental categories are shown in Table S2. Under decreased TF conditions, the $lnR_{moisture}$ values were significantly affected by the soil texture category (Table S2), with the smallest values shown by coarse-textured soils ($p < 0.05$) (Figure 4b). On average, the $lnR_{moisture}$ values upon TF addition became larger in broadleaf forests than in mixed forests ($p < 0.05$) (Figure 4a). As regards the climate type categories, the $lnR_{moisture}$ values upon TF addition were on average the largest in Mediterranean forests, whereas no significant differences arose within the experimental category (Figure 4c). The TFsize category did not significantly affect the $lnR_{moisture}$ values across all selected forest stands (Table S2 and Figure 4d).

The lnR_{roots} values upon TF reduction were significantly affected by the climate type category (Table S2). On average, the lnR_{roots} values under the altered TF conditions were the largest in the coarse-textured soils among the soil texture category (Figure S1c). The lnR_{NH4} values upon TF addition were significantly influenced by the climate type, forest type, and soil texture categories, respectively (Table S2), and the values became negative in coarse-textured soils and positive in medium-textured soils, respectively ($p < 0.01$) (Figure S1b). The lnR_{mbc} values upon TF reduction within the forest type category became negative on average, along with the smallest values seen in coniferous forests compared with the values in broadleaf forests and mixed forests (Figure S1a). Both lnR_{DOC} and lnR_{pH} values under decreased TF conditions were significantly affected by the climate type and soil texture categories, respectively, and the lnR_{pH} values under increased TF conditions were significantly affected by the climate type category (Table S2).

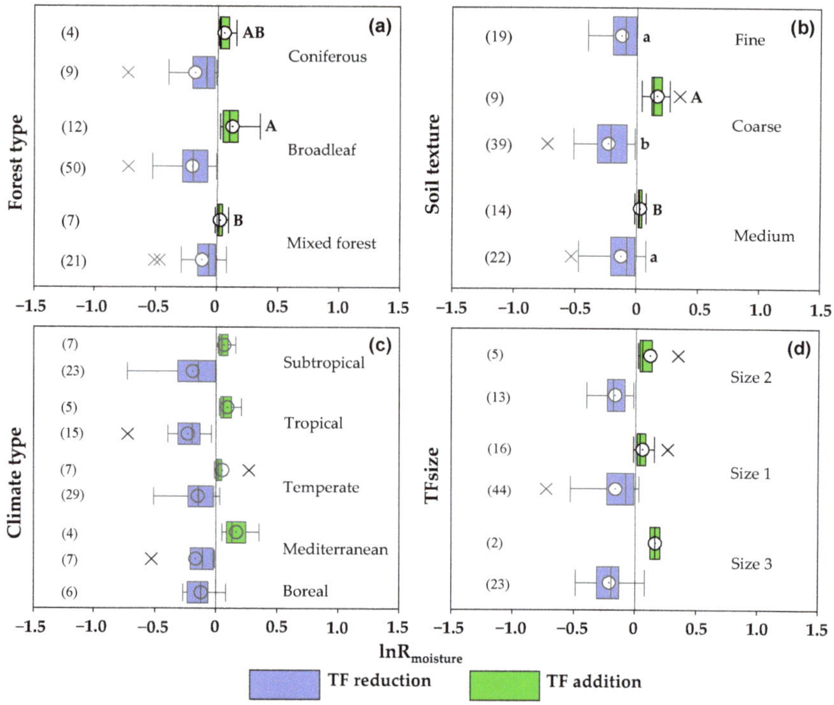

Figure 4. Box plots of sizes of effects on soil moisture ($lnR_{moisture}$) by forest type (**a**), soil textur (**b**), climate type (**c**), and TFsize (**d**) categories. The number of paired observations in brackets i shown according to relevant information from some studies. In (**b**), different small letters close t boxes represent statistical differences of the $lnR_{moisture}$ values upon TF reduction among the soi texture categories. In (**a**,**b**), different capital letters close to boxes represent statistical differences i the $lnR_{moisture}$ values upon TF addition among the forest type and soil texture categories, respectivel Boxes and vertical lines and circles within boxes represent interquartile ranges (IQRs), as well a median and mean values, respectively. Right and left whiskers represent 75 percentiles plus 1.5 IQI and 25 percentiles minus 1.5 IQR, respectively.

3.4. Relationships between lnR_{CO2} Values and Environmental Variables

Figure 5 shows the relationships among the effect sizes on the soil CO_2 emission versus soil moisture, microbial biomass C, and fine root biomass across the whole availabl dataset. The results of linear regression analysis strongly indicate that the lnR_{CO2} value significantly increased with the increase in the $lnR_{moisture}$, lnR_{mbc}, and lnR_{roots} values in a selected forest ecosystems around the globe, respectively.

The SEM analysis was performed to investigate the causal relationships betwee the lnR_{CO2} values and environmental variables such as soil properties, fine root bioma and hydrothermal conditions (Figures 6 and 7). The annual mean air temperature, annua rainfall, the ratio of soil total organic C to total N (SOCtoTN), and bulk density could explai 18% of the variability in the lnR_{CO2} values under the altered TF conditions (Figure 6a), wit a relatively large positive standardized total effect of annual mean air temperature an negative effects of bulk density and the SOCtoTN (Figure 6b). Considering the role of th effect sizes of soil properties in regulating the lnR_{CO2} values under altered TF condition the SEM analysis indicates that the percentage changes in throughfall (ΔTP), $lnR_{moisture}$ lnR_{roots}, lnR_{mbc}, and lnR_{DOC} could explain 61% of the variability in the lnR_{CO2} values fo the whole dataset (Figure 7a), with ΔTP and lnR_{roots} as well as lnR_{mbc} being the dominan factors in regulating the lnR_{CO2} (Figure 7b).

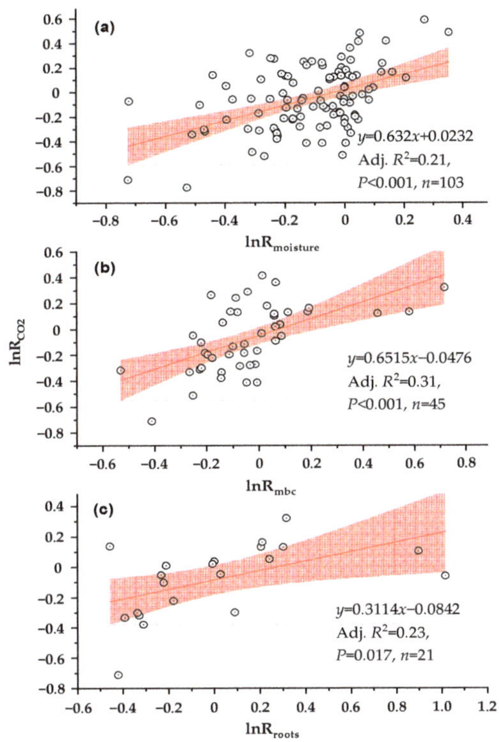

Figure 5. Relationships of the effect sizes on the soil CO_2 emissions ($\ln R_{CO2}$) against the sizes of effects on soil moisture ($\ln R_{moisture}$) (**a**), soil microbial biomass C ($\ln R_{mbc}$) (**b**), and fine root biomass ($\ln R_{roots}$) (**c**) across the whole available dataset. Shaded area represents the upper and lower 95% confidence intervals, respectively.

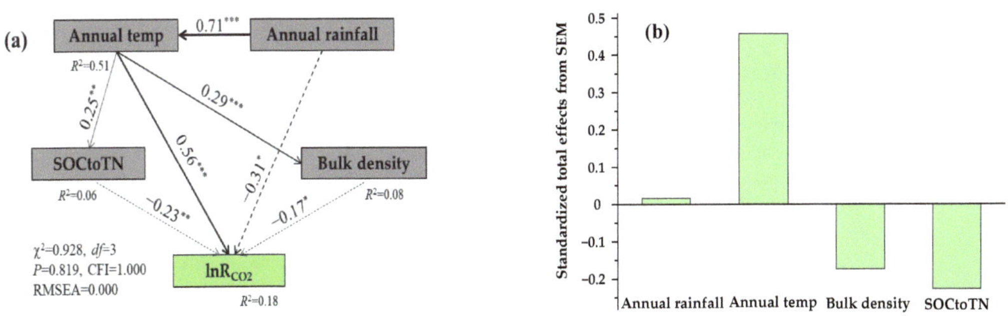

Figure 6. Structural equation modeling analysis explaining the direct and indirect effects of annual rainfall, annual mean air temperature (Annual temp), the ratio of soil total organic C to total N (SOCtoTN), and soil bulk density on the $\ln R_{CO2}$ values under the experimental altered throughfall conditions (**a**), and their standardized total effects on the $\ln R_{CO2}$ values (**b**). Single-headed arrows show the hypothesized direction of causation. Dashed and solid arrows show negative and positive relationships, respectively. The width of arrows is in proportion to the intensity of the relationship. The numbers close to the arrows are the standardized path coefficients, and the R^2 values show the proportions of variations interpreted by relationships with other variables. χ^2 stands for Chi-square, CFI for comparative fit index, and RMSEA for root-mean-square error of approximation. *df* and *p* represent degrees of freedom and probability level, respectively. Significant differences are indicated as *, $p \leq 0.05$; **, $p \leq 0.01$; and ***, $p \leq 0.001$.

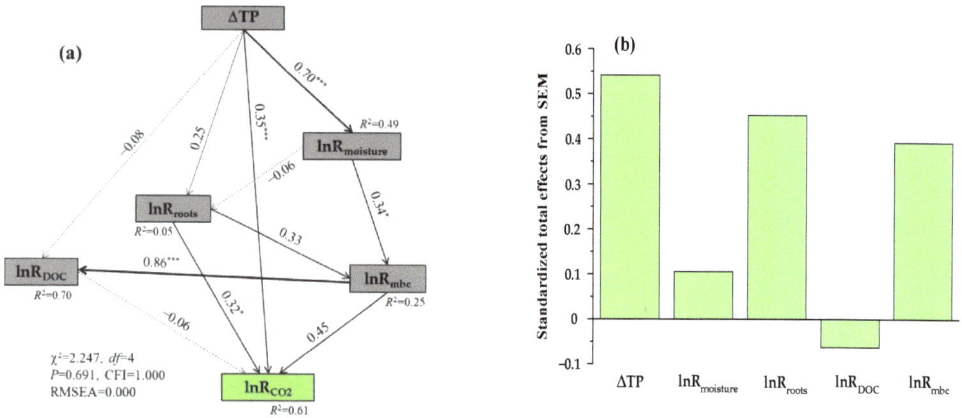

Figure 7. Structural equation modeling analysis of causal relationships among the percentage changes in throughfall (ΔTP), $\ln R_{CO2}$, $\ln R_{moisture}$, $\ln R_{roots}$, $\ln R_{mbc}$, and $\ln R_{DOC}$ (**a**), and their standardized total effects on the $\ln R_{CO2}$ values (**b**). Single-headed arrows show the hypothesized direction of causation. Dashed and solid arrows show negative and positive relationships, respectively. The width of arrows is in proportion to the intensity of the relationship. The numbers close to the arrows are the standardized path coefficients, and R^2 values show the proportion of variations interpreted by relationships with other variables. χ^2 stands for Chi-square, CFI for comparative fit index, and RMSEA for root-mean-square error of approximation. *df* and *p* represent degrees of freedom and probability level, respectively. Significant differences are indicated as *, $p \leq 0.05$; ***, $p \leq 0.001$.

4. Discussion

4.1. Throughfall Effects on Forest Soil CO_2 Emissions Vary with Climate Types

Global warming and land-use changes in the tropical region are predicted to result in drier conditions [51,52], which would affect soil CO_2 emissions from tropical forests. In humid tropical forests of south America and southwest China, many studies have reported an increase in soil respiration after precipitation reduction [24,25,46,53]. However, in a moist tropical forest such as the Brazilian Amazon, soil respiration did not change after throughfall exclusion, mainly due to the counteracting effects of fine root death and consequent increased microbial activity [44]. Additionally, the different responses in wet tropical soils at various sites could be partly due to fluctuating redox conditions and the interaction between soil iron mineral dynamics and dissolved organic C [54]. In a seasonally dry tropical forest of southern China, the simulation of a delayed wet season for two months via TF reduction did not change monthly measured soil respiration, which could be attributed to the increased microbial biomass offsetting the reduction in fine root biomass [45]. In a tropical forest of Indonesia, the simulated drought reduced soil CO_2 emission by 23% in the first 9 months and by 48% in the next 15.5 months, compared to the control, which was accompanied by significant reductions in both autotrophic and heterotrophic respiration [20,41]. The magnitude of the decrease in soil CO_2 emissions in Indonesia was much greater than that in the soil CO_2 emission observed in the eastern Amazon [19]. Furthermore, due to the differences in regular extended drought stress in tropical forests at various sites, the tropical forests in the eastern Amazon have developed deeper root systems than those in Southeast Asia [55–57]. A previous review documented a positive correlation between the length of dry season and rooting depth, and a negative correlation between annual precipitation and rooting depth in tropical forests [56]. In conclusion, the differences in the development of root systems of tropical forests and soil properties (e.g., microbial biomass, pH, dissolved organic C, and iron mineral dynamics) under fluctuating redox conditions most likely explain the above-mentioned different responses of the soil CO_2 emissions to the decreased precipitation. Probably, the different responses of soil CO_2 emissions to precipitation reduction in tropical forests at various sites lead to the apparently unchanged soil CO_2 emissions under TF reduction conditions (Figure 2a).

Compared with those from tropical, Mediterranean, and boreal forests, soil CO_2 emissions from subtropical and temperate forests became more sensitive to precipitation reduction (Figure 2a). Soil CO_2 emissions from temperate and Mediterranean forests showed significant stimulation responses to the increased precipitation compared with those from subtropical and tropical forests (Figure 2b). The results indicate that the altered precipitation-induced soil CO_2 emissions from forest ecosystems would vary with climate types. Compared with temperate, Mediterranean, subtropical, and tropical forests, there were a few field-paired observations showing soil CO_2 emissions from boreal forests under TF reduction [36,58]. As shown in Figure 2a, soil CO_2 emissions caused by TF reduction in boreal forests were not significantly different from the control. This is due to the changes in soil temperature and water availability in different boreal microtopography areas [36]. The boreal forests contain 16 of the world's soil carbon [59], and the soil carbon mostly stored in the permafrost of boreal landscapes, especially in young forests with low density, is becoming more vulnerable to the shifts in precipitation and increases in temperature [60,61]. Due to the lack of field experimental data, more field-paired observations will be needed to further explore the changes in the precipitation-induced soil CO_2 emissions of global boreal forests with different forest development stages, and the influencing mechanism involved.

A meta-analysis showed that soil respiration across all terrestrial ecosystems tends to be more sensitive to increased rainfall in more arid areas, and more responsive to decreased rainfall in more humid areas [62]. This result can partly support the stimulating responses of soil CO_2 emissions to TF addition in Mediterranean forests (Figure 2b). In the Mediterranean climate zone, small precipitation during the growing season [32,63] and mostly coarse-textured soils [32,33] normally result in relatively large $\ln R_{CO2}$ values upon increased TF in Mediterranean forests than in tropical and subtropical forests ($p < 0.05$) (Figure 2b). Furthermore, following TF addition, an increase in soil nitrogen availability [17] and an increase in soil pH [45] under seasonally dry forest ecosystem conditions would increase the activity of soil microorganisms, probably resulting in increased soil CO_2 emissions in Mediterranean forests, mostly with poor soil nutrients [33]. However, in tropical and subtropical forests, there were no significant differences in the $\ln R_{CO2}$ values upon TF addition (Figure 2b). This results from the relatively high annual precipitation, and the different responses of the development of root systems and soil properties (e.g., microbial biomass, pH, dissolved organic C, and iron mineral dynamics) to the varying soil moisture in tropical and subtropical forests at various sites around the globe [45,54–57].

To date, there has been no research reporting that the responses of annual soil CO_2 emissions to the altered precipitation became more sensitive in temperate forests than in tropical and subtropical forests at a global scale ($p < 0.05$) (Figure 2). Here, the following three reasons would explain this hypothesis. The changes in precipitation regimes can alter both belowground and aboveground C allocation in forest ecosystems [44]. Compared with tropical and subtropical forest soils, the carbon stored in temperate forest soils normally has relatively long residence times [64], and it becomes more sensitive to climate changes, particularly in the non-growing season, probably due to the relatively high levels of soil pH and substrate availability [37,43,45,65]. Unlike tropical and subtropical forests, temperate forests are most likely to experience variations in soil freezing and thawing events annually, which can be strengthened by the reduction in winter snowfall against the background of warming [66]. The release of dissolved organic matter and inorganic nitrogen into the soil, and changes in the activities of soil enzymes following soil freeze–thaw events [67–69], would, to some extent, aggravate the altered precipitation-induced soil CO_2 emissions from global temperate forests in the growing season. Compared with boreal, Mediterranean, and tropical terrestrial plants around the globe, a relatively high concentration of non-structural carbohydrates was found in the belowground part of temperate terrestrial plants [70]. This would partly account for the relatively high sensitivity of CO_2 emissions from temperate forest soils to altered precipitation (Figure 2). Together with the results of all the throughfall manipulation experiments available, it can be reasonably concluded that the responses of annual soil CO_2 emissions in forest ecosystems around the world to altered precipitation vary with climate types.

4.2. Throughfall Effects on Soil CO_2 Emissions Vary with Forest Types and Soil Properties

The forest type, a vital part of the stand structure, can affect the variability in the properties of aboveground litter and belowground root biomass and distribution [56,71], which could in turn affect the responses of forest soil respiration to precipitation changes [17,45,72–74]. Soil CO_2 emissions in coniferous forests appear to be insensitive to increased TF (Figure 2b) compared with the control, whereas the responses of the soil CO_2 emissions to TF reduction are more negative in coniferous and mixed forests than in broadleaf forests (Figure 2a). Compared with the control, soil microbial biomass C concentrations under coniferous forests and mixed forests were significantly reduced upon TF reduction ($p < 0.05$), whereas no significant difference was observed under broadleaf forests (Figure S1a). This would partly account for the relatively high inhibition of CO_2 emissions from coniferous forest soils and mixed forest soils upon TF reduction (Figure 2a). Taken together, the changes in the precipitation-induced soil CO_2 emissions varied with forest types, via the differences in the development of root systems and the decomposition of aboveground litter [20,55–57], as well as soil microbial biomass and dissolved organic C dynamics under the altered precipitation conditions [17,22,37,39,72]. This would be further supported by the results of the SEM analysis in this study (Figure 7).

Precipitation changes can influence forest ecosystems via dissolved organic matter (DOM) leaching [22,24,75]. When the ambient precipitation is high, the amount of DOM leached from broadleaf tree leaf litter normally becomes larger than that from coniferous tree leaf litter in temperate, subtropical, and boreal forests [76–78], and broadleaf trees show relatively larger litter-derived DOM biodegradation than coniferous trees [76,79]. The differences in litter decay under broadleaf and coniferous trees, and the associated litter-derived DOM dynamics [77,79,80], would, to some extent, explain the relatively large soil CO_2 emission response in broadleaf forests compared to that in coniferous forests following TF addition (Figure 2b). Besides DOM input into the soil, changes in precipitation regimes can normally alter both aboveground and belowground C allocations in forest ecosystems [44]. Under drought conditions, terrestrial plants can allocate more photosynthate to fine roots while directing less C to aboveground growth, and this phenomenon might become more prevalent in coniferous forests than that in broadleaf forests, probably due to the relatively low non-structural carbohydrate (NSC) concentrations in conifer tree roots [70,81,82]. This would partly explain why soil CO_2 emissions in broadleaf forests have become insensitive to TF reductions (Figure 2a). Additionally, compared with broadleaf trees, coniferous trees in boreal, temperate, and subtropical forests showed more drought vulnerability, especially in spring, mainly due to the low water-holding capacity and high NSC concentrations in conifer needles, and the poor water transport system in conifer foliage [70,83,84]. This would, to some extent, result in the significantly negative responses of soil CO_2 emissions to TF reductions in coniferous forests (Figure 2a).

Regardless of TF addition or TF reduction, the soil CO_2 emissions under broadleaf and coniferous mixed forests showed significant differences compared with the control (Figure 2). Litter mixtures of coniferous and broadleaf tree leaves have been reported to decompose faster than single-species leaf litter [85]. The differences in litter decay in coniferous and broadleaf mixed forests, and variations in soil biotic and abiotic properties (e.g., pH, substrate availability, fine root biomass, microbial biomass, and microbial community), as well as different C allocation patterns in the aboveground and belowground parts of mixed forests [70,76,86,87], would result in the significant responses of soil CO_2 emissions to altered precipitation. Although forest type may be uniform within the same site, significant differences in plant species can arise between different sites, which would affect the C allocation between the aboveground and belowground parts of plants, the changes in soil properties, and soil respiration under the altered precipitation conditions

Therefore, plant species should be included as a variable in future studies seeking to explore the mechanisms affecting lnR_{CO2} values (Figure 8).

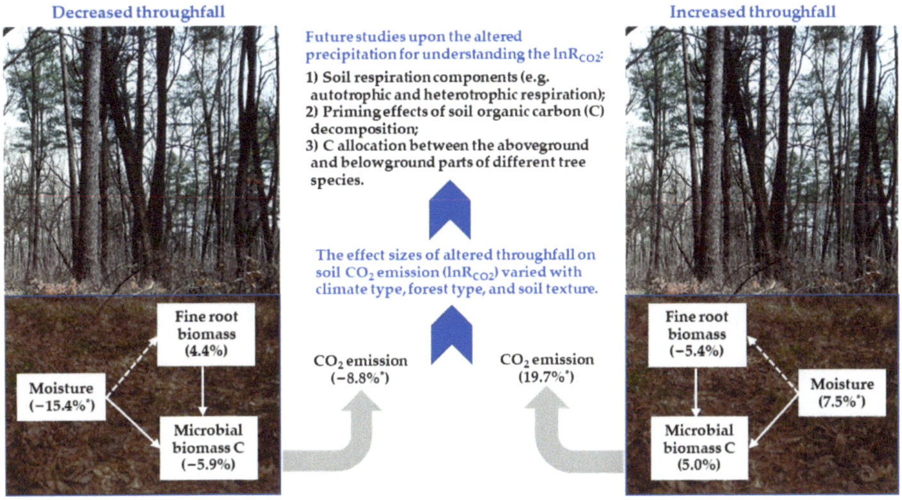

Figure 8. Conceptual model illustrating the potential effects of changes in throughfall on the soil CO_2 emission and future studies upon altered precipitation for understanding the lnR_{CO2}. Data in parenthesis represent the average percentage change of the altered throughfall treatments for each of the variables selected. Dotted and solid arrows show negative and positive effects, respectively. *, significant at the $a = 0.05$ level because the 95% CI of the effect size did not exceed zero.

Soil texture can normally influence the growth of plant roots and the activity of soil microorganisms, especially upon rewetting, mainly via the plant–soil water relationship characteristics and soil aeration, thus affecting soil respiration [88,89]. The smallest negative response of soil exchangeable NH_4-N content (lnR_{NH4}) and the largest positive response of fine root biomass (lnR_{roots}) following TF addition (Figure S1b,c) contribute to the largest CO_2 emission responses of coarse-textured soils (Figure 2b). On average, decreased throughfall increased fine root biomass by 4.4% and decreased soil microbial biomass C by 5.9% in the whole dataset, while increased throughfall reduced fine root biomass by 5.4% and increased soil microbial biomass C by 5.0% (Figure 8). Whether throughfall increases or decreases, the changes in soil moisture would affect fine root biomass and soil microbial biomass C mainly via the regulation of oxygen and the translocation of nutrients in the soil, and thus soil respiration (Figure 8). In a subtropical forest, it was reported that the variations in soil respiration responses to doubling annual precipitation were consistent with the changes in fine root biomass and soil microbial biomass [72]. In humid tropical forests, many studies have indicated that increased soil respiration following precipitation reduction results from enhanced soil microbial activity [24,46,53]. In the eastern Amazonia Basin, a decrease in precipitation was reported to decrease the rates of root growth and increase the amount of dead fine roots and soil nitrogen availability, which may affect soil respiration [17,19,20]. In subtropical and temperate forests, soil respiration was reported to be strongly correlated with fine root biomass [28,73]. The SEM analysis showed that the effect sizes of fine root biomass, soil microbial biomass and dissolved organic C content (e.g., lnR_{roots}, lnR_{mbc} and lnR_{DOC}) played a substantial role in regulating annual soil CO_2 emissions under the altered TF conditions (Figure 7). The changes in soil microbial biomass C and fine root biomass upon altered precipitation would affect soil heterotrophic and autotrophic respiration [28,73,74], which is likely related to the changes in the lnR_{CO2} values among the experimental categories. To date, only 10 paired observation data points are available showing the responses of soil heterotrophic respiration and autotrophic respiration to decreased precipitation in temperate ($n = 7$) and subtropical broadleaf forests ($n = 3$) [21,28,73,74,90,91]. On average, annual soil autotrophic respiration upon decreased precipitation was increased by 33.4% (95% CI: −24.0% to 134.1%) in temperate forests, while it was decreased by 54.0% (95% CI: −79.7% to 4.2%) in subtropical forests. Contary to soil autotrophic respiration, annual soil heterotrophic respiration in temperate and subtropical forests upon decreased precipitation was, on average, reduced by 5.9% (95% CI: −16.3% to 5.7%) and 8.8% (95% CI: −23.0% to 8.1%), respectively. The few paired observation data indicate that soil autotrophic respiration in forest ecosystems would become more sensitive to drought than soil heterotrophic respiration. Due to the scarcity of observation data, heterotrophic respiration and total soil respiration were not separated in this meta-analysis. Although some explanations involving the lnR_{CO2} are provided above, future studies should focus on the measurements of soil respiration components, the priming effects of soil organic C decomposition, and C allocation between the aboveground and belowground parts of different tree species under altered precipitation conditions, which is useful for further understanding changes in the lnR_{CO2} values and the key mechanisms involved (Figure 8).

5. Conclusions and Future Perspectives

Based on a review and data analysis of TF manipulation experiments in forest ecosystems around the globe, the variations in the altered precipitation-induced soil CO_2 emissions could vary with climate types, types of forests, and soil texture. Soil CO_2 emissions from temperate forests are more sensitive to altered precipitation than those from subtropical and tropical forests, and Mediterranean forests show a relatively large increase in soil CO_2 emissions upon increased rainfall. More importantly, the differences in the precipitation-induced annual soil CO_2 emission responses across different climate types can be partly masked by the natural interannual variability in rainfall. To effectively predict the effects of, and feedbacks to, climate change in forest ecosystems, a multi-year observation database is needed to understand the effects of changes in annual rainfall on soil CO_2 emissions from forest ecosystems, especially from tropical/subtropical and temperate forests [22,23]. In recent years, soil microorganism parameters such as microbial biomass have been effectively incorporated into soil C cycling-related models, such as Millennial and MIMICS models [92,93], to predict soil C dynamics in terrestrial ecosystems under future climate change scenarios. Based on the findings of this meta-analysis, both soil microbial biomass C and fine root biomass could be considered as the main factors in regulating the lnR_{CO2} under altered precipitation conditions, which should be incorporated into some terrestrial microbial-mediated C cycling models to reasonably predict the altered precipitation-induced soil CO_2 emissions from forest ecosystems over large scales. Over the past few decades, pulse-labeling experiments and the measurement of stable carbon isotopes have been widely carried out to study the C allocation in tree and soil compartments, and between forest ecosystems and the atmosphere [94–97]. To fully extend our understanding of the altered precipitation-induced soil CO_2 emissions in forest ecosystems, and the key influencing mechanisms involved, the measurement of soil respiration components (e.g., autotrophic and heterotrophic respiration), the priming effects of soil organic C decomposition, and C allocation between the aboveground and belowground parts of different tree species under altered precipitation conditions (e.g., drought and rewetting) should be given more attention in the future [66,98,99].

Supplementary Materials: The following supporting information can be downloaded at https://www.mdpi.com/article/10.3390/f14051037/s1, Table S1: Main message of studies looking at various forest stand sites around the globe; Table S2: Results of analysis variance (ANOVA) on the effects of experimental categories on the effect sizes of soil moisture ($lnR_{moisture}$), fine root biomass (lnR_{roots}), microbial biomass (lnR_{mbc}), KCl-exchangeable ammonium-N content (lnR_{NH4}), KCl-exchangeable nitrate-N content (lnR_{NO3}), dissolved organic carbon content (lnR_{DOC}), and pH values (lnR_{pH}) for the whole dataset; Figure S1: Box plots of sizes of effects on soil microbial biomass C (lnR_{mbc}), (a), exchangeable ammonium-N content (lnR_{NH4}) (b), and biomass of fine roots (lnR_{roots}) (c) by forest type and soil texture categories. References [100–110] are cited in the Supplementary Materials.

Funding: This research was funded by the National Natural Science Foundation of China (grant number 41175133, 41775163, 41975121 and 42275130).

Data Availability Statement: The datasets analyzed during the current study are available from the corresponding author on reasonable request.

Acknowledgments: The authors would like to thank all the researchers for the data used in this meta-analysis. The authors sincerely thank the anonymous reviewers for their insightful comments and suggestions on how to improve this manuscript.

Conflicts of Interest: The authors declare no conflict of interest.

eferences

FAO. *Global Forest Resources Assessment 2015: How Are the World's Forests Changing?* Food and Agriculture Organization of the United Nations: Rome, Italy, 2015.

Cathcart, J.F.; Kline, J.D.; Tilton, D.M. Carbon storage and Oregon's land-use planning program. *J. For.* **2007**, *105*, 167–172.

Zaehle, S.; Bondeau, A.; Carter, T.R.; Cramer, W.; Erhard, M.; Prentice, I.C.; Reginster, I.; Rounsevell, M.D.A.; Sitch, S.; Smith, B.; et al. Projected changes in terrestrial carbon storage in Europe under climate and land-use change, 1990–2100. *Ecosystems* **2007**, *10*, 380–401. [CrossRef]

Law, B.E.; Ryan, M.G.; Anthoni, P.M. Seasonal and annual respiration of a ponderosa pine ecosystem. *Glob. Chang. Biol.* **1999**, *5*, 169–182. [CrossRef]

Janssens, I.A.; Lankreijer, H.; Matteucci, G.; Kowalski, A.S.; Buchmann, N.; Epron, D.; Pilegaard, K.; Kutsch, W.; Longdoz, B.; Grunwald, T.; et al. Productivity overshadows temperature in determining soil and ecosystem respiration across European forests. *Glob. Chang. Biol.* **2001**, *7*, 269–278. [CrossRef]

Bolstad, P.V.; Davis, K.J.; Martin, J.; Cook, B.D.; Wang, W. Component and whole-system respiration fluxes in northern deciduous forests. *Tree Physiol.* **2004**, *24*, 493–504. [CrossRef]

Cox, P.M.; Betts, R.A.; Jones, C.D.; Spall, S.A.; Totterdell, I.J. Acceleration of global warming due to carbon-cycle feedbacks in a coupled climate model. *Nature* **2000**, *408*, 184–187. [CrossRef]

Valentini, R.; Matteucci, G.; Dolman, A.J.; Schulze, E.D.; Rebmann, C.; Moors, E.J.; Granier, A.; Gross, P.; Jensen, N.O.; Pilegaard, K.; et al. Respiration as the main determinant of carbon balance in European forests. *Nature* **2000**, *404*, 861–865. [CrossRef]

Lal, R.; Kimble, J.; Levine, E.; Stewart, B. *Soils and Global Change*; CRC Press: Boca Raton, FL, USA, 1995.

). Schimel, D.S.; House, J.I.; Hibbard, K.A.; Bousquet, P.; Ciais, P.; Peylin, P.; Braswell, B.H.; Apps, M.J.; Baker, D.; Bondeau, A.; et al. Recent patterns and mechanisms of carbon exchange by terrestrial ecosystems. *Nature* **2001**, *414*, 169–172. [CrossRef]

1. Bond-Lamberty, B.; Bailey, V.L.; Chen, M.; Gough, C.M.; Vargas, R. Globally rising soil heterotrophic respiration over recent decades. *Nature* **2018**, *560*, 80–83. [CrossRef]

2. Wentz, F.J.; Ricciardulli, L.; Hilburn, K.; Mears, C. How much more rain will global warming bring? *Science* **2007**, *317*, 233–235. [CrossRef]

3. Zhou, X.; Zhou, L.; Nie, Y.; Fu, Y.; Du, Z.; Shao, J.; Zheng, Z.; Wang, X. Similar responses of soil carbon storage to drought and irrigation in terrestrial ecosystems but with contrasting mechanisms: A meta-analysis. *Agric. Ecosyst. Environ.* **2016**, *228*, 70–81. [CrossRef]

4. Houghton, R.A. Balancing the global carbon budget. *Annu. Rev. Earth Planet. Sci.* **2007**, *35*, 313–347. [CrossRef]

5. Schindlbacher, A.; Wunderlich, S.; Borken, W.; Kitzler, B.; Zechmeister-Boltenstern, S.; Jandl, R. Soil respiration under climate change: Prolonged summer drought offsets soil warming effects. *Glob. Chang. Biol.* **2012**, *18*, 2270–2279. [CrossRef]

6. Mazza, G.; Amorini, E.; Cutini, A.; Manetti, M.C. The influence of thinning on rainfall interception by Pinus pinea L. in Mediterranean coastal stands (Castel Fusano-Rome). *Ann. For. Sci.* **2011**, *68*, 1323–1332. [CrossRef]

7. Cattânio, J.; Davidson, E.; Nepstad, D.; Verchot, L.; Ackerman, I. Unexpected results of a pilot throughfall exclusion experiment on soil emissions of CO_2, CH_4, N_2O, and NO in eastern Amazonia. *Biol. Fertil. Soils* **2002**, *36*, 102–108.

8. Borken, W.; Savage, K.; Davidson, E.A.; Trumbore, S.E. Effects of experimental drought on soil respiration and radiocarbon efflux from a temperate forest soil. *Glob. Chang. Biol.* **2006**, *12*, 177–193. [CrossRef]

9. Sotta, E.D.; Veldkamp, E.; Schwendenmann, L.; Guimarães, B.R.; Paixão, R.K.; Lourdes, M.D.; Ruivo, P.; Costa, A.C.L.D.; Meir, P. Effects of an induced drought on the soil CO_2 production and soil CO_2 efflux in an eastern Amazonian rainforest, Brazil. *Glob. Chang. Biol.* **2007**, *13*, 2218–2229. [CrossRef]

0. van Straaten, O.; Veldkamp, E.; Corre, M.D. Simulated drought reduces soil CO_2 efflux and production in a tropical forest in Sulawesi, Indonesia. *Ecosphere* **2011**, *2*, 119. [CrossRef]

1. Hinko-Najera, N.; Fest, B.; Livesley, S.J.; Arndt, S.K. Reduced throughfall decreases autotrophic respiration, but not heterotrophic respiration in a dry temperate broadleaved evergreen forest. *Agric. For. Meteorol.* **2015**, *200*, 66–77. [CrossRef]

2. Xu, X.K.; Duan, C.T.; Wu, H.H.; Luo, X.B.; Han, L. Effects of changes in throughfall on soil GHG fluxes under a mature temperate forest, northeastern China. *J. Environ. Manag.* **2021**, *294*, 112950. [CrossRef]

23. Davidson, E.A.; Nepstad, D.C.; Ishida, F.Y.; Brando, P.M. Effects of an experimental drought and recovery on soil emission of carbon dioxide, methane, nitrous oxide, and nitric oxide in a moist tropical forest. *Glob. Chang. Biol.* **2008**, *14*, 2582–259 [CrossRef]
24. Cleveland, C.C.; Wieder, W.R.; Reed, S.C.; Townsend, A.R. Experimental drought in a tropical rain forest increases soil carbo dioxide losses to the atmosphere. *Ecology* **2010**, *91*, 2313–2323. [CrossRef]
25. Zhang, X.; Zhang, Y.P.; Sha, L.Q.; Wu, C.S.; Tan, Z.H.; Song, Q.H.; Liu, Y.T.; Dong, L.Y. Effects of continuous drought stress on soi respiration in a tropical rainforest in southwest China. *Plant Soil* **2015**, *394*, 343–353. [CrossRef]
26. Zhou, S.X.; Huang, C.D.; Xiang, Y.B.; Xiao, Y.X.; Tang, J.D.; Han, B.H. Effects of throughfall exclusion on soil respiration in natura evergreen broadleaved forest in rainy area of Western China. Chin. *J. Appl. Environ. Biol.* **2016**, *22*, 499–504.
27. Xiang, Y.B.; Huang, C.D.; Hu, T.X.; Tu, L.H.; Zhou, S.X.; Xiao, Y.X.; Gao, B.D. Effects of simulated nitrogen deposition an precipitation changes on soil respiration in an evergreen broadleaved forest in the rainy area of western China. *Acta Ecol. Sin* **2016**, *36*, 5227–5235.
28. Liu, Y.C.; Liu, S.R.; Wan, S.Q.; Wang, J.X.; Luan, J.W.; Wang, H. Differential responses of soil respiration to soil warming an experimental reduction in a transitional oak forest in central China. *Agric. For. Meteorol.* **2016**, *226–227*, 186–198. [CrossRef]
29. Chen, L.; Liu, S.R.; Zeng, J.; Wang, H.; Lu, L.H. Effect of precipitation reduction on soil greenhouse gas emission in *Castanopsi hystrix* and *Pinus massoniana* plantations. *J. Cent. South Univ. For. Tech.* **2019**, *39*, 86–92.
30. Joffre, R.; Qurcival, J.M.; Rambal, S.; Rocheteau, A. The Key-role of topsoil moisture on CO_2 efflux from a Mediterranean *Quercu ilex* forest. *Annu. For. Sci.* **2003**, *60*, 519–526. [CrossRef]
31. Mission, L.; Rocheteau, A.; Rambal, S.; Ourcival, J.M.; Limousin, J.M.; Rodriguez, R. Functional changes in the control of carbo fluxes after 3 years of increased drought in a Mediterranean evergreen forest? *Glob. Chang. Biol.* **2010**, *16*, 2461–2475. [CrossRef
32. Cotrufo, M.F.; Alberti, G.; Inglima, I.; Marjanovic, H.; LeCain, D.; Zaldei, A.; Peressotti, A.; Miglietta, F. Decreased summe drought affects plant productivity and soil carbon dynamics in a Mediterranean woodland. *Biogeosciences* **2011**, *8*, 2729. [CrossRef
33. Shvaleva, A.; Lobo-do-Vale, R.; Cruz, C.; Castaldi, S.; Rosa, A.P.; Chaves, M.M.; Pereira, J.S. Soil-atmosphere greenhouse gase (CO_2, CH_4 and N_2O) exchange in evergreen oak woodland in southern Portugal. *Plant Soil Environ.* **2011**, *57*, 471–477. [CrossRef
34. Wu, Z.; Dijkstra, P.; Koch, G.W.; Penuelas, J.; Hungate, B.A. Responses of terrestrial ecosystems to temperature and precipitatio change: A meta-analysis of experimental manipulation. *Glob. Chang. Biol.* **2011**, *17*, 927–942. [CrossRef]
35. Deng, Q.; Hui, D.; Zhang, D.; Zhou, G.; Liu, J.; Liu, S.; Chu, G.; Li, J. Effects of precipitation increase on soil respiration: A three-year field experiment in subtropical forests in China. *PLoS ONE* **2012**, *7*, e41493. [CrossRef]
36. Billings, S.A.; Richter, D.D.; Yarie, J. Soil carbon dioxide fluxes and profile concentrations in two boreal forests. *Can. J. For. Re* **1998**, *28*, 1773–1783. [CrossRef]
37. Xu, X.K.; Han, L.; Wang, Y.S.; Inubushi, K. Influence of vegetation types and soil properties on microbial biomass carbon an metabolic quotients in temperate volcanic and tropical forest soils. *Soil Sci. Plant Nutr.* **2007**, *53*, 430–440. [CrossRef]
38. Wood, T.E.; Silver, W.L. Strong spatial variability in trace gas dynamics following experimental drought in a humid tropical fores *Glob. Biogeochem. Cycles* **2012**, *26*, GB3005. [CrossRef]
39. Xu, X.K.; Luo, X.B. Effect of wetting intensity on soil GHG fluxes and microbial biomass under a temperate forest floor durin dry season. *Geoderma* **2012**, *170*, 118–126. [CrossRef]
40. Fender, A.C.; Gansert, D.; Jungkunst, H.F.; Fiedler, S.; Beyer, F.; Schützenmeister, K.; Thiele, B.; Valtanen, K.; Polle, A.; Leuschne C. Root-induced tree species effects on the source/sink strength for greenhouse gases (CH_4, N_2O and CO_2) of a temperat deciduous forest soil. *Soil Biol. Biochem.* **2013**, *57*, 587–597. [CrossRef]
41. van Straaten, O.; Veldkamp, E.; Kohler, M.; Anas, I. Spatial and temporal effects of drought on soil CO_2 efflux in a caca agroforestry system in Sulawesi, Indonesia. *Biogeosciences* **2010**, *7*, 1223–1235. [CrossRef]
42. Suseela, V.; Dukes, J.S. The responses of soil and rhizosphere respiration to simulated climatic changes vary by season. *Ecolog* **2013**, *94*, 403–413. [CrossRef]
43. Bréchet, L.M.; Lopez-Sangil, L.; George, C.; Birkett, A.J.; Baxendale, C.; Trujillo, B.C.; Sayer, E.J. Distinct responses of so respiration to experimental litter manipulation in temperate woodland and tropical forest. *Ecol. Evol.* **2018**, *8*, 3787–379 [CrossRef]
44. Davidson, E.A.; Ishida, F.Y.; Nepstad, D.C. Effects of an experimental drought on soil emissions of carbon dioxide, methan nitrous oxide, and nitric oxide in a moist tropical forest. *Glob. Chang. Biol.* **2004**, *10*, 718–730. [CrossRef]
45. Yu, S.Q.; Mo, Q.F.; Chen, Y.Q.; Li, Y.W.; Li, Y.X.; Zou, B.; Xia, H.P.; Jun, W.; Li, Z.A.; Wang, F.M. Effects of seasonal precipitatio change on soil respiration processes in a seasonally dry tropical forest. *Ecol. Evol.* **2020**, *10*, 467–479. [CrossRef]
46. Wood, T.E.; Detto, M.; Silver, W.L. Sensitivity of soil respiration to variability in soil moisture and temperature in a humid tropica forest. *PLoS ONE* **2013**, *8*, e80965. [CrossRef]
47. Post, W.M.; Kwon, K.C. Soil carbon sequestration and land-use change: Processes and potential. *Glob. Chang. Biol.* **2000**, 317–327. [CrossRef]
48. Rosenberg, M.S.; Adams, D.C.; Gurevitch, J. *MetaWin: Statistical Software for Meta-Analysis, Version 2*; Sinauer Associate Sunderland, MA, USA, 1999.
49. Liu, L.; Estiarte, M.; Peñuelas, J. Soil moisture as the key factor of atmospheric CH_4 uptake in forest soils under environment change. *Geoderma* **2019**, *355*, 113920. [CrossRef]

50. Li, L.F.; Zheng, Z.Z.; Wang, W.J.; Biederman, J.A.; Xu, X.L.; Ran, Q.W.; Qian, R.Y.; Xu, C.; Zhang, B.; Wang, F.; et al. Terrestrial N_2O emissions and related functional genes under climate changes: A global meta-analysis. *Glob. Chang. Biol.* **2020**, *26*, 931–943. [CrossRef]
51. Nepstad, D.C.; Veríssimo, A.; Alencar, A.; Nobre, C.; Lima, E.; Lefebvre, P.; Schlesinger, P.; Potter, C.; Moutinho, P.; Mendoza, E.; et al. Large-scale impoverishment of Amazonian forests by logging and fire. *Nature* **1999**, *398*, 505–508. [CrossRef]
52. Costa, M.H.; Foley, J.A. Combined effects of deforestation and doubled atmospheric CO_2 concentrations on the climate of Amazonia. *J. Clim.* **2000**, *13*, 18–34. [CrossRef]
53. Waring, B.G.; Hawkes, C.V. Short-term precipitation exclusion alters microbial responses to soil moisture in a wet tropical forest. *Microb. Ecol.* **2015**, *69*, 843–854. [CrossRef]
54. Bhattacharyya, A.; Campbell, A.N.; Tfaily, M.M.; Lin, Y.; Kukkadapu, R.K.; Silver, W.L.; Nico, P.S.; Pett-Ridge, J. Redox fluctuations control the coupled cycling of iron and carbon in tropical forest soils. *Environ. Sci. Technol.* **2018**, *52*, 14129–14139. [CrossRef] [PubMed]
55. Nepstad, D.C.; Decarvalho, C.R.; Davidson, E.A.; Jipp, P.H.; Lefebvre, P.A.; Negreiros, G.H.; Dasilva, E.D.; Stone, T.A.; Trumbore, S.E.; Vieira, S. The role of deep roots in the hydrological and carbon cycles of Amazonian forests and pastures. *Nature* **1994**, *372*, 666–669. [CrossRef]
56. Schenk, H.J.; Jackson, R.B. The global biogeography of roots. *Ecol. Monogr.* **2002**, *72*, 311–328. [CrossRef]
57. Hertel, D.; Moser, G.; Culmsee, H.; Erasmi, S.; Horna, V.; Schuldt, B.; Leuschner, C. Below- and above-ground biomass and net primary production in a paleotropical natural forest (Sulawesi, Indonesia) as compared to neotropical forests. *For. Ecol. Manag.* **2009**, *258*, 1904–1912. [CrossRef]
58. Martins, C.S.C.; Nazaries, L.; Delgado-Baquerizo, M.; Macdonald, C.A.; Anderson, I.C.; Hobbie, S.E.; Venterea, R.T.; Reich, P.B.; Singh, B.K. Identifying environmental drivers of greenhouse gas emissions under warming and reduced rainfall in boreal-temperate forests. *Funct. Ecol.* **2017**, *31*, 2356–2368. [CrossRef]
59. Post, W.M.; Emanuel, W.R.; Zinke, P.J.; Stangenberger, A.G. Soil carbon pools and world life zones. *Nature* **1982**, *298*, 156–159. [CrossRef]
60. Thom, D.; Golivets, M.; Edling, L.; Meigs, G.W.; Gourevitvh, D.J.; Sonter, L.J.; Galford, G.L.; Keeton, W.S. The climate sensitivity of carbon, timber, and species richness covaries with forest age in boreal-temperate North America. *Glob. Chang. Biol.* **2019**, *25*, 2446–2458. [CrossRef]
61. Stuenzi, S.; Boike, J.; Gädeke, A.; Herzschuh, U.; Kruse, S.; Pestryakova, L.A.; Westermann, S.; Langer, M. Sensitivity of ecosystem-protected permafrost under changing boreal forest structures. *Environ. Res. Lett.* **2021**, *16*, 084045. [CrossRef]
62. Liu, L.; Wang, X.; Lajeunesse, M.J.; Miao, G.; Piao, S.; Wan, S.; Wu, Y.; Wang, Z.; Yang, S.; Li, P. A cross-biome synthesis of soil respiration and its determinants under simulated precipitation changes. *Glob. Chang. Biol.* **2016**, *22*, 1394–1405. [CrossRef]
63. Darenova, E.; Čater, M. Different structure of sessile oak stands affects soil moisture and soil CO_2 efflux. *For. Sci.* **2018**, *64*, 340–348.
64. Amundson, R. The carbon budget in soils. *Annu. Rev. Earth Planet. Sci.* **2001**, *29*, 535–562. [CrossRef]
65. Hayakawa, C.; Funakawa, S.; Fujii, K.; Kadono, A.; Kosaki, T. Effects of climatic and soil properties on cellulose decomposition rates in temperate and tropical forests. *Biol. Fertil. Soils* **2014**, *50*, 633–643. [CrossRef]
66. Xu, X.K. Effect of freeze-thaw disturbance on soil C and N dynamics and GHG fluxes of East Asia forests: Review and future perspectives. *Soil Sci. Plant Nutr.* **2022**, *68*, 15–26. [CrossRef]
67. Xu, X.K.; Duan, C.T.; Wu, H.H.; Li, T.S.; Cheng, W.G. Effect of intensity and duration of freezing on soil microbial biomass, extactable C and N pools, and N_2O and CO_2 emissions from forest soils in cold temperate region. *Sci. China Earth Sci.* **2016**, *59*, 156–169. [CrossRef]
68. Wu, H.H.; Xu, X.K.; Cheng, W.G.; Han, L. Dissolved organic matter and inorganic N jointly regulate greenhouse gases fluxes from forest soils with different moistures during a freeze-thaw period. *Soil Sci. Plant Nutr.* **2020**, *66*, 163–176. [CrossRef]
69. Wu, H.H.; Xu, X.K.; Fu, P.Q.; Cheng, W.G.; Fu, C.S. Responses of soil WEOM quantity and quality to freeze-thaw and litter manipulation with contrasting soil water content: A laboratory experiment. *Catena* **2021**, *198*, 105058. [CrossRef]
70. Martínez-Vilalta, J.; Sala, A.; Asensio, D.; Galiano, L.; Hoch, G.; Palacio, S.; Piper, F.I.; Lloret, F. Dynamics of non-structural carbohydrates in terrestrial plants: A global synthesis. *Ecol. Monogr.* **2016**, *86*, 495–516. [CrossRef]
71. Harteveld, M.; Hertel, D.; Wiens, M.; Leuschner, C. Spatial and temporal variability of fine root abundance and growth in tropical moist forests and agroforestry systems (Sulawesi, Indonesia). *Ecotropica* **2007**, *13*, 111–120.
72. Jiang, H.; Deng, Q.; Zhou, G.; Hui, D.; Zhang, D.; Liu, S.; Chu, G.; Li, J. Responses of soil respiration and its temperature/moisture sensitivity to precipitation in three subtropical forests in southern China. *Biogeosciences* **2013**, *10*, 3963–3982. [CrossRef]
73. Huang, S.D.; Ye, G.F.; Lin, J.; Chen, K.T.; Xu, X.; Ruan, H.H.; Tan, F.L.; Chen, H.Y.H. Autotrophic and heterotrophic soil respiration responds asymmetrically to drought in a subtropical forest in the Southeast China. *Soil Biol. Biochem.* **2018**, *123*, 242–249. [CrossRef]
74. Zhang, J.L.; Liu, S.R.; Liu, C.J.; Wang, H.; Luan, J.W.; Liu, X.J.; Guo, X.W.; Niu, B.L. Different mechanisms underlying divergent responses of autotrophic and heterotrophic respiration to long-term throughfall reduction in a warm-temperate oak forest. *For. Ecosyst.* **2021**, *8*, 41. [CrossRef]
75. Deng, Q.; Zhang, D.; Han, X.; Chu, G.; Zhang, Q.; Hui, D. Changing rainfall frequency rather than drought rapidly alters annual soil respiration in a tropical forest. *Soil Biol. Biochem.* **2018**, *121*, 8–15. [CrossRef]

76. Kiikkilä, O.; Kitunen, V.; Spetz, P.; Smolander, A. Characterization of dissolved organic matter in decomposing Norway spruce and silver birch litter. *Eur. J. Soil Sci.* **2012**, *63*, 476–486. [CrossRef]
77. Hensgens, G.; Lechtenfeld, O.J.; Guillemette, F.; Laudon, H.; Berggren, M. Impacts of litter decay on organic leachate composition and reactivity. *Biogeochemistry* **2020**, *154*, 99–117. [CrossRef]
78. Xu, J.W.; Ji, J.H.; Hu, D.N.; Zheng, Z.; Mao, R. Tree fresh leaf- and twig-leached dissolved organic matter quantity and biodegradability in subtropical plantations in China. *Forests* **2022**, *13*, 833. [CrossRef]
79. Wu, P.P.; Ding, Y.D.; Li, S.L.; Sun, X.X.; Zhang, Y.; Mao, R. Carbon, nitrogen and phosphorus stoichiometry controls interspecific patterns of leaf litter-derived dissolved organic matter biodegradation in subtropical plantations of China. *iForest* **2021**, *14*, 80–85 [CrossRef]
80. Xu, X.K.; Yin, L.; Duan, C.T.; Jing, Y.S. Effect of N addition, moisture, and temperature on soil microbial respiration and microbial biomass in forest soil at different stages of litter decomposition. *J. Soils Sediments* **2016**, *16*, 1421–1439. [CrossRef]
81. Oberhuber, W.; Gruber, A.; Lethaus, G.; Winkler, A.; Wieser, G. Stem gridling indicates prioritized carbon allocation to the root system at the expense of radial stem growth in Norway spruce under drought conditions. *Environ. Exp. Bot.* **2017**, *138*, 109–118 [CrossRef]
82. Smith, M.G.; Miller, R.E.; Arndt, S.K.; Kasel, S.; Bennett, L.T. Whole-tree distribution and temporal variation of non-structural carbohydrates in broadleaf evergreen trees. *Tree Physiol.* **2018**, *38*, 570–581. [CrossRef]
83. Peltier, D.M.P.; Ogle, K. Legacies of more frequent drought in ponderosa pine across the western United States. *Glob. Chang. Biol.* **2019**, *25*, 3803–3816. [CrossRef]
84. Li, S.L.; Xu, J.W.; Ding, Y.D.; Mao, R. Litter water-holding and water-loss characteristics of trees and ferns in the water conservation forests at the middle reaches of the Gan River. *J. Soil Water Conserv.* **2021**, *35*, 170–176.
85. Gartner, T.B.; Cardon, Z.G. Decomposition dynamics in mixed-species leaf litter. *Oikos* **2004**, *104*, 230–246. [CrossRef]
86. Rousk, J.; Brookes, P.C.; Bååth, E. Investigating the mechanisms for the opposing pH relationships of fungal and bacterial growth in soil. *Soil Biol. Biochem.* **2010**, *42*, 926–934. [CrossRef]
87. Smolander, A.; Kitunen, V. Comparison of tree species effects on microbial C and N transformations and dissolved organic matter properties in boreal forest floors. *Appl. Soil Ecol.* **2011**, *49*, 224–233. [CrossRef]
88. Bouma, T.J.; Bryla, D.R. On the assessment of root and soil respiration for soils of different textures: Interactions with soil moisture contents and soil CO_2 concentrations. *Plant Soil* **2000**, *227*, 215–221. [CrossRef]
89. Dilustro, J.J.; Collins, B.; Duncan, L.; Crawford, C. Moisture and soil texture effects on soil CO_2 efflux components in southeastern mixed pine forests. *For. Ecol. Manag.* **2005**, *204*, 85–95. [CrossRef]
90. Wang, Y.; Liu, S.R.; Luan, J.W.; Chen, C.; Cai, C.J.; Zhou, F.; Di, Y.P.; Gao, X.M. Nitrogen addition exacerbates the negative effect of throughfall reduction on soil respiration in a Bamboo Forest. *Forests* **2021**, *12*, 724. [CrossRef]
91. Lu, H.B.; Liu, S.R.; Wang, H.; Luan, J.W.; Schindlbacher, A.; Liu, Y.C.; Wang, Y. Experimental throughfall reduction barely affects soil carbon dynamics in a warm-temperate oak forest, central China. *Sci. Rep.* **2017**, *7*, 15099. [CrossRef]
92. Abramoff, R.; Xu, X.F.; Hartman, M.; O'Brien, S.; Feng, W.T.; Davidson, E.; Finzi, A.C.; Moorhead, D.; Schimel, J.; Torn, M. The Millennial model: In search of measurable pools and transformations for modeling soil carbon in the new century. *Biogeochemistry* **2018**, *137*, 51–71. [CrossRef]
93. Zhang, H.C.; Goll, D.S.; Wang, Y.P.; Ciais, P.; Wieder, W.R.; Abramoff, R.; Haung, Y.Y.; Guenet, B.; Preschner, A.K.; Rossel, R.A.V. et al. Microbial dynamics and soil physicochemical properties explain large-scale variations in soil organic carbon. *Glob. Chang. Biol.* **2020**, *26*, 2668–2685. [CrossRef]
94. Epron, D.; Bahn, M.; Derrien, D.; Lattanzi, F.A.; Pumpanen, J.; Gessler, A.; Högberg, P.; Maillard, P.; Dannoura, M.; Gérant, D. et al. Pulse-labelling trees to study carbon allocation dynamics: A review of methods, current knowledge and future prospects *Tree Physiol.* **2012**, *32*, 776–798. [CrossRef]
95. McCarroll, D.; Whitney, M.; Young, G.H.F.; Loader, N.J.; Gagen, M.H. A simple stable carbon isotope method for investigating changes in the use of recent versus old carbon in oak. *Tree Physiol.* **2017**, *37*, 1021–1027. [CrossRef]
96. Diao, H.Y.; Wang, A.Z.; Yuan, F.H.; Guan, D.X.; Dai, G.H.; Wu, J.B. Environmental effects on carbon isotope discrimination from assimilation to respiration in a coniferous and broadleaf-leaved mixed forest of Northeast China. *Forests* **2020**, *11*, 1156. [CrossRef]
97. Wen, T.; Qu, Y.X.; Lu, K.L.; Guan, C.; Zhao, C.M. Combining tree-ring width and carbon isotope data to investigate stem carbon allocation in an evergreen coniferous species. *Agric. For. Meteorol.* **2022**, *316*, 108845. [CrossRef]
98. Trumbore, S. Age of soil organic matter and soil respiration: Radiocarbon constraints on belowground C dynamics. *Ecol. Appl* **2000**, *10*, 399–411. [CrossRef]
99. Moyes, A.B.; Gaines, S.J.; Siegwolf, R.T.W.; Bowling, D.R. Diffusive fractionation complicates isotopic partitioning of autotrphic and heterotrophic sources of soil respiration. *Plant Cell Environ.* **2010**, *33*, 1804–1819. [CrossRef]
100. Dařenová, E.; Crabbe, R.A.; Knott, R.; Uherková, B.; Kadavý, J. Effect of coppicing, thinning and throughfall reduction on soil water content and soil CO_2 efflux in a sessile oak forest. *Silva Fennica* **2018**, *52*, 9927. [CrossRef]
101. Muhr, J.; Borken, W. Delayed recovery of soil respiration after wetting of dry soil further reduced C losses from a Norway spruce forest soil. *J. Geophys. Res.* **2009**, *114*, G04023. [CrossRef]
102. Borken, W.; Davidson, E.A.; Savage, K.; Sundquist, E.T.; Steudler, P. Effect of summer throughfall exclusion, summer drought and winter snow cover on methane fluxes in a temperate forest soil. *Soil Biol. Biochem.* **2006**, *38*, 1388–1395. [CrossRef]

3. Liu, Y.C.; Shang, Q.; Wang, L.; Tian, Y.; Ju, Y.X.; Gan, J.B. Responses of soil respiration to changing precipitation regimes in an oak forest at a climate transitional zone. *Acta Ecol. Sin.* **2016**, *36*, 8054–8061.
4. Peng, X.H.; Xu, X.F.; Han, H.R.; Cheng, X.Q.; Cai, M.K.; Bai, Y.C. Effects of throughfall exclusion on soil respiration in *Larix principis-rupprechtii* plantation. *Ecol. Environ. Sci.* **2017**, *26*, 1310–1316.
5. Vasconcelos, S.S.; Zarin, D.J.; Capanu, M.; Littell, R.; Davidson, E.A.; Ishida, F.Y.; Santos, E.B.; Araújo, M.M.; Aragão, D.V.; Rangel-Vasconcelos, L.G.T.; et al. Moisture and substrate availability constrain soil trace gas fluxes in an eastern Amazonian regrowth forest. *Global Biogeochem. Cycles* **2004**, *18*, GB2009. [CrossRef]
6. Bréchet, L.; Courtois, E.A.; Saint-Germain, T.; Janssens, I.A.; Asensio, D.; Ramirez-Rojas, I.; Soong, J.L.; Langenhove, L.V.; Verbruggen, E.; Stahl, C. Disentangling drought and nutrient effects on soil carbon dioxide and methane fluxes in a tropical forest. *Front. Environ. Sci.* **2019**, *7*, 180. [CrossRef]
7. Xu, J. Response of Soil Carbon Processes to Throughfall Exclusion in Two Subtropical Plantations in Southern China. Ph.D. Thesis, Chinese Academy of Forestry, Beijing, China, 2014.
8. Liu, W.; Wang, L.H.; Fu, R.; Zhou, S.X.; Liu, L.; Wu, X.H.; Huang, C.D. Effects of precipitation increase on soil respiration of a *Pinus yunnanensis* plantation in a dry river valley area. *Acta Ecol. Sin.* **2017**, *37*, 1391–1400.
9. Li, S.J. Effects of Precipitation Exclusion on Soil Microbial Community Structure and Greenhouse Gas Fluxes in a Natural *Castanopsis carlesii* Forest. Master's Thesis, Fujian Normal University, Fuzhou, China, 2018.
0. Ju, H. The Characteristic of Soil Activated Carbon and Nitrogen and Greenhouse Gas Fluxes of the Different Types of Forests in Shennongjia. Ph.D. Thesis, Inner Mongolia Agricultural University, Hohhot, China, 2016.

forests

MDPI

Article

Seasonal Dynamics of Soil Enzymatic Activity under Different Land-Use Types in Rocky Mountainous Region of North China

Yuhua Kong, Anran Qu, Erpeng Feng, Rui Chen, Xitian Yang and Yong Lai *

College of Forestry, Henan Agricultural University, Zhengzhou 450046, China
* Correspondence: yonglai@henau.edu.cn

Abstract: To reveal the effects of different land-use types on soil enzyme activities, soil samples wer collected from 0–10, 10–20 and 20–30 cm soil layers to compare and analyze soil β-glucosidase (BG urease (URE), protease (PROT) and catalase (CAT) activities in farmland (FL), abandoned land (AL and three plantation forests: *Platycladus orientalis* (L.) Franco (PO), *Robinia pseudoacacia* L. (RP) an *Quercus variabilis* Bl. (QV) in the rocky mountainous region of North China. The results showed tha the soil enzyme activities varied significantly under different land-use types, and the interannua mean values of FL and PO were remarkably higher than those of other land uses in the 0–30 cm soi layer, in which the soil BG and URE activities of FL were 22% and 12% higher than those of AL, an 428% and 179% higher than those of QV, respectively; the soil PROT and CAT activities of PO wer 66% and 23% higher than those of AL, and 479% and 113% higher than those of QV, respectivel Soil BG, URE and PROT activities were all higher in June and lower in December, while soil CA activity was slightly lower in June. The soil enzymatic activities all showed a notable decrease wit the depth of the soil layer. Soil BG, URE, PROT and CAT activities were remarkably ($p < 0.01$) o significantly ($p < 0.05$) positively correlated with available nitrogen, available phosphorus, availabl potassium, dissolved organic carbon (DOC), $NO_3{}^-$-N, soil organic carbon, water content, clay and si volume fraction, and significantly negatively correlated with sand volume fraction. Soil DOC and pH were important factors influencing soil enzymatic activity, implying that changes in soil enzymati activity under different land-use types may be the result of a combination of temperature, moistur and plant type. In conclusion, PO plantations are conducive to improving the physicochemical an biological properties of soil and enhance soil fertility, which is a reasonable land-use method t achieve sustainable development in the rocky mountainous region of North China.

Keywords: land-use types; soil enzymatic activity; rocky mountainous region of North China

check for updates

Citation: Kong, Y.; Qu, A.; Feng, E.; Chen, R.; Yang, X.; Lai, Y. Seasonal Dynamics of Soil Enzymatic Activity under Different Land-Use Types in Rocky Mountainous Region of North China. *Forests* **2023**, *14*, 536. https://doi.org/10.3390/f14030536

Academic Editor: Richard D. Bowden

Received: 27 December 2022
Revised: 1 March 2023
Accepted: 6 March 2023
Published: 9 March 2023

1. Introduction

As the most active among soil organic components, soil enzymes take part in all bi chemical processes in the soil environment, and can reflect the quality and health of the so ecosystem and environment [1–3]. They are mainly derived from plant and animal residue plant roots and microorganisms [4–6]. As an indicator of microorganisms, soil enzym activity is an important factor in controlling organic matter decomposition and nutrien movement, and enzyme responses can serve as an early alert about soil physicochemica and biological variations [7]. Soil enzyme activities change with the nutrient requirement and metabolic activity of soil microorganisms [8,9]. For example, β-glucosidase (BG) an urease (URE) take part in the carbon and nitrogen cycles, respectively [10], and catalas (CAT) and protease (PROT) are essential for the redox processes of macromolecules co tained in soil organic matter [11]. The critical roles of soil enzymes in different nutrien cycles are well demonstrated [12,13]. However, there are still some blind spots in th response of soil enzymatic activity to seasonal and land-use changes.

Land-use types and histories can alter soil microbial activity and its communit structure [9] by influencing soil organic matter content [14]; soil physical structure [15 as well as the type, amount and residue of vegetation apoplast, and all these changes govern soil enzymatic activity [16,17]. It has been shown that soil enzymatic activity varies notably among land-use types. For instance, microbial communities can be governed by different land-use practices by altering soil environmental factors [18]. The content of soil mineral elements, soil enzymatic activity and soil fungal communities in karst areas are obviously controlled by land-use practices and soil depth [19]. It was also noted that the impact of different land-use types on soil enzymatic activity in the Minjiang River Basin showed the pattern secondary forest > plantation forest > scrub > sloping land [20]. Additionally, the biological and geochemical cycles of soil elements in terrestrial ecosystems may be influenced by seasonal dynamics, particularly temperature and precipitation [21]. It has been reported that temperature and precipitation are the main factors regulating soil enzymatic activity [22]. Temperature is a decisive factor for the physicochemical dissolution of soil organic matter and the depolymerization of macromolecules [23], and higher air temperatures lead to higher soil temperatures and lower soil moisture [24]. Precipitation, as another vital climate change driver, can alter the structure and composition of soil microbial communities, and its variation can alter soil enzymatic activity by altering nutrient inputs to soil from plants [25]. Therefore, temperature, precipitation and their interaction profoundly affect soil enzymatic activity. However, little is known about the regulation mechanisms of land-use types, seasonal dynamics and their interactions on soil active C and N turnover. Clearly revealing the response of soil enzyme activities to different land-use types would not only help to understand the changes of soil enzyme activities, but would also be favorable for further elucidating the function and stability of plantation ecosystems.

The rocky mountainous region of North China, as one of the key areas of forest ecological engineering in China, is an important ecological barrier of the North China Plain [26]. This area has poor soils, drought and water shortage, severe soil erosion and numerous unplanted or barren limestone hills [21]. The land-use types in this region mainly include farmland, abandoned land and artificial forest, and it is a typical area to study land-use change. At present, the research reports in this region mainly focus on the effects of different plantation types and tree ages on soil active carbon and nitrogen and greenhouse gas emissions, while little is known about the effects of land-use types and seasonal changes on the pivotal enzyme activities of soil carbon and nitrogen cycles. We hypothesized that land-use types would significantly affect soil enzyme activities by influencing inputs of litter and root exudates, as well as microclimate, and that seasonal dynamics coupled with land-use types would have complex effects on soil enzyme activities. We studied the seasonal characteristics of soil BG, URE, PROT and CAT activities under three typical land-use types; the results of this study are significant for revealing the relationship between enzyme activities and soil physicochemical properties as well as identifying the controlling factors. Furthermore, our aims are to improve soil quality, select reasonable land-use modes and provide a basis for sustainable land development in this region.

2. Materials and Methods

2.1. Study Area and Soil Sampling

In Jiyuan City of Henan Province, the study area is part of the Yellow River Xiaolangdi Forest Ecosystem Location Research Station (35°01′ N, 112°28′ E) of the Chinese Forest Ecosystem Network, which is located in the middle reaches of the Yellow River and close to the Taihang Mountains. It has a warm temperate continental monsoon climate with hot and rainy summers and cold and dry winters. The annual average temperature is 13.1 °C, the highest is 32 °C and the lowest is −4 °C, and the annual sunshine hours are above 2300 h. The annual average rainfall is 641.7 mm, which is unevenly distributed during the year, mostly concentrated in July to September, accounting for more than 60% of the annual rainfall. The soil types are brown loam and Alfisol.

Three typical land-use types were selected as farmland (FL), abandoned land (AL) and artificial forest (i.e., *Platycladus orientalis* (L.) Franco (PO), *Robinia pseudoacacia* L. (RP) and *Quercus variabilis* Bl. (QV)). The farmland in this region is dominated by winter rape and summer maize rotation, treated with compound fertilizer and farm organic manure Compound fertilizer containing 15% nitrogen (N), 15% phosphorus (P) and 15% potassium (K) is inputted at ca. 500 and 200 kg ha^{-1} before sowing (i.e., at the middle of October and April, respectively). The main raw material composing farm organic manure is animal manure, which is spread according to farmers' habits and crop growth situation. The AL had been abandoned for 1 year, adjacent to the Locust plantation. All land-use types are in close proximity (5 km apart), having similar slope position and aspect. Details of sampling lands are given in Table 1.

Table 1. The basic details of the sampling plots.

Land-Use Types	Stand Age (Years)	Elevation (m)	Mean Tree Height (m)	Mean DBH (cm)	Crown Density
FL	/	410	/	/	/
AL	/	380	/	/	/
PO	35	410	8.5	12.90	0.85
RP	35	370	13.5	15.61	0.86
QV	35	380	13.1	7.32	0.84

FL, farmland; AL, abandoned land; PO, *Platycladus orientalis* artificial forest; RP, *Robinia pseudoacacia* artificial forest; QV, *Quercus variabilis* artificial forest; DBH, diameter at breast height.

According to the principle of proximity arrangement, three 10 m × 10 m fixed sample squares were set up in each sampling plot, totaling 15 sampling squares. Five sampling points were set by the grid-based sampling method in each square, and soil samples were collected from three soil layers (0–10, 10–20 and 20–30 cm). Fresh soil was sieved through a 2 mm sieve and refrigerated at 4 °C until analysis.

2.2. Analysis of Soil Enzymatic Activity

Soil BG activity was evaluated by the nitro-phenol-colorimetric method and expressed as mg ρ-nitrophenol per g of dry soil after 1 h of incubation [27]. Soil URE activity was determined according to the method described by Paudel et al. [28], reported as mg NH_4^+-N per g dry soil after 1 d of incubation. Soil PROT activity was detected by the ninhydrin colorimetric method and expressed as milligrams of glycine produced after enzymatic reaction per gram of soil after 1 day of incubation [29]. Soil CAT activity was conducted by $KMnO_4$ titration method with H_2O_2 as substrate, expressed in cm^3 of 0.1 mol $KMnO_4$ dm^{-3} solution titrated per g of dry soil after 1 h of incubation [30].

2.3. Analysis of Soil Physicochemical and Biological Properties

Soil dissolved organic carbon (DOC), dissolved organic nitrogen (DON), ammonia nitrogen (NH_4^+-N) and nitrate nitrogen (NO_3^--N) contents were determined by leaching 5 g of fresh soil (soil:liquid = 1:5) with 1 mol dm^{-3} KCL solution using a SKALAR flow analyzer (HQ-Skalar Analytical B.V., Breda, Netherlands). Soil TN content was determined by automatic elemental analyzer EuroEA3000 (Eurovector, Pavia, Italy). Soil available nitrogen (AN), available phosphorus (AP) and available potassium (AK) contents were determined by the alkaline diffusion method, sodium bicarbonate leaching-molybdenum antimony anti-colorimetric method and ammonium acetate leaching-flame photometer FP6450 (Hinotek, Ningbo, China) method, respectively. Soil microbial biomass carbon (MBC) and nitrogen (MBN) contents were determined by chloroform fumigation and extraction method [31]. Soil pH, moisture content and bulk density were determined as described by Bao Shidan [32]. The soil basic physical and chemical properties were given in Gong et al. [21].

2.4. Statistical Analysis

Multivariate analysis of variance (MANOVA) was used to analyze the influences of land-use type, season, soil layer and their interactions on soil BG, URE, PROT and

CAT activity. The relationship between soil enzymatic activity and other soil properties was evaluated by Pearson correlation analysis. Stepwise multiple-regression analysis was conducted to determine the main parameters influencing the soil enzymatic activity. All statistical analysis was conducted in SPSS 20.0 software (IBM, New York, NY, USA).

3. Results

3.1. Seasonal Dynamics of Soil β-Glucosidase and Urease Activities under Different Land-Use Types

Soil BG and URE activities in the 0–30 cm soil layer varied remarkably ($p < 0.05$) under different land-use types, both showing the lowest level in QV (Figure 1a) (Table 2). The variation of soil BG activity under the five sampling sites ranged from 0.02 to 2.32 mg ρ-nitrophenol g^{-1} h^{-1}, and its interannual mean values from the 0–30 cm layer showed an overall variation pattern of FL > PO > AL > RP > QV (Figure 1a), among which the activity in FL soil was 8%, 22%, 107% and 428% higher than in PO, AL, RP and QV, respectively. Under the five sampling sites, the variation of soil URE activity ranged from 0.11 to 0.70 mg NH_4^+-N g^{-1} h^{-1}, and the mean values showed an overall pattern of PO (0.44 mg NH_4^+-N g^{-1} h^{-1}) > FL (0.33 mg NH_4^+-N g^{-1} h^{-1}) > AL (0.29 mg NH_4^+-N g^{-1} h^{-1}) > QV (0.12 mg NH_4^+-N g^{-1} h^{-1}) ≈ RP (0.11 mg NH_4^+-N g^{-1} h^{-1}) from the 0–30 cm layer, and the soil URE activity of PO was 26%, 36%, 74% and 75% higher than that of FL, AL, QV and RP, respectively ($p < 0.05$) (Figure 1b).

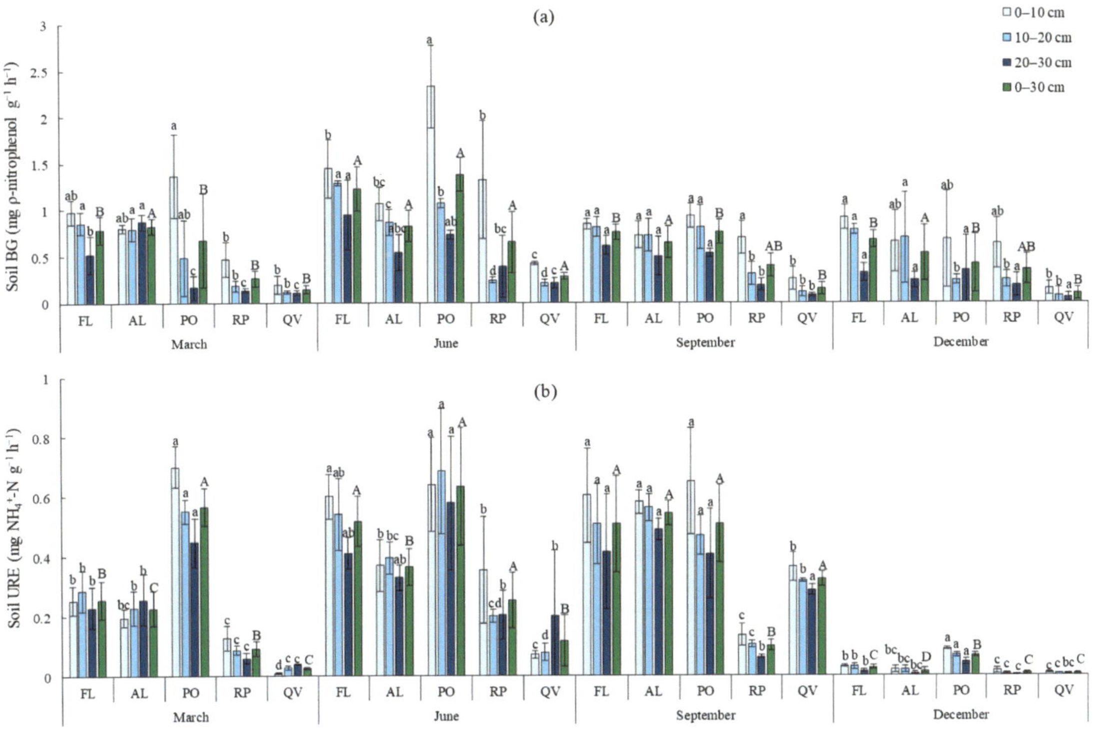

Figure 1. Effects of different land-use types on the soil BG (**a**) and URE (**b**) activities. Different lowercase letters indicate that different land-use types are significantly different in the same soil layer in the same season ($p < 0.05$). Different capital letters indicate that different seasons are significantly different in the 0–30 cm soil layer ($p < 0.05$).

Table 2. Variance analysis of effects of land-use types, season and soil layer on soil enzymatic activity

	Source	Land-Use Types (LUT)	Season (S)	Soil Layer (SL)	LUT × S	LUT × SL	S × SL	LUT × S × SL
p	BG	0.000	0.000	0.000	0.000	0.000	0.004	0.107
	URE	0.000	0.000	0.000	0.000	0.039	0.269	0.636
	PROT	0.000	0.000	0.000	0.000	0.000	0.224	0.396
	CAT	0.000	0.000	0.012	0.015	0.337	0.006	0.408

There was an obvious seasonal pattern in soil BG and URE activities under differen land-use types ($p < 0.05$) (Table 2). Soil BG activity showed a unimodal trend under all fiv sites, with the highest activity in June and the lowest in December, while soil URE activit showed higher levels in June or September and the lowest level in December (Figure 1).

With the increase of soil depth, most of the soil BG and URE activities tended t decrease, showing enrichment in the 0–10 cm soil layer (Figure 1) ($p < 0.05$). For exampl in June, soil BG activity in the 0–10 cm soil layer was 1.5–3.4 times that in the 20–30 cm soi layer. In September, soil URE activity in the 0–10 cm soil layer was 1.2–2.2 times than in th 20–30 cm soil layer (Figure 1).

3.2. Seasonal Dynamics of Soil Protease and Catalase Activities under Different Land-Use Types

Different land-use types significantly influenced soil PROT and CAT activities in th 0–30 cm layer ($p < 0.05$) (Table 2), with both showing the lowest values in QV (Figure 2 Soil PROT activity ranged from 1.72 to 155.70 μg glycine g^{-1} h^{-1}, and its interannual mea values showed an overall variation pattern of PO > FL > RP > AL > QV from the 0 to 30 cr soil layer (Figure 2a). Among them, the PROT activity of PO was 31%, 39%, 40% and 83% higher than that in FL, RP, AL and QV, respectively ($p < 0.05$) (Figure 2a). Soil CAT activit ranged from 0.32 to 1.51 cm^3 of (0.1 mol $KMnO_4$ dm^{-3}) g^{-1} h^{-1}, and its interannual mea values in the 0–30 cm layer showed the trend of RP > PO > FL > AL > QV, where CA activities in RP, PO, AL and FL soils were 116%, 113%, 72% and 10% higher than that in Q respectively (Figure 2b).

Soil PROT and CAT activities showed notable seasonal patterns under different lanc use types ($p < 0.05$) (Table 2). Soil PROT activity was higher in March or June and lowe in September and December. The CAT activity of FL soil in March was significantl higher than that in other months, while that in the RP soil showed the lowest level i June. However, soil CAT activity had no significant seasonal variation in AL, PO and Q (Figure 2). With increasing soil depth, soil PROT and CAT activities of all land-use type showed a decreasing trend in December, while FL, AL and QV had no remarkable chang in other months.

3.3. Correlation Analysis of Soil Enzymatic Activity and Soil Physicochemical Properties

As shown in Figure 3, there was a highly significant positive correlation among soi BG, URE, PROT and CAT ($p < 0.01$). Soil BG, URE, PROT and CAT activities were positivel remarkable correlated with AN, AP, AK, DOC, NO_3^--N, SOC, water content, clay and sil volume fractions ($p < 0.05$), and notably negatively correlated with sand volume fractio ($p < 0.05$). In addition, soil BG, PROT and CAT activities were all positively significan correlated with DON, MBC and TN contents ($p < 0.05$). Soil BG, URE and PROT activitie were all positively correlated with NH_4^+-N ($p < 0.05$).

In order to further analyze the main influencing factors of four soil enzyme activitie we conducted stepwise regression analysis. Soil BG, URE, PROT and CAT activities wer influenced by different factors (Table 3). DOC was the key factor affecting the activities c soil BG and PROT, accounting for 51% and 62% of the variations, respectively. Secondly, th activities of BG and PROT were affected by AK and MBC, respectively, with the contributio rates of 10% and 5%. Soil URE activity was mainly regulated by soil silt volume fraction an pH, which contributed 29% and 8% of the variations, respectively. Soil pH could explai

47% of the variation in soil CAT activity. In addition, soil water content was the main factor influencing the activities of soil BG (2%), URE (7%), PROT (5%) and CAT (9%) activities.

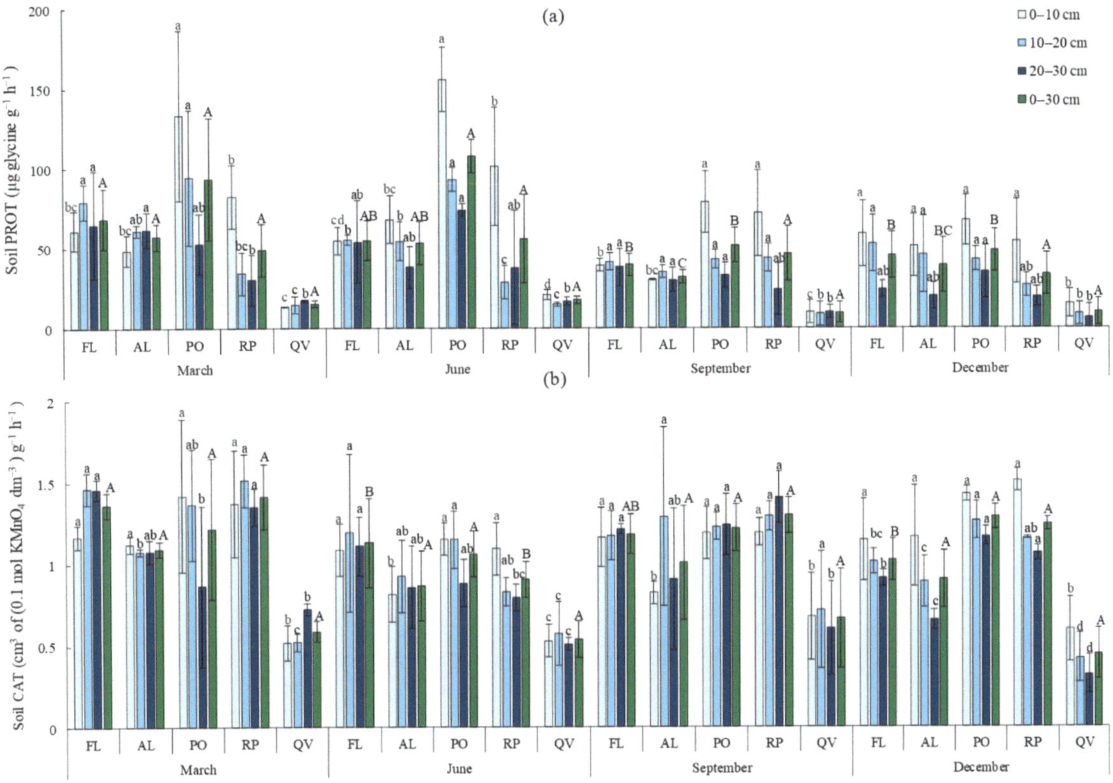

Figure 2. Effects of different land-use types on the soil PROT (**a**) and CAT (**b**) activities. Different lowercase letters indicate that different land-use types are significantly different in the same soil layer in the same season ($p < 0.05$). Different capital letters indicate that different seasons are significantly different in the 0–30 cm soil layer ($p < 0.05$).

Table 3. Multiple-regression analysis model of soil enzyme activities and other soil properties.

Equations	Adjusted R^2	P
$Y_1 = 0.483X_1 + 2.831X_2 + 3.608X_3 + 27.942X_4 - 67.521X_5 + 5.964X_6 + 0.143X_7 - 57.761$	0.673	0.046
$Y_2 = 0.007X_8 - 0.121X_5 + 0.003X_3 + 0.022X_4 + 0.374$	0.509	0.000
$Y_3 = 0.058X_1 + 0.035X_9 + 2.067X_4 + 0.144X_3 - 0.084X_{10} + 0.375X_6 - 20.87$	0.745	0.032
$Y_4 = 0.184X_5 + 0.028X_4 + 0.006X_6 - 0.824$	0.591	0.009

Y_1, soil BG activity; Y_2, soil URE activity; Y_3, soil PROT activity; Y_4, soil CAT activity; X_1, DOC; X_2, AK; X_3, AN; X_4, water content; X_5, pH; X_6, NO_3^--N content; X_7, SOC; X_8, silt volume fractions; X_9, MBC; X_{10}, MBN.

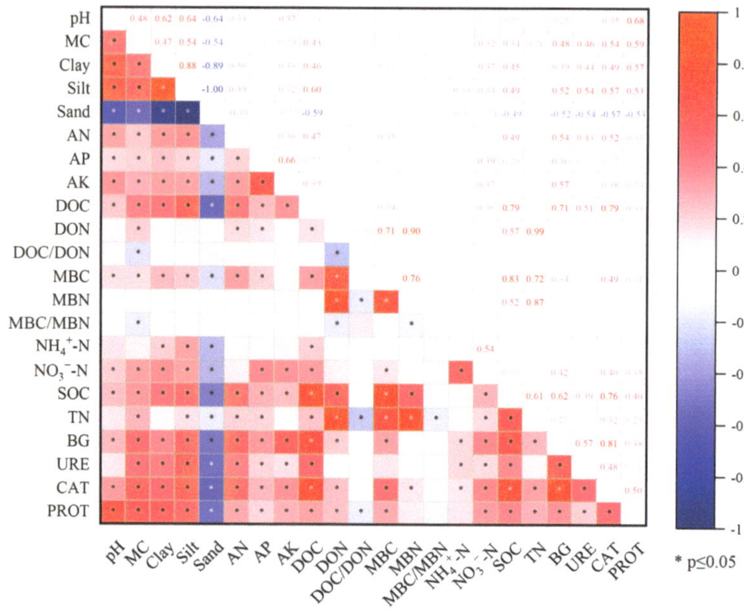

Figure 3. Correlation coefficients between soil enzyme activities and other soil physical and chemical properties.

4. Discussion

Soil enzymes, as biologically active indicators of soil quality, not only reflect the basic condition of the soil, but are also influenced by soil physicochemical properties temperature, moisture and vegetation type [33–35]. Therefore, changes in land-use types seasons and soil layers are able to induce differences in soil enzymatic activity. In the present study, soil BG, URE, PROT and CAT activities differed remarkably among land use types (Figures 1 and 2), and all four soil enzyme activities were higher in FL, AL and PO plantations and the lowest in QV plantations, which is not consistent with the results of Silva et al. [36]. In addition to different climate type and soil conditions, artificial fertilization as well as plant characteristics were the main reasons for this phenomenon [37] FL may have indirectly increased its soil enzymatic activity due to the artificial fertilizer application that inputs large amounts of nutrients to the soil and promotes soil microbial colonization and plant root growth, which are the main sources of soil enzymes released by soil microorganisms and secreted by plant roots [38]. The higher enzymatic activity in the AL soil may have been due to the fact that some crop stubble still remained in the soil, and the stubble decomposition rate reaches its highest value after one year of abandonment, prompting microorganisms to release large amounts of nutrients, enzymes and other active substances into the soil [39]. All four soil enzyme activities were higher in the PO plantation, in contrast to other studies [40], mainly due to their well-developed root system and large annual apoplastic volume [26], lower soil bulk density and adequate source of soil nutrients [21], thus indirectly increasing soil enzyme activities. Meanwhile the four soil enzyme activities under the QV artificial forest were all the lowest, mainly because of its poor physical structure, such as lower soil bulk density and higher sand volume fraction, which lead to poor water and fertilizer retention abilities [21], and thus reduced soil microbial activities and the production of enzymes.

In this study, soil BG, URE, PROT and CAT activities were characterized by notable seasonal dynamic changes. Soil BG and URE activities were higher in summer and lower in winter under all land-use types, and all showed a highly positive correlation with soil water content, probably because the higher temperatures in summer enhanced the activity of soil microorganisms and frequent rainfall promoted enzyme mobility [23]. In addition, the rapid plant growth and increased root secretion in summer [41] may have significantly increased soil BG and URE activities. In winter, low temperature and dry climate lead to very slow plant growth and low microbial activity, which notably reduced soil organic matter turnover and mineralization rates [21], indirectly reducing soil BG and URE activities. However, the seasonal patterns of the four soil enzyme activities in this study were not entirely consistent, implying that they may be regulated by different factors. Soil PROT activity was mainly influenced by soil DOC, and was higher in March and lower in September. In spring, when the temperature rose, frozen soil melted and released large amounts of C and N nutrients to the soil, which stimulated microbial growth and thus indirectly increased soil PROT activity; in autumn, plants still needed to absorb nutrients from the soil, which made it easy for soil microorganisms to form a competitive relationship [42]. Soil CAT activities were affected by pH and water content, with persistent high temperature and frequent rainfall in summer causing a decrease in activity, while soil pH did not vary significantly throughout the seasons, so there were no notable changes in soil CAT activities between March, September and December.

With the increase of soil depth, soil BG, URE, PROT and CAT activities showed a decreasing trend under the five land-use types (Figure 1), showing superficial enrichment, consistent with the results of Song et al. [43], which is mainly due to the long-term deposition of highly polymerized plant material (especially leaves) and its decomposition, that increases the content of humus and organic matter in the superficial soil and improves the superficial soil structure. In the present study, it was found that the surface soil had less bulk, its soil aeration and water retention were good, and its soil organic matter content was high [21], which provided a suitable living environment and sufficient nutrient sources for microbial growth and enhanced microbial metabolism, resulting in higher enzymatic activity in the surface soil. In addition, all four enzyme activities of the soils in our paper were notably positively correlated with soil AN, AP and AK contents, while the AN, AP and AK contents in top soil were higher than those in the deep soil [21], indicating that the top soil can provide more substrates for microbial metabolism and thus higher enzyme activities. In the vertical spatial variation of the soil profile, soil enzymatic activities are more susceptible to the variations of soil physical properties and available substrates.

5. Conclusions

Different land-use types had remarkable influences on soil enzyme activities, showing higher values in FL cultivation and PO plantation. However, the mechanism behind this phenomenon is not entirely consistent. The high activity in farmland is mainly attributable to high fertilizer inputs and annual agricultural production, while that in PO plantation is mainly caused by its large amount of litter and well-developed root system. Soil BG, URE and PROT activities were found to be very sensitive to the changes of soil temperature and moisture, and all showed high activities in summer. The activities of these enzymes were closely related to soil physicochemical and biological properties. This suggests that, in general, soil enzyme activities can be used as an important indicator of changes in soil quality. In view of its high quality and strong adaptability, *Platycladus orientalis* can be used as the dominant tree species in the barren rocky mountainous region of North China.

Author Contributions: Conceptualization, Y.K. and Y.L.; methodology, A.Q. and Y.K.; software, investigation and data curation, R.C. and E.F.; formal analysis, X.Y. and Y.L.; writing—original draft preparation, Y.K. and A.Q.; writing—review and editing, Y.K. and Y.L.; funding acquisition, Y.L. and X.Y. All authors have read and agreed to the published version of the manuscript.

Funding: This research was supported by the Research and Demonstration of Key Technologies of Afforestation in Barren Lands of Henan Province (Grant No. YLK202209) and the Training Program of Young Key Teachers in Colleges and Universities of Henan Province (Grant No. 2021GGJS033).

Institutional Review Board Statement: Not applicable.

Informed Consent Statement: Not applicable.

Data Availability Statement: Not applicable.

Acknowledgments: We would like to express our sincere thanks to Jia Changrong for his help in th selection and setting up of experimental plots in the forestry farm.

Conflicts of Interest: The authors declare no conflict of interest.

References

1. Duan, C.; Fang, L.; Yang, C.; Chen, W.; Cui, Y.; Li, S. Reveal the response of enzyme activities to heavy metals through in sit zymography. *Ecotoxicol. Environ. Saf.* **2018**, *156*, 106–115. [CrossRef]
2. Qu, Y.; Tang, J.; Li, Z.; Zhou, Z.; Wang, J.; Wang, S.; Cao, Y. Soil enzyme activity and microbial metabolic function diversity i soda saline–alkali rice paddy fields of northeast China. *Sustainability* **2020**, *12*, 10095. [CrossRef]
3. Xiao, L.; Liu, G.; Li, P.; Li, Q.; Xue, S. Ecoenzymatic stoichiometry and microbial nutrient limitation during secondary successio of natural grassland on the loess plateau, China. *Soil Tillage Res.* **2020**, *200*, 104605. [CrossRef]
4. Fierer, N. Embracing the unknown: Disentangling the complexities of the soil microbiome. *Nat. Rev. Microbiol.* **2017**, *15*, 579–59 [CrossRef]
5. Chen, Y.; Wei, T.; Sha, G.; Zhu, Q.; Liu, Z.; Ren, K. Soil enzyme activities of typical plant communities after vegetation restoratio on the loess plateau, China. *Appl. Soil Ecol.* **2022**, *170*, 104292. [CrossRef]
6. Huang, H.; Di, T.; Zhou, L.; Su, H.; Ma, S.; Feng, Y.; Tang, Z.; Zhu, J.; Ji, C.; Fang, J. Effects of afforestation on soil microbia diversity and enzyme activity: A meta-analysis. *Geoderma* **2022**, *423*, 115961. [CrossRef]
7. Pausch, J.; Kuzyakov, Y. Carbon input by roots into the soil: Quantification of rhizodeposition from root to ecosystem scale. *Glo Chang. Biol.* **2018**, *24*, 1–12. [CrossRef]
8. Xu, G.R.; Ma, W.W.; Song, L.C.; Tang, Y.M.; Zhou, X.L.; Shang, Y.X.; Yang, X. Characteristics of soil nitrogen content enzym activity in Gahai wetland under different vegetation degradation conditions. *Acta Ecol. Sin.* **2020**, *40*, 8917–8927.
9. Li, Y.; Han, C.; Sun, S.; Zhao, C. Effects of tree species and soil enzyme activities on soil nutrients in dryland plantations. *Forest* **2021**, *12*, 1153. [CrossRef]
10. Kang, H.; Gao, H.; Yu, W.; Yi, Y.; Wang, Y.; Ning, M. Changes in soil microbial community structure and function after afforestatio depend on species and age: Case study in a subtropical alluvial island. *Sci. Total Environ.* **2018**, *625*, 1423–1432. [CrossRef]
11. Zhao, F.Z.; Ren, C.J.; Han, X.H.; Yang, G.H.; Wang, J.; Doughty, R. Changes of soil microbial and enzyme activities are linked t soil C, N and P stoichiometry in afforested ecosystems. *For. Ecol. Manag.* **2018**, *427*, 289–295. [CrossRef]
12. Zhang, W.; Qiao, W.; Gao, D.; Dai, Y.; Deng, J.; Yang, G.; Han, X.; Ren, G. Relationship between soil nutrient properties an biological activities along a restoration chronosequence of *Pinus tabulaeformis* plantation forests in the Ziwuling Mountains, Chin *Catena* **2018**, *161*, 85–95. [CrossRef]
13. Li, Q.; Chen, J.; Feng, J.; Wu, J.; Zhang, Q.; Jia, W.; Lin, Q.; Cheng, X. How do biotic and abiotic factors regulate soil enzym activities at plot and microplot scales under afforestation? *Ecosystems* **2020**, *23*, 1408–1422. [CrossRef]
14. Di Iorio, E.; Napoletano, P.; Circelli, L.; Memoli, V.; Santorufo, L.; De Marco, A.; Colombo, C. Comparison of natural an technogenic soils developed on volcanic ash by Vis-NIR spectroscopy. *Catena* **2022**, *216*, 106369. [CrossRef]
15. Ji, L.; Yang, Y.; Yang, L.; Zhang, D. Effect of land uses on soil microbial community structures among different soil depths i northeastern China. *Eur. J. Soil Biol.* **2020**, *99*, 103205. [CrossRef]
16. Manoharan, L.; Kushwaha, S.K.; Ahrén, D.; Hedlund, K. Agricultural land use determines functional genetic diversity of soi microbial communities. *Soil Biol. Biochem.* **2017**, *115*, 423–432. [CrossRef]
17. Vazquez, C.; Verdenelli, R.A.; Merlo, C.; Brandan, C.P.; Kowaljow, E.; Meriles, J.M. Influence of land-use changes on microbia community structure and diversity in a semiarid region. *Land Degrad. Dev.* **2022**, *33*, 3690–3702. [CrossRef]
18. Guo, X.; Zhou, Y. Effects of land use patterns on the bacterial community structure and diversity of wetland soils in the Sanjian Plain. *J. Soil Sci. Plant Nutr.* **2021**, *21*, 1–12. [CrossRef]
19. Gong, J.Y.; Hou, W.P.; Liu, J.; Malik, K.; Kong, X.; Wang, L.; Chen, X.L.; Tang, M.; Zhu, R.Q.; Cheng, C.; et al. Effects of differen land use types and soil depths on soil mineral elements, soil enzyme activity, and fungal community in karst area of Southwes China. *Int. J. Environ. Res. Public Health* **2022**, *19*, 3120. [CrossRef]
20. Yao, H.U.; Li, Y.; Hou, Y. The variation of soil organic carbon fractions and soil enzyme activity of different land use types i Minjiang river valley. *Ecol. Environ. Sci.* **2018**, *27*, 1617–1624.
21. Gong, S.S.; Feng, Z.P.; Qu, A.R.; Sun, J.H.; Xu, X.K.; Lai, Y.; Kong, Y.H. Effects of land-use types on the temporal dynamics of so active carbon and nitrogen in the rocky mountainous of North China. *Soil Sci. Plant Nutr.* **2022**, *68*, 72–80. [CrossRef]
22. Xu, H.W.; Qu, Q.; Chen, Y.H.; Liu, G.B.; Xue, S. Responses of soil enzyme activity and soil organic carbon stability over time afte cropland abandonment in different vegetation zones of the loess plateau of China. *Catena* **2021**, *196*, 104812. [CrossRef]
23. Lee, M.H.; Park, J.H.; Matzner, E. Sustained production of dissolved organic carbon and nitrogen in forest floors during continuou leaching. *Geoderma* **2018**, *310*, 163–169. [CrossRef]
24. Han, C.; Kang, Y.M.; Yu, H.L. Effects of precipitation on soil enzyme activities during litter decomposition in a desert steppe o Northwestern China. *Ecol. Environ. Sci.* **2022**, *31*, 1802–1812.

25. Li, G.; Kim, S.; Han, S.H.; Chang, H.; Du, D.L.; Son, Y. Precipitation affects soil microbial and extracellular enzymatic responses to warming. *Soil Biol. Biochem.* **2018**, *120*, 212–221. [CrossRef]
26. Kong, Y.; Ma, N.L.; Yang, X.; Lai, Y.; Feng, Z.; Shao, X.; Xu, X.; Zhang, D. Examining CO_2 and N_2O pollution and reduction from forestry application of pure and mixture forest. *Environ. Pollut.* **2020**, *265*, 114951. [CrossRef]
27. Eivazi, F.; Tabatabai, M.A. Glucosidases and galactosidases in soils. *Soil Biol. Biochem.* **1988**, *20*, 601–606. [CrossRef]
28. Kandeler, E.; Gerber, H. Short-term assay of soil urease activity using colorimetric determination of ammonium. *Biol. Fertil. Soils* **1988**, *6*, 68–72. [CrossRef]
29. Ladd, J.N.; Butler, J.H.A. Short-term assays of soil proteolytic enzyme activities using proteins and dipeptide derivatives as substrates. *Soil Biol. Biochem.* **1972**, *4*, 19–30. [CrossRef]
30. Guan, S. *Soil Enzyme and Its Research Methods*; China Agriculture Press: Beijing, China, 1986.
31. Jenkinson, D.S.; Brookes, P.C.; Powlson, D.S. Measuring soil microbial biomass. *Soil Biol. Biochem.* **2004**, *36*, 5–7. [CrossRef]
32. Bao, S.D. *Soil Agrochemical Analysis*, 3rd ed.; China Agriculture Press: Beijing, China, 2000.
33. Wu, J.; Wang, H.; Li, G.; Ma, W.; Wu, J.; Gong, Y.; Xu, G. Vegetation degradation impacts soil nutrients and enzyme activities in wet meadow on the Qinghai-Tibet Plateau. *Sci. Rep.* **2020**, *10*, 21271. [CrossRef]
34. Zhu, Y.; Guo, B.; Liu, C.; Lin, Y.; Fu, Q.; Li, N.; Li, H. Soil fertility, enzyme activity, and microbial community structure diversity among different soil textures under different land use types in coastal saline soil. *J. Soils Sediments* **2021**, *21*, 2240–2252. [CrossRef]
35. Li, B.B.; Shen, X.J.; Zhao, Y.J.; Cong, P.J.; Wang, H.Y.; Wang, A.J.; Chang, S.W. Sloping Farmlands Conversion to Mixed Forest Improves Soil Carbon Pool on the Loess Plateau. *Int. J. Environ. Res. Public Health* **2022**, *19*, 5157. [CrossRef]
36. Silva, E.D.; de Medeiros, E.V.; Duda, G.P.; Lira, M.A.; Brossard, M.; de Oliveira, J.B.; dos Santos, U.J.; Hammecker, C. Seasonal effect of land use type on soil absolute and specific enzyme activities in a Brazilian semi-arid region. *Catena* **2019**, *172*, 397–407. [CrossRef]
37. Leeuwen, J.P.; Djukic, I.; Bloem, J.; Lehtinen, T.; Hemerik, L.; de Ruiter, P.C.; Lair, G.J. Effects of land use on soil microbial biomass, activity and community structure at different soil depths in the Danube floodplain. *Eur. J. Soil Biol.* **2017**, *79*, 14–20. [CrossRef]
38. Vaidya, B.P.; Hagmann, D.F.; Haramuniz, J.; Krumins, J.A.; Goodey, N.M. Artificial root exudates restore microbial functioning in a metal contaminated, barren, inactive soil. *Environ. Pollut.* **2022**, *312*, 120007. [CrossRef]
39. Chu, H.Y.; Hosen, Y.; Yagi, K.; Okada, K.; Ito, O. Soil microbial biomass and activities in a Japanese Andisol as affected by controlled release and application depth of urea. *Biol. Fertil. Soils* **2005**, *42*, 89–96. [CrossRef]
40. Pan, M.; Zhu, Q.; Gong, S.; Zhang, Z.; Lei, L.; Kong, Y. Effects of different land-use types on soil biological and physicochemical properties. *Sci. Soil Water Conserv.* **2021**, *19*, 24–33.
41. Wan, X.H.; Yu, Z.P.; Wang, M.J.; Zhang, Y.; Huang, Z.Q. Litter and root traits control soil microbial composition and enzyme activities in 28 common subtropical tree species. *J. Ecol.* **2022**, *110*, 3012–3022. [CrossRef]
42. Bardgett, R.D.; Mommer, L.D.; De Vries, F.T. Going underground: Root traits as drivers of ecosystem processes. *Trends Ecol. Evol.* **2014**, *29*, 692–699. [CrossRef]
43. Wang, J.J.; Shu, K.L.; Wang, S.Y.; Zhang, C.; Feng, Y.C.; Gao, M.; Li, Z.H.; Cai, H.G. Soil enzyme activities affect SOC and TN in aggregate fractions in sodic-alkali soils, Northeast of China. *Agronomy* **2022**, *12*, 2549. [CrossRef]

forests

Article

Distribution Characteristics of Active Soil Substances along Elevation Gradients in the Southern of Taihang Mountain, China

Erpeng Feng [1], Liwei Zhang [2], Yuhua Kong [1,*], Xingkai Xu [3,4], Ting Wang [1] and Caifeng Wang [2]

1 College of Forestry, Henan Agricultural University, Zhengzhou 450046, China
2 China Construction Seventh Engineering Division Corp. Ltd., Zhengzhou 450004, China
3 State Key Laboratory of Atmospheric Boundary Layer Physics and Atmospheric Chemistry, Institute of Atmospheric Physics, Chinese Academy of Sciences, Beijing 100029, China
4 Department of Atmospheric Chemistry and Environmental Science, College of Earth and Planetary Sciences, University of Chinese Academy of Sciences, Beijing 100049, China
* Correspondence: y.kong@henau.edu.cn

Abstract: Active soil substances, which can indicate environmental changes sensitively, have the fastest turnover rate. Vegetation diversity and soil bio-physicochemical properties according to five elevations classes (800 m, 1000 m, 1100 m, 1200 m, and 1500 m a.s.l.) in the Southern Taihang Mountain were investigated. Soil ammonium-N (NH_4^+—N), nitrate-N (NO_3^-—N), microbial biomass carbon (MBC), and nitrogen (MBN), as well as soil urease (URE) and sucrose (SUC) activities were determined. The results showed that elevation gradients, soil layers, and their interaction had significant effects on most of the active soil substances. With the rise of elevation, soil NO_3^-—N, inorganic N, MBC, and MBN contents, as well as SUC activity and SUC/MBC ratio basically showed an inverted V-shaped distribution trend and reached the peak value at 1100 m ($p < 0.05$). Soil URE showed a fluctuating upward trend and reached the peak value at 1500 m ($p < 0.05$), and the URE/MBC ratio showed a straight upward trend. With the depth of soil layer, the contents of active soil substances tended to decrease, showing a phenomenon of surface aggregation. Active soil substances were correlated with plant community diversity indexes, soil water content, pH, available N, and available phosphorus contents, and available N was the main factor affecting active soil substances, which could explain 34.4% of the variation. In summary, at the intermediate slope of 1100 m, soil moisture and tightness were suitable for soil microbial activity and plant growth, the highest contents of active soil substances, indicating a fast turnover of soil carbon and nitrogen. The present study enriched our understanding of soil carbon and nitrogen turnover mechanisms in the mountain ecosystem.

Keywords: active soil substances; elevation gradients; plant diversity indexes; Taihang mountain

Citation: Feng, E.; Zhang, L.; Kong, Y.; Xu, X.; Wang, T.; Wang, C. Distribution Characteristics of Active Soil Substances along Elevation Gradients in the Southern of Taihang Mountain, China. *Forests* **2023**, *14*, 370. https://doi.org/10.3390/f14020370

Academic Editor: Choonsig Kim

Received: 31 December 2022
Revised: 7 February 2023
Accepted: 8 February 2023
Published: 12 February 2023

1. Introduction

Global warming causes ecological imbalances, leading to the decline of biodiversity, and mountain ecosystems, as one of the most climate-sensitive ecosystems in the world, are particularly vulnerable to the effects of climate warming [1–3]. Forests, the largest carbon pool in terrestrial ecosystems, play an important role in the carbon cycle and also affect nitrogen dynamics [4,5]. Soil active carbon (C) and nitrogen (N) are the most active components underground, and their dynamics are of great significance to the C and N cycle processes within and between ecosystems [6]. In addition, altitude has been identified as a potential factor affecting soil nutrient cycling. With the increase of altitude, microclimates will affect the distribution and composition of vegetation, soil conditions, and nutrient cycling [7,8].

Active soil substances, including soil inorganic nitrogen, soil microbial biomass, and soil enzymes, have a fast turnover rate and are very sensitive to micro-environmental changes. Soil ammonium-N (NH_4^+—N), nitrate-N (NO_3^-—N), soil microbial biomass carbon (MBC), and nitrogen (MBN) are important components of soil active C and N, which are sensitive to environmental changes [9]. NH_4^+—N and NO_3^-—N are the material basis of important biochemical processes in the soil N cycle and are important indicators of soil N biological effectiveness [10]. MBC and MBN are involved in the soil-atmosphere C and N cycle and are also one of the indexes of soil microbial indicators. Soil urease (URE) activity is often used as an indicator of N mineralization. URE is catalyzing the hydrolysis of urea to ammonia and CO_2, one of the many organic substrates of nitrogen in the soil [11]. Sucrase (SUC) decomposes soil carbohydrates and increases the soil soluble nutrient content, which is the main energy source of soil microorganisms. Its activity affects the decomposition, conversion and accumulation efficiency of soil organic carbon. Therefore, SUC activity is often used as an indicator of the carbon cycle [11]. Soil enzyme activity can be expressed as absolute enzyme activity and relative enzyme activity [12], and most of studies in mountain ecosystems mainly focused on the variations of absolute enzyme activity [13,14], which is greatly affected by environmental heterogeneity. Moreover, relative soil enzyme activity is commonly used to characterize soil quality status [15,16], which can better reflect the catalytic ability of soil enzymes, making the comparison of enzyme activities under different environments more meaningful. Soil enzyme activity can be an important indicator of soil quality and microbial activity due to its functional properties and sensitivity to the environment [17]. The analysis of active soil substances along the altitude gradient is helpful to improve our understanding of spatial changes of active soil substances and provide a scientific basis for vegetation restoration in vulnerable mountain ecosystems.

Elevation gradient, which can significantly change soil properties, vegetation types and climate characteristics in a small geographical scale, is an important topographic factor for mountain ecosystems [18]. In addition, elevation gradient affects biological processes and plant community composition and growth by affecting water distribution and temperature changes, and indirectly changes litter loss rate [19]. However, with altitude gradient, the variation characteristics of soil nutrient content are not consistent. For example, soil available P, total nitrogen (TN), and soil organic C (SOC) increased with elevation in the Segila Mountains of Southeast Tibet along an elevation gradient (3500–4300 m a.s.l.) [20]. SOC and pH increased and available P decreased significantly according to elevation, whereas other soil factors had no significant variations on the south slope of Qinling Mountains according to elevations (1500–1900 m a.s.l.) [21]. Soil MBC content increased and then decreased with the elevation gradient in a broadleaf forest [22], soil enzyme activity increased with elevation in the mountain bamboo forest [13], while Hofmann et al. found that the soil MBC linearly decreased along the altitudinal gradient in the Austrian Central Alps [23]. These similar or opposite results indicate that the effects of elevation gradient and vegetation type on soil properties are very complex. It is of great significance to comprehensively understand the distribution characteristics of soil active carbon and nitrogen, available nutrients and enzyme activities under different vegetation types along with altitudes.

Taihang Mountain, located in the center of North China, is an important ecological barrier of the North China Plain and a key area of forestry ecological engineering construction in China [24]. However, human overexploitation and the unique continental monsoon climatic conditions lead to the degradation of the ecosystem, the reduction of vegetation richness and soil fertility, which seriously limits the sustainable development of this region [25]. Since the 1950s, a series of afforestation projects have gradually improved the soil conditions and the local ecological environment [26]. At present, a large number of scientific studies have been carried out in this region, but most of them focus on the effects of tree species or age on soil properties, while ignoring the effects of elevation change, which is an important topographic factor in mountain ecosystems, on active soil substances. We hypothesized that the change of elevation gradients would influence soil carbon and nitrogen turnover by governing plant diversity and the microenvironment, and that different types of active soil substances may have different responses to the rise of the elevation gradient. Therefore, we investigated vegetation and soil according to five elevation classes (800–1500 m a.s.l.) in this region, and analyzed plant community composition and soil biotic and abiotic factors. We aimed to explore the vertical distribution patterns of active soil substances and vegetation and their relationships, so as to give some new insights into the carbon and nitrogen turnover mechanisms in mountain ecosystems.

2. Materials and Methods

2.1. Study Area Description

The study site is located in the Wangwu Mountain scenic spot in the south of Taihang Mountain, northwest of Jiyuan City, Henan Province (35°8′45″–35°13′43″ N, 112°14′24″–112°18′17″ E) with a total area of 265 km^2. The study area belongs to a temperate monsoon climate, with four distinct seasons. The annual sunshine is 1727.6 h, and the annual precipitation is 860 mm. The precipitation is mainly concentrated from July to September, accounting for 70% of the annual precipitation. The geographical conditions of Wangwu Mountain area are complex, the terrain is high in the north and low in the south, the main mountain in the area is the Tiantan peak with an altitude of 1715 m, while the altitude of the southern mountain is only 400 m. Due to the large height difference, the climate has a large vertical change. The soil type in the study area is mainly cambisols. The vegetation in the study area is mainly secondary forest dominated by *Quercus variabilis Blume*, *Quercus aliena Blume*, etc. The shrub layer is dominated by *Forsythia suspensa*, *Vitex negundo L.* var. *heterophylla*, *Viburnum dilatatum*, and the herbaceous layer includes *Carex rigescens*, *Oplismenus undulatifolius*, *Circaea lutetiana*, and *Thalictrum aquilegiifolium*.

2.2. Soil Sampling and Plant Investigation

In August 2020, five sampling sites with elevation gradients of 800 m, 1000 m, 1100 m, 1200 m, and 1500 m were set up in the research area (Figure 1). Three sample plots with an area of 20 × 20 m was set up in each elevation using a total station, and the spacing of each two quadrants is greater than 50 m. The elevation, latitude, and longitude coordinates, slope direction, slope, and other environmental factors at the center of the sample plot were also recorded, and the basic conditions of the sample plots are shown in Table 1.

Table 1. Summary of site characteristics along the elevation gradients.

Elevation/m	Canopy Density	Mean Diameter at Breast Height/cm	Litter	
			Thickness/cm	Biomass/(t·ha^{-1})
800	0.75 ± 0.01 B	12.47 ± 2.27 B	3.37 ± 0.64 B	2.93 ± 0.24 D
1000	0.73 ± 0.03 B	11.53 ± 1.51 B	3.7 ± 1.13 B	2.94 ± 0.26 D
1100	0.81 ± 0.04 A	18.03 ± 2.61 A	7 ± 1.00 A	5.39 ± 0.44 B
1200	0.78 ± 0.02 AB	14.4 ± 1.41 B	7.3 ± 0.58 A	6.73 ± 0.50 A
1500	0.81 ± 0.04 A	12.87 ± 1.87 B	4.7 ± 0.58 B	4.52 ± 0.32 C

Different capital letters indicate significant differences at $p < 0.05$ with elevation gradients.

In each plot, five 2 × 2 m shrub subplots and five 1 × 1 m herb subplots were surveyed, respectively. Plant species names of all trees, shrubs, and herb layer species in the plots and tree sizes were recorded in detail. For trees, the diameters at breast height, tree height, and tree density were measured. Shrub surveyed items included shrub density, shrub height, and shrub coverage. In the case of the herb layers, relative coverage of each species and herb height were recorded.

In each sample plot, five sample points were selected according to the grid-based sampling method. Following the removal of the surface dead leaves and small shrubs and other herbs, the soil profile was excavated for a total of 75, and soil samples were collected from 0–10 cm, 10–20 cm, and 20–30 cm, respectively. Each soil sample was about 500 g and a total of 225 samples were collected. At the same time, soil core samples were also collected. The soil samples were taken back to the laboratory, and the gravel, plants, roots, stems, leaves, insects, and other animal and plant remains were picked out. All samples were sieved through a 2 mm sieve and divided into two parts, one part was refrigerated at 4 °C and the other part was dried at room temperature for the determination of relevant indicators.

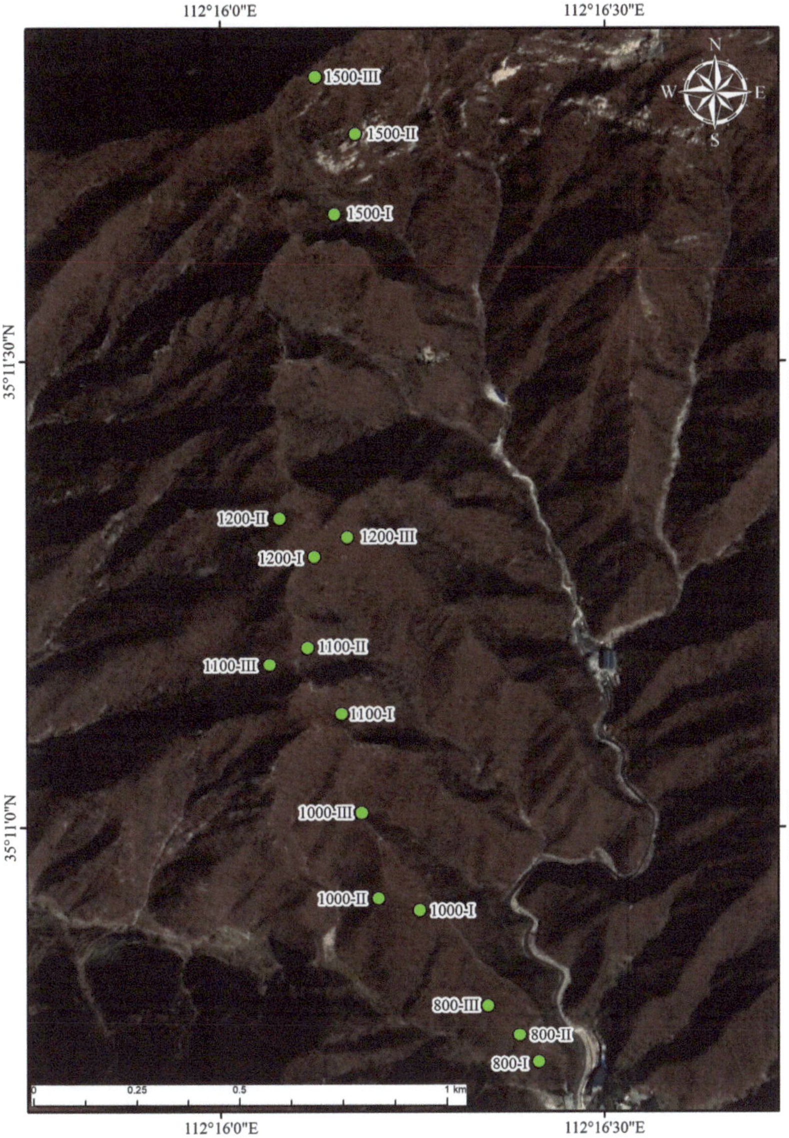

Figure 1. Map of the sampling sites in the Wangwu Mountain scenic spot.

2.3. Chemical and Biochemical Analyses

Air-dried soil (10 g) was mixed with 25 mL distilled water to perform the soil pH by a glass electrode (Leici PHS-3C, Shanghai INESA Scientific Instrument Co., Ltd., Shanghai, China). Soil bulk density and soil porosity were determined by a soil triple-phase meter (Daiki-1130, Daiki Rika Kogyo Co., Ltd., Saitama, Japan) combined with the drying method for core cutter samples. The soil water content determination was made by drying in an oven at 105 °C. The soil available P contents were determined by the Olsen method, the soil available K was determined by flame photometry after extraction with 1 M CH_3COONH_4 [16].The soil available N content was determined by the alkaline diffusion method. The soil total carbon (TC) and total nitrogen (TN) were measured using the elemental analyzer (Euro EA3000, Euro Vector, Pavia, Italy).

The soil NH_4^+—N and NO_3^-—N contents were determined by 1M KCl extracts by a leaching-flow analyzer (SAN++, HQ-Skalar Analytical B.V., Breda, Netherlands). The soil inorganic nitrogen (SIN) was the sum of NH_4^+—N and NO_3^-—N. The soil MBC and MBN were determined by the fumigation–extraction protocol [22]. The soil urease (URE) activity was determined by sodium phenol-sodium hypochlorite colorimetric method, expressed as the amount of NH_3—N contained per g of soil for 1 h after 24 h incubation (mg). The soil sucrase (SUC) activity was determined by the 3, 5-dinitrosalicylic acid colorimetric method expressed as milligrams of glucose produced per g of soil for 1 h after incubation [27].

2.4. Plant Diversity and Litter Biomass

Based on previous studies, this paper adopts the following four indicators to indicate the number of species contained in the community and the distribution of species in the community, avoiding the disadvantage of a single selection. The main four formulas are described as follows:

$$\text{Shannon-Wiener index: } H = -\Sigma(P_i \ln P_i)$$

$$\text{Simpson index: } D = 1 - \Sigma P_i^2$$

$$\text{Pielou index: } J = H/\ln S$$

$$\text{Margalef index: } R = (S - 1)/\ln N$$

where $P_i = N_i/N$ is the relative abundance of a certain species; N_i is the number of individuals of species i; S is the total number of plants of all species in the sample plot where species i resides.

Litter biomass: The withering samples brought back to the laboratory were dried in an oven at 65 °C to a constant weight, and the dry weight was obtained and calculated.

2.5. Statistical Analysis

A one-way ANOVA was used to examine the significant differences of the soil physicochemical properties, such as NH_4^+—N, NO_3^-—N, MBC, MBN, URE, and SUC at different altitude gradients, and a multi-way ANOVA was used to investigate the effects of the altitude gradient, soil layer changes, and their interaction on active soil substances; CANOCO 5.0 software was used to investigate the effects of altitude on soil NH_4^+—N, NO_3^-—N MBC, MBN, URE, and SUC. The relationship between active soil substances and other soil parameters was explored by a Pearson correlation analysis (Biometris-Plant research international, Wageningen, The Netherlands). All data were statistically analyzed using SPSS 20 software (IBM, New York, NY, USA) and plotted using Origin 2022 (Origin Lab Northampton, MA, USA).

3. Results

3.1. Variation Characteristics of Plant Diversity along Altitudinal Gradients

The altitude gradient has a significant effect on Shannon–Wiener, Simpson, and Pielou indexes ($p < 0.05$), whereas it has no remarkable effect on the Margalef index (Table 2). With the increase of altitude gradient, the Shannon–Wiener, Simpson, and Pielou indexes all showed an increasing trend and reached the peak value at 1500 m ($p < 0.05$), while the Margalef index showed a slight increase, but did not reach a notable change. In addition with the increase of altitude gradient, the litter thickness and litter biomass increased first and then decreased, reaching the maximum values at 1200 m (Table 1).

Table 2. Diversity index of the plant community at different elevation gradients.

Elevation/m	Shannon–Wiener Index	Simpson Index	Pielou Index	Margalef Index
800	1.51 ± 0.31 B	0.66 ± 0.13 B	0.64 ± 0.12 B	2.34 ± 0.16 A
1000	1.90 ± 0.33 AB	0.73 ± 0.10 AB	0.67 ± 0.09 B	3.34 ± 0.83 A
1100	1.81 ± 0.51 AB	0.75 ± 0.11 AB	0.74 ± 0.09 AB	2.68 ± 1.01 A
1200	1.85 ± 0.31 AB	0.77 ± 0.05 AB	0.78 ± 0.01 AB	2.59 ± 0.86 A
1500	2.47 ± 0.22 A	0.89 ± 0.02 A	0.86 ± 0.02 A	3.93 ± 0.91 A

Different capital letters indicate significant differences at $p < 0.05$ with elevation gradients.

3.2. Characteristics of the Soil's Physical and Chemical Properties

The altitudinal gradient significantly affected the soil's physical and chemical properties ($p < 0.05$), but they showed different trends with the increase of the altitudinal gradient (Table 3). Soil porosity and soil water contents SWC showed a fluctuating upward trend and reached the peak value at 1500 m, while soil bulk density had an opposite trend and reached the maximum value at 800 m. The soil pH ranged from 4.67 to 6.09, which was slightly acidic and showed a unimodal trend with the increase of altitude. The average contents of the soil available N, available K, and available P in the 0–30 cm layer fluctuated with the increase of altitude, and the highest values occurred at 1100 m or 1500 m, and their maximum value was 2.49–3.51 times the minimum value. The soil's TN and TC content also showed a fluctuating trend and reached the maximum value at 1500 m.

Table 3. Basic physical and chemical properties of the soils at different elevation gradients.

Soil Variables	Soil Layer	800 m	1000 m	1100 m	1200 m	1500 m
Soil porosity/%	0–10 cm	44.67 ± 5.59 a	54.87 ± 6.79 a	62.01 ± 5.56 a	66.73 ± 9.83 a	78.06 ± 10.86 a
	10–20 cm	45.54 ± 7.43 a	48.26 ± 6.98 a	61.32 ± 5.72 a	58.75 ± 7.94 a	67.71 ± 6.65 a
	20–30 cm	48.12 ± 4.46 a	50.63 ± 5.47 a	65.33 ± 4.39 a	56.28 ± 9.22 a	76.31 ± 6.23 a
	0–30 cm	46.11 ± 5.80 C	51.25 ± 6.36 BC	62.89 ± 5.19 AB	60.59 ± 8.99 B	74.03 ± 7.91 A
Soil bulk density /(g·cm^{-3})	0–10 cm	1.78 ± 0.10 a	1.53 ± 0.09 a	1.33 ± 0.09 a	1.18 ± 0.13 b	1.39 ± 0.11 a
	10–20 cm	1.71 ± 0.15 a	1.63 ± 0.11 a	1.39 ± 0.18 a	1.40 ± 0.10 ab	1.45 ± 0.10 a
	20–30 cm	1.65 ± 0.09 a	1.58 ± 0.16 a	1.24 ± 0.12 a	1.49 ± 0.13 a	1.43 ± 0.13 a
	0–30 cm	1.71 ± 0.11 A	1.58 ± 0.12 AB	1.32 ± 0.13 C	1.36 ± 0.12 C	1.42 ± 0.11 BC
Soil water contents/%	0–10 cm	26.5 ± 0.08 a	33.78 ± 0.05 a	39.82 ± 0.04 a	32.51 ± 0.03 a	36.77 ± 0.02 a
	10–20 cm	15.16 ± 0.05 b	19.35 ± 0.03 b	34.16 ± 0.05 a	18.07 ± 0.01 b	32.54 ± 0.01 b
	20–30 cm	13.31 ± 0.02 b	15.36 ± 0.02 b	21.71 ± 0.06 b	15.53 ± 0.01 b	19.69 ± 0.01 c
	0–30 cm	18.32 ± 3.72 B	22.83 ± 2.29 B	31.90 ± 3.65 A	22.04 ± 1.69 B	33.00 ± 0.48 A
pH	0–10 cm	4.67 ± 0.39 a	5.63 ± 0.25 a	5.72 ± 0.11 b	5.57 ± 0.31 a	4.93 ± 0.09 b
	10–20 cm	4.90 ± 0.38 a	5.67 ± 0.40 a	5.87 ± 0.12 b	5.73 ± 0.48 a	5.11 ± 0.19 ab
	20–30 cm	5.13 ± 0.10 a	5.75 ± 0.29 a	6.09 ± 0.09 a	5.64 ± 0.49 a	5.37 ± 0.10 a
	0–30 cm	4.90 ± 0.28 B	5.68 ± 0.25 A	5.89 ± 0.03 A	5.65 ± 0.39 A	5.14 ± 0.11 B
AvailableN /(mg·kg^{-1})	0–10 cm	97.31 ± 12.17 a	169.17 ± 23.76 a	170.41 ± 19.99 a	113.67 ± 18.23 a	145.46 ± 9.67 a
	10–20 cm	29.78 ± 9.17 b	74.08 ± 14.32 b	100.97 ± 15.13 b	71.37 ± 11.69 b	105.29 ± 15.66 b
	20–30 cm	18.42 ± 4.73 b	42.58 ± 15.55 b	91.74 ± 12.39 b	42.25 ± 8.61 c	69.26 ± 11.17 c
	0–30 cm	48.50 ± 8.68 D	95.28 ± 5.42 BC	121.04 ± 15.82 A	75.76 ± 12.84 C	106.67 ± 12.13 AB
Available P /(mg·kg^{-1})	0–10 cm	1.40 ± 0.44 a	3.06 ± 0.95 a	3.85 ± 1.34 a	2.62 ± 0.38 a	3.67 ± 1.06 a
	10–20 cm	0.76 ± 0.59 ab	0.85 ± 0.23 b	2.11 ± 0.77 ab	1.27 ± 0.03 b	1.86 ± 0.72 b
	20–30 cm	0.21 ± 0.03 b	0.22 ± 0.24 b	1.83 ± 0.36 b	0.99 ± 0.09 b	1.31 ± 0.60 b
	0–30 cm	0.79 ± 0.29 C	1.37 ± 0.35 C	2.60 ± 0.44 A	1.63 ± 0.10 BC	2.28 ± 0.77 AB
Available K /(mg·kg^{-1})	0–10 cm	98.79 ± 27.34 a	204.70 ± 35.66 a	238.42 ± 53.37 a	202.89 ± 20.95 a	230.95 ± 42.47 a
	10–20 cm	38.83 ± 32.92 b	108.30 ± 16.63 b	224.84 ± 33.14 a	118.03 ± 56.93 b	160.80 ± 34.75 b
	20–30 cm	13.71 ± 11.22 b	98.12 ± 16.20 b	58.74 ± 11.57 b	85.44 ± 35.17 b	141.34 ± 20.24 b
	0–30 cm	50.44 ± 20.71 B	137.04 ± 22.53 A	174.00 ± 29.97 A	135.45 ± 37.41 A	177.70 ± 32.29 A

Table 3. *Cont.*

Soil Variables	Soil Layer	800 m	1000 m	1100 m	1200 m	1500 m
TN /(g·kg^{-1})	0–10 cm	4.04 ± 0.67 a	6.08 ± 0.20 a	5.89 ± 1.13 a	4.63 ± 0.15 a	8.92 ± 0.90 a
	10–20 cm	1.32 ± 0.59 b	3.32 ± 1.44 b	5.73 ± 0.97 a	3.16 ± 0.75 b	6.36 ± 1.31 b
	20–30 cm	0.87 ± 0.05 b	1.04 ± 0.28 c	3.47 ± 0.75 b	1.50 ± 0.07 c	5.58 ± 0.49 b
	0–30 cm	2.08 ± 0.15 D	3.48 ± 0.44 C	5.03 ± 0.95 B	3.09 ± 0.32 CD	6.95 ± 0.89 A
TC /(g·kg^{-1})	0–10 cm	47.90 ± 6.80 a	78.13 ± 5.39 a	74.27 ± 5.75 a	55.53 ± 3.51 a	95.42 ± 8.30 a
	10–20 cm	11.92 ± 5.51 b	46.42 ± 10.01 b	63.56 ± 4.70 a	40.51 ± 3.15 b	68.56 ± 6.75 b
	20–30 cm	9.58 ± 1.53 b	11.03 ± 5.32 c	43.07 ± 7.51 b	15.70 ± 1.53 c	66.30 ± 6.20 b
	0–30 cm	23.13 ± 3.60 D	45.20 ± 2.08 C	60.30 ± 5.98 B	37.25 ± 2.72 C	76.76 ± 7.08 A

Values are the mean ± standard error ($n = 3$). Different lowercase letters indicate significant differences between the different soil layers at the same elevation gradient ($p < 0.05$). Different capital letters indicate the significant difference along different elevation gradients ($p < 0.05$).

The soil layer had no significant influence on soil porosity, soil bulk density and soil pH, while remarkably, it controlled the distribution of the soil water contents, available N, available P, available K, TN, and TC content in the soil ($p < 0.05$). With the depth of the soil layer, the contents of the 0–10 cm soil layer were notably higher than those of the subsurface layer or deep layer ($p < 0.05$), showing the phenomenon of surface aggregation (Table 3).

3.3. Variation Characteristics of Active Soil Substances along the Elevation Gradients

The soil's NH_4^+—N, NO_3^-—N and SIN contents were significantly affected by altitude, soil layer, and their interaction ($p < 0.0001$, Table 4). With the increase of elevation the soil NH_4^+—N content showed a U-shaped trend, and the NO_3^-—N and SIN content showed an inverted V-shaped trend. The highest content of NH_4^+—N is at 800 m ($p < 0.05$) which was 1.1–1.5 times that of other altitudes. The trend of the soil's SIN was basically consistent with that of NO_3^-—N, peaked at 1100 m, indicating that the SIN in the study area was dominated by NO_3^-—N in the study area ($p < 0.05$, Figure 2). Except for 1500 m the soil's NO_3^-—N and SIN contents decreased remarkably with the deepening of the soil layer, while the soil's NH_4^+—N content only reached a prominent difference at 800 m and 1500 m.

With the rise of elevation, both of the soil's MBC and MBN content showed an inverted V-shaped distribution trend and reached the peak value at 1100 m ($p < 0.05$, Figure 3). The mean content of MBC in the 0–30 cm soil layer varied from 218.25 to 400.29 mg·kg^{-1}. The maximum content of MBN was 81.39 mg·kg^{-1} at 1100 m, which was 1.97–4.29 times that that of other elevations. In addition, the soil's MBC/MBN was not chiefly governed by elevation and soil layer, and exhibited a fluctuating trend along with the elevation rise.

Table 4. Analysis of variance between the altitude and soil layer on active soil substances.

Parameters	Elevation		Soil Layer		Elevation × Soil Layer	
	F	P	F	P	F	P
NH_4^+—N	8.436	<0.001 ***	24.922	<0.001 ***	6.636	<0.001 ***
NO_3^-—N	22.315	<0.001 ***	38.646	<0.001 ***	6.845	<0.001 ***
SIN	19.393	<0.001 ***	46.36	<0.001 ***	6.146	<0.001 ***
MBC	3.99	0.01 *	21.452	<0.001 ***	1.867	0.103
MBN	12.443	<0.001 ***	6.883	0.003 **	0.665	0.718
MBC/MBN	2.506	0.063	0.763	0.475	0.674	0.71
URE	46.826	<0.001 ***	52.583	<0.001 ***	9.032	<0.001 ***
SUC	58.87	<0.001 ***	66.812	<0.001 ***	5.497	<0.001 ***
URE/MBC	7.149	<0.001 ***	2.641	0.088	5.814	<0.001 ***
SUC/MBC	4.969	0.003 **	0.628	0.054	1.269	0.296

*, **, and *** indicate significant effects at $p < 0.05$, $p < 0.01$, and $p < 0.001$, respectively.

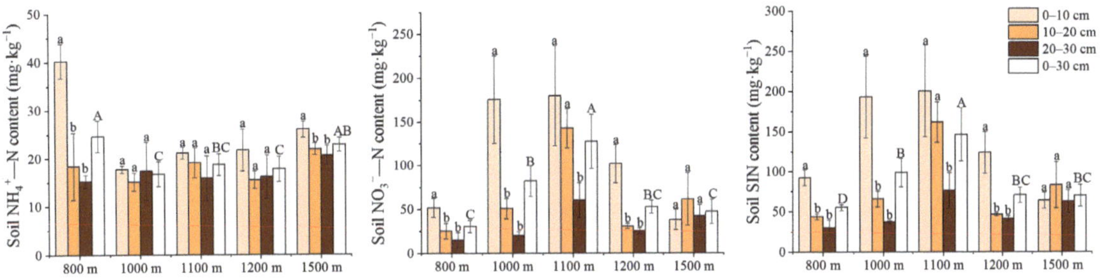

Figure 2. Variation characteristics of the soil's NH_4^+—N, NO_3^-—N and SIN contents at different elevations. Different lowercase letters indicate significant differences between different soil layers at the same elevation gradient ($p < 0.05$). Different capital letters indicate significant differences among the different elevation gradients ($p < 0.05$).

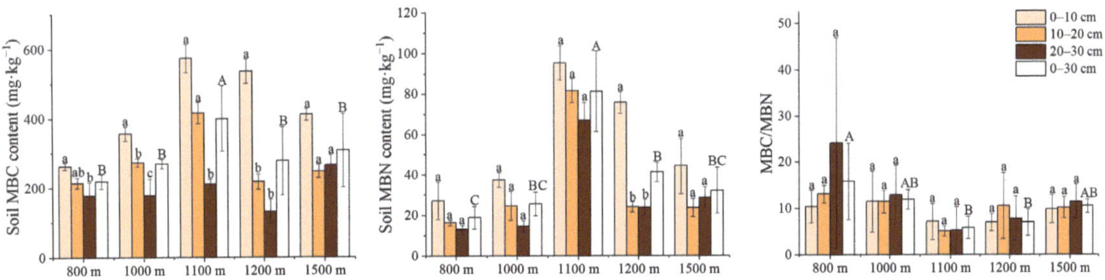

Figure 3. Variation characteristics of the soil's MBC, MBN contents and MBC/MBN at different elevations. Different lowercase letters indicate significant differences between different soil layers at the same elevation gradient ($p < 0.05$). Different capital letters indicate significant differences along different elevation gradients ($p < 0.05$).

Soil URE and SUC activities were significantly influenced by altitude, soil layer, and their interaction ($p < 0.0001$, Table 4). With the elevation increasing, the soil's URE activity in 0–30 cm soil layer showed a fluctuating pattern and reached its peak value at 1500 m, which was 173%, 83%, 23%, and 104% higher than that of 800 m, 1000 m, 1100 m, and 1200 m, respectively (Figure 4). The largest soil SUC activity was found at 1100 m, which was 209%, 112%, 322%, and 123% higher than that of 800 m, 1000 m, 1200 m, and 1500 m, respectively (Figure 4). The soil's URE/MBC ranged from 68.85 to 170.71 g NH_3-N·g^{-1}·h^{-1}·MBC^{-1}, and showed a linear increase with the elevation. The soil's SUC/MBC was remarkably affected by elevation, peaked at 1100 m, which was 1.56–1.94 times higher than other elevations (Figure 4).

The soil layer had no remarkable influence on MBC/MBN, URE/MBC, and SUC/MBC, but had remarkable influences on the soil's NH_4^+—N, NO_3^-—N, SIN MBC, MBN, URE, and SUC ($p < 0.05$, Table 4). With the depth of the soil layer, the content of the 0–10 cm soil layer was notably higher than those of the subsurface layer or deep layer, showing the phenomenon of surface aggregation.

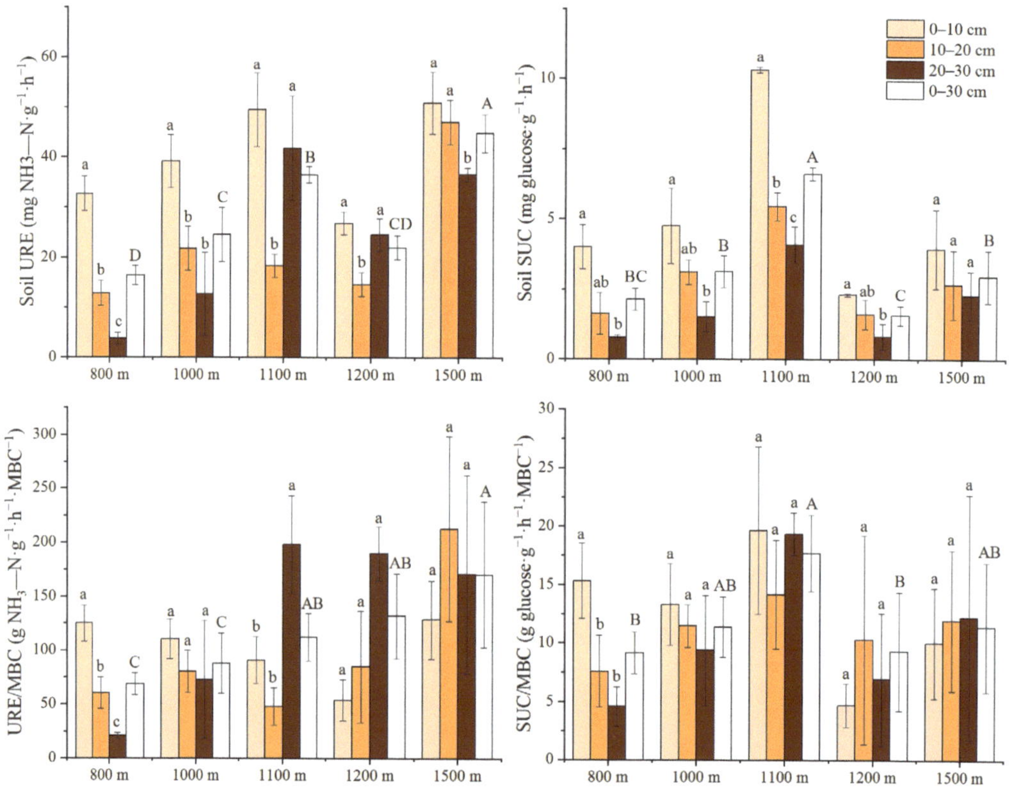

Figure 4. Absolute and relative activities of soil enzymes at different elevations. Different lowercase letters indicate significant differences between different soil layers at the same elevation gradient ($p < 0.05$). Different capital letters indicate significant differences along different elevation gradients ($p < 0.05$).

3.4. Correlation between Active Soil Substances and Environmental Factors

A Pearson's correlation analysis showed that soil NH_4^+—N, NO_3^-—N, and SIN were significantly correlated with pH ($p < 0.05$, Figure 5). Interestingly, only NO_3^-—N and SIN were positively correlated with pH, which may be due to the predominance of NO_3^-—N in SIN in this study area. The soil's MBC and MBN were significant positively correlated with available N and available P content, the soil's MBC/MBN ratio was notably negatively correlated with the soil water contents, available P, available K, and the Pielou index ($p < 0.05$, Figure 5). URE was considerably positively correlated with the soil porosity, soil water contents, available N, available P, available K, TN, TC, Shannon–Wiener, Simpson and Pielou indexes. While the SUC was only remarkably influenced by the soil water contents, available N, and available P contents ($p < 0.05$, Figure 5). The URE/MBC was prominent positively correlated with the available K, TN, TC, and Pielou indexes, while SUC/MBC was only notably influenced by the soil's available P content ($p < 0.05$, Figure 5) indicating that the URE is more susceptible to environmental influences.

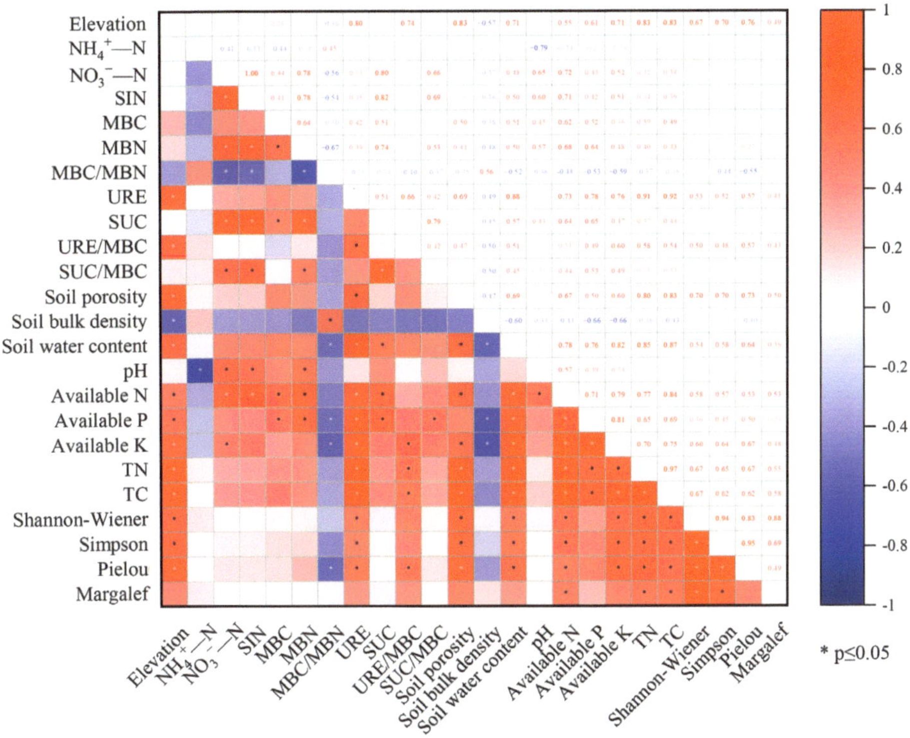

Figure 5. Pearson correlation analysis of the active soil substances with environmental factors and soil physicochemical properties at different elevation gradients.

To further understand the main influencing factors of active soil substances, a redundancy analysis (RDA) of active soil substances and soil physicochemical properties was performed. The first axis was closely related to NO_3^-—N, MBN, URE, SUC, MBC/MBN, and SUC/MBC, which explained the 50.96% variation. While the second axis was mainly related to NH_4^+—N and URE/MBC, explained the 18.46% variation (Figure 6). The role of available N was the most pronounced, explaining the 34.4% of the variation for active soil substances, followed by elevation, soil bulk density, and TN (Table 5), which together they accounted for 67.4% of the variation.

Table 5. Explanation of the impact factors in the redundancy analysis.

Impact Factors	Explains/%	F	P	Impact Factors	Explains/%	F	P
Available N	34.4	6.8	0.002	Available P	2.6	1	–
Elevation	14.5	3.4	0.004	Soil water contents	1.3	0.4	0.73
Soil bulk density	10.3	2.8	0.03	Shannon–Wiener	1.8	0.6	0.678
TN	8.2	2.5	–	Simpson	5	2	0.256
Margalef	5.9	2	0.098	Pielou	1.5	0.5	–
TC	4.9	1.8	0.14	Available K	1	0.2	0.818
Soil porosity	3.6	1.4	0.256	pH	5.1	<0.1	–

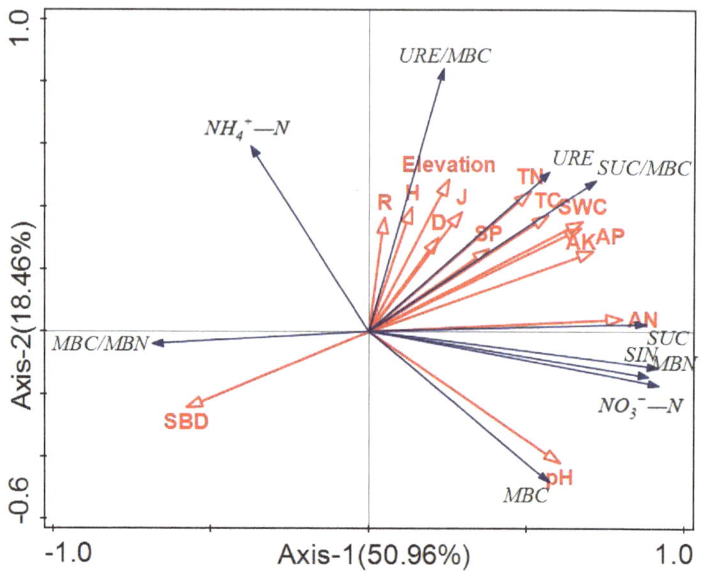

Figure 6. Redundancy analysis of the active soil substances along the elevation. H, Shannon–Wiene index; D, Simpson index; J, Pielou index; R, Margalef index; SP, soil porosity; SBD, soil bulk densit SWC, soil water contents; AN, available N; AP, available P; AK, available K.

4. Discussion

The elevation gradient is one of the important factors of mountain ecosystems. Wit the increase of altitude, the average annual temperature decreases and the average ar nual precipitation increases, thus causing climate change in mountain systems [16,28 Changes in environmental factors caused by elevation gradients directly or indirectly lea to changes in plant species composition and community structure. Species richness of plan communities in alpine meadows decreases with the elevation increase [29], and vegetatio community diversity and biomass present a unimodal distribution with the elevatio gradient [30]. However, in this study, the plant diversity indexes were positively correlate with the elevation gradient (Table 2), probably due to the differences in climatic condition and soil properties. Furthermore, with the increase of the altitude gradient, higher wate content, soil porosity, and rich nutrient conditions (Table 3), it promoted the growth of th dominant understory vegetation, and thus increased the diversity indexes. In additio we also found that the Shannon–Wiener, Simpson, and Pielou indexes are significantl positively correlated with the URE activity, and the Margalef index is also significantl positively correlated with URE/MBC (Figure 5). This may be due to the abundant an diverse plant roots, which produce more root exudates, thus providing substrate for th microbial growth, leading to increased soil enzyme activity [31]. Previous studies have als shown that a higher diversity of understory vegetation is conducive to litter decompositio and nutrient cycling [32,33], and the decomposition rate of litter is regulated by the soi enzyme activity [34], and abundant soil nutrients and higher soil enzyme activity in tur promote plant growth [35]. The relationship between the plant diversity index and soi nutrients and enzyme activities in this study could also indirectly reflect this view.

Active soil substances are greatly affected by the environment, among which the soil' SIN is closely related to N mineralization, nitrification, and denitrification [36]. NH_4^+—N and NO_3^-—N are fast-acting N which can be directly absorbed and utilized by plants [37 In this study, the soil's NH_4^+—N and NO_3^-—N contents were significantly differen according to elevations ($p < 0.05$). At the elevation of 1100 m, the soil's SIN and NO_3^-—N showed a higher content while NH_4^+—N showed a lower content, probably because of more active soil nitrifying bacteria at 1100 m, accelerating the nitrification reaction. The change in the altitude gradient causes the change of temperature and water, which influences nitrogen mineralization and inorganic nitrogen production [38,39]. Furthermore, in this study, soil moisture and SIN contents showed a highly consistent variation trend (Table 2, Figure 2). It has been generally accepted that the soil nitrification rate depends largely on the soil pH [40,41], and the result of this study also supports this view (Figure 5).

The soil's MBC and MBN are closely related to many biotic and abiotic parameters, and they have a fast turnover and high sensitivity for a rapid response to changes of elevation and soil environment [20,42]. Some studies on altitudes below 3000 m indicated that the soil microbial biomass was positively correlated with altitude [43,44], which is not consistent with the results of this study. In this paper, the soil's MBC and MBN contents showed a unimodal trend with increasing altitude (Figure 3), and were notably and positively correlated with the soil's available N and available P contents (Figure 5). On one hand, the surface temperature at a low altitude was higher and water evaporation was intense, which reduced the soil moisture content and thus the activity of soil microorganisms. On the other hand, at the altitude of 1100 m and 1200 m, a higher vegetation diversity resulted in the remarkable turnover rate of soil organic matter. Soil microorganisms benefited from this since high quantities of dead root and leaf litter, as well as higher soil water contents and available nutrients, primarily provide an ideal environment for their growth and development [45]. Thus, the accumulation of soil microbial carbon and nitrogen was promoted.

Soil enzyme activities were significantly affected by the soil water content [46], and positively correlated with most soil nutrients [13,14,23]. In this study, the soil's URE and SUC activities were positively correlated with the soil water contents, available N, available P, and available K contents (Figure 5). The soil URE and SUC participate in soil N and C cycling and different stages of organic matter degradation, so they do not respond identically to changes in elevation gradient [4,13]. Previous studies under similar elevation gradients have shown that soil enzyme activity increases with elevation [13]. It has been shown that the relative activity of soil enzymes can characterize the metabolic activity of microbial communities and, to some extent, can reflect the catalytic efficiency of enzymes [47], which can be used as a tool to stabilize changes in enzyme activity. In this study, URE/MBC and SUC/MBC were significantly affected by elevation changes (Table 4). Interestingly, SUC/MBC showed the same trend as MBC with elevation increasing, while URE/MBC showed an increasing change with elevation increasing. In general, enzymes are derived from animal, plant, and microbial secretions and the decomposition of their residues, and an increase in MBC increases the enzyme activity, while a lower MBC environment improves the efficiency of the enzyme utilization of nutrients, maintains soil metabolic activity, and indirectly increases the proportion of enzymes in MBC. URE/MBC was remarkably correlated with TC and TN in the study area, and SUC/MBC was only remarkably correlated with available P, indicating that URE/MBC was mainly limited by soil carbon and nitrogen, while SUC/MBC was more influenced by the soil's available P content, and a more pronounced response of relative enzyme activity to elevation changes in the study area.

With the deepening of the soil layer, the soil SIN, MBC, and MBN, as well as the soil enzyme activities all showed decreasing trends, which were consistent with the results of previous studies [20,27,48]. This result could be explained by a smaller soil bulk density, good hydrothermal, and aeration conditions in the surface soil layer. In addition, the poor ventilation and nutrient content of the deep soil weaken the microbial activity and affect the active matter content. At the same time, an abundant litter return in the surface soil layer, which provides a suitable living environment and sufficient nutrient sources for microbial growth, and the microbial metabolism is strong, resulting in a higher active matter content.

However, the present experiment did not take into account the effect of elevation gradients on soil microbial community structure. It has been shown that soil bacterial and fungal composition is closely related to soil organic matter and total nitrogen content [13]. Therefore, it is necessary to further investigate the effects of microbial community structures on soil carbon and nitrogen content along elevation gradients in mountain ecosystems.

5. Conclusions

In the Southern Taihang Mountains, active soil substances contents were significantly influenced by elevation gradients and soil layer. With the rise of elevation, the soil's NO_3^-—N, MBC and MBN content as well as SUC activity basically showed an inverted V-shaped distribution trend and reached the peak value at 1100 m, while soil URE activity showed a fluctuating pattern. With the deepening of the soil layer, the content of active soil substances tended to decrease, showing a phenomenon of surface aggregation. The soil's available N, elevation, and the soil bulk density and TN were important factors driving the changes of active soil substances, and together they accounted for 67.4% of the variation. The soil microbial community structure is mostly probably to explain the changes of soil active substance contents with altitude more deeply. It is suggested to further strengthen the study of soil microbial community structure in mountain ecosystem.

Author Contributions: Conceptualization, E.F. and Y.K.; Methodology and investigation, C.W. and T.W.; Software and data curation, E.F. and L.Z.; Formal analysis, X.X. and L.Z.; Writing—original draft preparation, E.F. and Y.K.; Writing—review and editing, Y.K. and E.F.; Supervision, Y.K. and X.X.; Funding acquisition, Y.K. and T.W. All authors have read and agreed to the published version of the manuscript.

Funding: This research was supported by the Research and Demonstration of Key Technologies of Afforestation in Barren Lands of Henan Province (Grant No. YLK202209) and the pilot project for ecological protection and restoration of mountains, water, forests, fields, lakes, and grasses in South Taihang, Henan (JGZJ—Grant—2019125).

Data Availability Statement: Not applicable.

Conflicts of Interest: The authors declare no conflict of interest.

References

1. Fang, J.; Chen, A.; Peng, C.; Zhao, S.; Ci, L. Changes in forest biomass carbon storage in China between 1949 and 1998. *Science* **2001**, *292*, 2320–2322. [CrossRef] [PubMed]
2. Yu, G.; Song, X.; Wang, Q.; Liu, Y.; Guan, D.; Yan, J.; Sun, X.; Zhang, L.; Wen, X. Water-use efficiency of forest ecosystems in eastern China and its relations to climatic variables. *New Phytol.* **2008**, *177*, 927–937. [CrossRef]
3. Verrall, B.; Pickering, C.M. Alpine vegetation in the context of climate change: A global review of past research and future directions. *Sci. Total Environ.* **2020**, *748*, 141344. [CrossRef] [PubMed]
4. Nottingham, A.T.; Turner, B.L.; Whitaker, J.; Ostle, N.J.; McNamara, N.P.; Bardgett, R.D.; Salinas, N.; Meir, P. Soil microbial nutrient constraints along a tropical forest elevation gradient: A belowground test of a biogeochemical paradigm. *Biogeosciences* **2015**, *12*, 6071–6083. [CrossRef]
5. Ndossi, E.M.; Becker, J.N.; Hemp, A.; Dippold, M.A.; Kuzyakov, Y.; Razavi, B.S. Effects of land use and elevation on the functional characteristics of soil enzymes at Mt. Kilimanjaro. *Eur.J.Soil Biol.* **2020**, *97*, 103167. [CrossRef]
6. Kaiser, K.; Guggenberger, G. The role of DOM sorption to mineral surfaces in the preservation of organic matter in soils. *Org. Geochem.* **2000**, *31*, 711–725. [CrossRef]
7. Johnson, B.G.; Verburg, P.S.J.; Arnone, J.A., III. Effects of climate and vegetation on soil nutrients and chemistry in the Great Basin studied along a latitudinal-elevational climate gradient. *Plant Soil* **2014**, *382*, 151–163. [CrossRef]
8. Sundqvist, M.K.; Wardle, D.A.; Vincent, A.; Giesler, R. Contrasting nitrogen and phosphorus dynamics across an elevational gradient for subarctic tundra heath and meadow vegetation. *Plant Soil* **2014**, *383*, 387–399. [CrossRef]
9. Parfitt, R.L.; Yeates, G.W.; Ross, D.J.; Mackay, A.D.; Budding, P.J. Relationships between soil biota, nitrogen and phosphorus availability, and pasture growth under organic and conventional management. *Appl. Soil Ecol.* **2005**, *28*, 1–13. [CrossRef]
10. Michel, K.; Matzner, E. Nitrogen content of forest floor Oa layers affects carbon pathways and nitrogen mineralization. *Soil Biol. Biochem.* **2002**, *34*, 1807–1813. [CrossRef]
11. Sardans, J.; Penuelas, J.; Estiarte, M. Changes in soil enzymes related to C and N cycle and in soil C and N content under prolonged warming and drought in a Mediterranean shrubland. *Appl. Soil Ecol.* **2008**, *39*, 223–235. [CrossRef]
12. Wang, B.; Xue, S.; Liu, G.B.; Zhang, G.H.; Li, G.; Ren, Z.P. Changes in soil nutrient and enzyme activities under different vegetations in the Loess Plateau area, Northwest China. *Catena* **2012**, *92*, 186–195. [CrossRef]

3. Chang, E.-H.; Chen, T.-H.; Tian, G.; Chiu, C.-Y. The effect of altitudinal gradient on soil microbial community activity and structure in moso bamboo plantations. *Appl. Soil Ecol.* **2016**, *98*, 213–220. [CrossRef]

4. Siles, J.A.; Cajthaml, T.; Minerbi, S.; Margesin, R. Effect of altitude and season on microbial activity, abundance and community structure in Alpine forest soils. *FEMS Microbiol. Ecol.* **2016**, *92*, fiw008. [CrossRef]

5. Liu, X.; Guo, K.; Huang, L.; Ji, Z.; Jiang, H.; Li, H.; Zhang, J. Responses of absolute and specific enzyme activity to consecutive application of composted sewage sludge in a Fluventic Ustochrept. *PLoS ONE* **2017**, *12*, e0177796. [CrossRef]

6. Xiao, L.; Li, P.; Shi, P.; Liu, Y. Soil nutrient stoichiometries and enzymatic activities along an elevational gradient in the dry-hot valley region of southwestern China. *Arch. Agron.Soil Sci.* **2019**, *65*, 322–333. [CrossRef]

7. Luo, L.; Meng, H.; Gu, J.-D. Microbial extracellular enzymes in biogeochemical cycling of ecosystems. *J.Environ.Manage.* **2017**, *197*, 539–549. [CrossRef]

8. Khan, S.; Chen, N.; Zhang, C.; Wang, L.; Han, C.; Lu, K.; Li, Y.; Rafiq, M.; Iqbal, A.; Zhao, C. Soil fungal taxonomic diversity along an elevation gradient on the semi-arid Xinglong Mountain, Northwest China. *Arch. Microbiol.* **2020**, *202*, 2291–2302. [CrossRef]

9. Cardelli, V.; De Feudis, M.; Fornasier, F.; Massaccesi, L.; Cocco, S.; Agnelli, A.; Weindorf, D.C.; Corti, G. Changes of topsoil under Fagus sylvatica along a small latitudinal-altitudinal gradient. *Geoderma* **2019**, *344*, 164–178. [CrossRef]

0. Li, J.; Chen, L.; Wang, H.; Ouyang, S.; Liu, X.; Lu, J. Pattern and drivers of soil fungal community along elevation gradient in the Abies georgei forests of Segila mountains, Southeast Tibet. *Glob.Ecol.Conserv.* **2022**, *39*, e02291. [CrossRef]

1. Zhou, Y.; Jia, X.; Han, L.; Liu, Z.; Kang, S.; Zhao, Y. Fungal community diversity in soils along an elevation gradient in a *Quercus aliena var. acuteserrata* forest in Qinling Mountains, China. *Appl. Soil Ecol.* **2021**, *167*, 104104. [CrossRef]

2. Massaccesi, L.; De Feudis, M.; Leccese, A.; Agnelli, A. Altitude and vegetation affect soil organic carbon, basal respiration and microbial biomass in Apennine forest soils. *Forests* **2020**, *11*, 710. [CrossRef]

3. Hofmann, K.; Lamprecht, A.; Pauli, H.; Illmer, P. Distribution of prokaryotic abundance and microbial nutrient cycling across a high-alpine altitudinal gradient in the Austrian Central Alps is affected by vegetation, temperature, and soil nutrients. *Microb. Ecol.* **2016**, *72*, 704–716. [CrossRef]

4. Tong, X.; Zhang, J.; Meng, P.; Li, J.; Zheng, N. Ecosystem water use efficiency in a warm-temperate mixed plantation in the North China. *J. Hydrol.* **2014**, *512*, 221–228. [CrossRef]

5. Gong, S.; Feng, Z.; Qu, A.; Sun, J.; Xu, X.; Yong, L.; Kong, Y. Effects of land-use types on the temporal dynamics of soil active carbon and nitrogen in the rocky mountainous of North China. *Soil Sci. Plant Nutr.* **2022**, *68*, 72–80. [CrossRef]

6. Kong, Y.; Ma, N.L.; Yang, X.; Lai, Y.; Feng, Z.; Shao, X.; Xu, X.; Zhang, D. Examining CO_2 and N_2O pollution and reduction from forestry application of pure and mixture forest. *Environ. Pollut.* **2020**, *265*, 114951. [CrossRef]

7. Guan, S. *Soil Enzymes and their Research Methods*; Agriculture Press: Beijing, China, 1986; pp. 274–301.

8. Zuo, Y.; Li, J.; Zeng, H.; Wang, W. Vertical pattern and its driving factors in soil extracellular enzyme activity and stoichiometry along mountain grassland belts. *Biogeochemistry* **2018**, *141*, 23–39. [CrossRef]

9. Han, W.; Chen, L.; Su, X.; Liu, D.; Jin, T.; Shi, S.; Li, T.; Liu, G. Effects of soil physico-chemical properties on plant species diversity along an elevation gradient over alpine grassland on the Qinghai-Tibetan Plateau, China. *Front. Plant Sci.* **2022**, *13*, 822268. [CrossRef]

0. Deng, Q.; Zhang, X.; Niu, J.; Qin, H. Species diversity of plant communities along an altitude gradient in Yinmachi Mountain, northwestern Shanxi, China. *Ecol. Environ.* **2019**, *28*, 865–872.

1. Gong, S.; Zhang, T.; Guo, R.; Cao, H.; Shi, L.; Guo, J.; Sun, W. Response of soil enzyme activity to warming and nitrogen addition in a meadow steppe. *Soil Res.* **2015**, *53*, 242–252. [CrossRef]

2. Xiong, Y.; Xia, H.; Li Za Cai Xa Fu, S. Impacts of litter and understory removal on soil properties in a subtropical Acacia mangium plantation in China. *Plant Soil* **2008**, *304*, 179–188. [CrossRef]

3. Qiao, Y.; Miao, S.; Silva, L.C.R.; Horwath, W.R. Understory species regulate litter decomposition and accumulation of C and N in forest soils: A long-term dual-isotope experiment. *For. Ecol. Manag.* **2014**, *329*, 318–327. [CrossRef]

4. Smart, K.A.; Jackson, C.R. Fine scale patterns in microbial extracellular enzyme activity during leaf litter decomposition in a stream and its floodplain. *Microb. Ecol.* **2009**, *58*, 591–598. [CrossRef]

5. Pauli, H.; Gottfried, M.; Dullinger, S.; Abdaladze, O.; Akhalkatsi, M.; Benito Alonso, J.L.; Coldea, G.; Dick, J.; Erschbamer, B.; Fernandez Calzado, R.; et al. Recent plant diversity changes on Europe's mountain summits. *Science* **2012**, *336*, 353–355. [CrossRef]

6. Kong, Y.; Watanabe, M.; Nagano, H.; Watanabe, K.; Yashima, M.; Inubushi, K. Effects of land-use type and nitrogen addition on nitrous oxide and carbon dioxide production potentials in Japanese Andosols. *Soil Sci. Plant Nutr.* **2013**, *59*, 790–799. [CrossRef]

7. Liu, Y.; Chen, L.; Duan, W.; Bai, Y.; Li, X. Effects of litter decomposition on soil N in Picea mongolica forest at different forest ages. *Forests* **2022**, *13*, 520. [CrossRef]

8. Cookson, W.R.; Cornforth, I.S.; Rowarth, J.S. Winter soil temperature (2–15 °C) effects on nitrogen transformations in clover green manure amended or unamended soils: A laboratory and field study. *Soil Biol. Biochem.* **2002**, *34*, 1401–1415. [CrossRef]

9. Wang, C.; Han, X.; Xing, X. Effects of grazing exclusion on soil net nitrogen mineralization and nitrogen availability in a temperate steppe in northern China. *J. Arid Environ.* **2010**, *74*, 1287–1293. [CrossRef]

0. Cheng, Y.; Wang, J.; Mary, B.; Zhang, J.-B.; Cai, Z.-C.; Chang, S.X. Soil pH has contrasting effects on gross and net nitrogen mineralizations in adjacent forest and grassland soils in central Alberta, Canada. *Soil Biol. Biochem.* **2013**, *57*, 848–857. [CrossRef]

41. Yao, H.; Campbell, C.D.; Qiao, X. Soil pH controls nitrification and carbon substrate utilization more than urea or charcoal i some highly acidic soils. *Biol. Fertil. Soils* **2011**, *47*, 515–522. [CrossRef]
42. Shen, R.; Xu, M.; Li, R.; Zhao, F.; Sheng, Q. Spatial variability of soil microbial biomass and its relationships with edaphi vegetational and climatic factors in the Three-River Headwaters region on Qinghai-Tibetan Plateau. *Appl. Soil Ecol.* **2015**, *9* 191–203. [CrossRef]
43. Huang, C.-Y.; Jien, S.-H.; Chen, T.-H.; Tian, G.; Chiu, C.-Y. Soluble organic C and N and their relationships with soil organic C an N and microbial characteristics in moso bamboo (*Phyllostachys edulis*) plantations along an elevation gradient in Central Taiwa *J. Soil Sediments* **2014**, *14*, 1061–1070. [CrossRef]
44. Whitaker, J.; Ostle, N.; Nottingham, A.T.; Ccahuana, A.; Salinas, N.; Bardgett, R.D.; Meir, P.; McNamara, N.P. Microbial communit composition explains soil respiration responses to changing carbon inputs along an Andes-to-Amazon elevation gradient. *J. Ecc* **2014**, *102*, 1058–1071. [CrossRef] [PubMed]
45. Prommer, J.; Walker, T.W.N.; Wanek, W.; Braun, J.; Zezula, D.; Hu, Y.; Hofhansl, F.; Richter, A. Increased microbial growt biomass, and turnover drive soil organic carbon accumulation at higher plant diversity. *Glob Chang Biol.* **2019**, *26*, 669–68 [CrossRef]
46. Zuccarini, P.; Asensio, D.; Ogaya, R.; Sardans, J.; Penuelas, J. Effects of seasonal and decadal warming on soil enzymatic activit in a P-deficient Mediterranean shrubland. *Global Change Biol.* **2020**, *26*, 3698–3714. [CrossRef]
47. Raiesi, F.; Beheshti, A. Soil specific enzyme activity shows more clearly soil responses to paddy rice cultivation than absolut enzyme activity in primary forests of northwest Iran. *Appl. Soil Ecol.* **2014**, *75*, 63–70. [CrossRef]
48. Wang, N.; Zhang, M.; Zhao, N.; Feng, F.; Zhao, M. Season-dependence of soil extracellular enzyme activities in a Pinus koraiensi forest on Changbai Mountain. *J. For. Res.* **2021**, *32*, 1713–1722. [CrossRef]

forests

Article

Biochar-Based Fertilizer Decreased Soil N_2O Emission and Increased Soil CH_4 Uptake in a Subtropical Typical Bamboo Plantation

Enhui Wang [1,2,3], Ning Yuan [1,2,3], Shaofeng Lv [1,2,3], Xiaoping Tang [1,2,3], Gang Wang [1,2,3], Linlin Wu [1,2,3], Yufeng Zhou [1,2,3], Guomo Zhou [1,2,3], Yongjun Shi [1,2,3] and Lin Xu [1,2,3,*]

1 State Key Laboratory of Subtropical Silviculture, Zhejiang A&F University, Hangzhou 311300, China
2 Key Laboratory of Carbon Cycling in Forest Ecosystems and Carbon Sequestration of Zhejiang Province, Zhejiang A&F University, Hangzhou 311300, China
3 School of Environmental and Resources Science, Zhejiang A&F University, Hangzhou 311300, China
* Correspondence: xul1210@zafu.edu.cn

Abstract: Soil is a crucial contributor to greenhouse gas (GHG) emissions from terrestrial ecosystems to the atmosphere. The reduction of GHG emissions in plantation management is crucial to combating and mitigating global climate change. A 12-month field trial was conducted to explore the effects of different fertilization treatments (control, without fertilizer (CK); biochar-based fertilizer treatment (BFT); chemical fertilizer treatment (CFT); and mixture of 50% BFT and 50% CFT (MFT)) on the soil GHG emissions of a typical bamboo (*Pleioblastus amarus* (Keng) Keng f.) plantation. The results demonstrated that compared with the CK, BFT reduced the annual cumulative soil N_2O emission by 16.3% ($p < 0.01$), while CFT and MFT significantly increased it by 31.0% and 23.3% ($p < 0.01$), respectively. Meanwhile, BFT and MFT increased the annual cumulative soil CH_4 uptake by 5.8% ($p < 0.01$) and 7.5% ($p < 0.01$), respectively, while there was no statistically significant difference between CFT and the control. In addition, BFT, CFT, and MFT significantly increased the annual cumulative soil CO_2 emission by 9.4% ($p < 0.05$), 13.0% ($p < 0.01$), and 26.5% ($p < 0.01$). The global warming potential (GWP) of BFT did not change significantly, while CFT and MFT increased the GWP by 13.7% ($p < 0.05$) and 28.6% ($p < 0.05$), respectively, compared with the control. Structural equation modeling revealed different treatments affected soil N_2O and CH_4 emission by changing soil labile carbon and labile nitrogen pools. This study suggests utilizing BFT new ideas and strategies for mitigating GHG emissions from soils in subtropical *Pleioblastus amarus* plantations.

Keywords: *Pleioblastus amarus* plantation; fertilizer; greenhouse gas; soil carbon pool; soil nitrogen pool; management strategy

Citation: Wang, E.; Yuan, N.; Lv, S.; Tang, X.; Wang, G.; Wu, L.; Zhou, Y.; Zhou, G.; Shi, Y.; Xu, L. Biochar-Based Fertilizer Decreased Soil N_2O Emission and Increased Soil CH_4 Uptake in a Subtropical Typical Bamboo Plantation. *Forests* **2022**, *13*, 2181. https://doi.org/10.3390/f13122181

Academic Editor: Xingkai Xu

Received: 20 September 2022
Accepted: 16 December 2022
Published: 19 December 2022

1. Introduction

Since the Industrial Revolution, the planet's population has been expanding quickly. The development of the world's major economies and the increased demands for human activities for natural resources have led to relatively high emissions of greenhouse gases (GHG), including methane (CH_4), carbon dioxide (CO_2), and nitrous oxide (N_2O). The relationship between humans and nature has experienced great tension under extreme weather events with high intensity and frequency [1]. Forests play a vital and unique position in lowering the concentration of CO_2 in the atmosphere and slowing down global warming. Forest ecosystems have a huge capacity to absorb and store CO_2. The annual carbon sequestration of world forests accounts for about 2/3 of the total terrestrial carbon sequestration, which profoundly affects the source and sink dynamics of CO_2 in the atmosphere [2–4]. Therefore, forest management technology can serve as one of the critical scientific approaches to mitigate global climate warming.

The bamboo forest serves an essential and constructive role in the world's forest resources, known as the "second forest" [5]. Nowadays, when the global forest area continues to decline, the bamboo forest area keeps increasing by about 3% per year on average, which suggests the bamboo forest is a potential expanding carbon sink [6]. *Pleioblastus amarus* (Keng) Keng f. is a lignified tree of the Poaceae and Bamboo genus, which has a great utilization value and is a crucial species for garden landscaping in east Asia. *Pleioblastus amarus* has become an excellent multi-purpose, economical bamboo species in southern China due to its strong adaptability, cold resistance, and barren resistance [7]. *Pleioblastus amarus* not only grows fast and has high economic value, but studies have also illustrated the carbon sequestration capacity of the *Pleioblastus amarus* plantation ecosystem was stronger than that of the Moso bamboo [8].

Regarding the ecological function and carbon sink capacity of *Pleioblastus amarus* plantations, Shen et al. [8] illustrated the carbon density, spatial distribution pattern, and carbon storage of the Sichuan *Pleioblastus amarus* plantation ecosystem were estimated by the biomass method. The results showed the carbon storage of the aboveground part is 2.19 times that of the underground part. He et al. [9] found rational management can effectively improve the carbon sink function of the *Pleioblastus amarus* plantation. Shen [10] also found the *Pleioblastus amarus* forest plays a greater role in carbon sequestration in coping with climate change in Sichuan. In plantation operation activities, farmers manage to improve the yield of *Pleioblastus amarus* by intensive management measures, such as removing underlayers vegetation, tilling soil, and applying chemical fertilizer. However, the above measures, especially the input of traditional fertilizers, usually increase soil GHG emissions from plantations [11,12], reduce active organic carbon storage, and change the chemical composition of soil organic carbon [13]. A recent paper reported from 2007 to 2016, 54% of total soil N_2O emission were caused by the application of nitrogen fertilizer [14]. Therefore, finding a management measure to reduce the soil GHG emissions and further improve the carbon sequestration of the *Pleioblastus amarus* plantation ecosystems is of great significance.

Because of its unique chemical structure, biochar has higher chemical stability and adsorption than other organic matters. Many have suggested biochar input into soil significantly affects soil physicochemical properties [15,16], increases the soil carbon pool [17,18] and reduces soil GHG emissions [19]. The study in a subtropical Chinese bamboo plantation for two consecutive years showed applying only a small amount of biochar can significantly reduce the total soil CO_2 emission and positively promoted soil organic carbon sequestration [20]. Xu et al. [21] revealed the input of different gradients of biochar could significantly improve the carbon sequestration capacity in the Moso bamboo forest and reduce the soil N_2O emission. Similarly, in a two-year field survey, Song et al. [22] demonstrated using biochar reduced soil N_2O emission by decreasing soil total denitrification and nitrification rates and by reducing NH_4^+-N and NO_3^--N concentrations. However, some studies reported different views on the impact of biochar application on GHG in forest soil. Hawthorne et al. [23] found CO_2 and N_2O emissions would increase following application of biochar at relatively high rates in forest soils. Lin et al. [24] indicated there was no significant difference in N_2O emission and CH_4 uptake of soil by applying different kinds of biochar in a subtropical forest of China. The reasons for the above different results may be due to the different types and application rates of biochar. In conclusion, it is necessary to study the impact mechanism of biochar application on soil GHG emission. However, the relatively low mineral nutrient content of biochar is frequently inadequate as a single supplement to support plantations' rapid growth [25,26].

Biochar-based fertilizer is a new type of material utilizing biomass charcoal as a carrier and mixed with a certain proportion of chemical fertilizers according to demand of soil for nutrient elements [27]. The ratio of each nutrient can be adjusted according to soil fertility levels in different regions. Recent studies revealed adding biochar as a base fertilizer improved soil fertility, increased the utilization ratio of fertilizers, reduced fertilizer loss [28,29], improved soil fertilizer retention capacity, and achieved a long-term balance of water and fertilizers [30]. At the same time, it achieved long-term, slow-release effects on water and fertilizer and improved soil ecological environment [31,32]. Compared with the traditional way biochar is applied to the soil in tons as a soil conditioner, the consumption of biochar-based fertilizer can save costs and practically mix chemical fertilizers and biochar dependent on the demand for the soil nutrient in the region. Studying the benefit of biochar-based fertilizers on soil GHG emissions has far-reaching significance for mitigating global climate warming, improving soil environment and fertility, and enhancing crop yields. However, in the past, most of the studies on the effectiveness of biochar on greenhouse gas emissions of soils were focused on farmlands and grasslands, while there were few reports on forest ecosystems [33]. Research on how biochar-based fertilizers can affect soil GHG emissions from *Pleioblastus amarus* ecosystems was rare.

The purpose of this study is to study the effects of applying biochar-based fertilizer and chemical fertilizer on soil GHG emission of the *Pleioblastus amarus* plantation compared to that without fertilizer. We will analyze the relationship between soil GHG and environmental factors, such as soil carbon pool, nitrogen pool, soil temperature, and soil moisture. The following two hypotheses will be tested. First, the addition of biochar-based fertilizers can reduce soil GHG emissions, whereas the application of chemical fertilizers can increase soil GHG emissions. Second, the decrease in soil GHG emissions following application of biochar-based fertilizers can result from the changes in soil labile carbon and labile nitrogen pools. This proposed study is anticipated to contribute to the possible management methods to mitigate the impact of soil GHG emissions in *Pleioblastus amarus* plantations.

2. Materials and Methods

2.1. Study Area

Our field experiment was conducted in Yuhang (30°11′12.29″ N, 119°51′06.58″ E), Zhejiang Province, China (Figure 1). With an annual average temperature of 15.3–16.2 °C, an annual sunlight duration of 1834 h, and an annual frost-free period of 230–260 days, the climate is classified as a subtropical monsoon climate. The study area's terrain is made up of hills and low mountains, and the altitude of the experimental region is 144 m above sea level, and the slope is about 30°. The main soil type in the experimental area is classified as slightly acidic laterite in the Chinese soil classification system [34].

Through the field investigations, a typical *Pleioblastus amarus* plantation was chosen to launch this field test in March 2021. The planting density of *Pleioblastus amarus* is about 13,000 culms ha^{-1}, and the average breast height diameter is 2.5 cm. Before applying the treatments to the experimental plots, samples (0–20 cm) of soil surface were collected to examine their physical and chemical characteristics. These results were: silt 31.50%, sand 46.70%, clay 19.60%, soil bulk density 1.14 g cm^{-3}, available K 64.5 mg kg^{-1}, pH value 4.77, available P 8.3 mg kg^{-1}, total N 1.96 g kg^{-1}, and the organic C 18.1 g kg^{-1}.

2.2. Experimental Design

In April 2021, a *Pleioblastus amarus* pure plantation site with a similar growth history, site conditions, and slope was selected at the Yuhang *Pleioblastus amarus* demonstration base. Four treatments with four repetitions each were employed in a fully randomized block design. The size of each experimental plot is 10 m × 10 m. A 3-m wide buffer zone was set up for each plot to eliminate the interference of the underground root whip of *Pleioblastus amarus* on the adjacent plots.

Four treatments were set up in this experiment: (1) control (CK, without any fertilizer application); (2) biochar-based fertilizer treatment (BFT, the application rate of biochar-based fertilizer was 133.33 g m^{-2}, the N, P_2O_5, and K_2O contents in the biochar-based fertilizer were 150 g kg^{-1}, 150 g kg^{-1}, and 100 g kg^{-1}, respectively); (3) chemical fertilizer treatment (CFT, 20 g N m^{-2}, 20 g P_2O_5 m^{-2}, and 13.3 g K_2O m^{-2} were applied in CFT to achieve the BFT level of nutrients, and the fertilizers in the CFT were provided by conventional chemical fertilizers); and (4) 50% BFT and 50% CFT mixed (MFT, 50% of N, P and K came from BFT (66.5 g m^{-2}), and 50% of nutrition came from CFT).

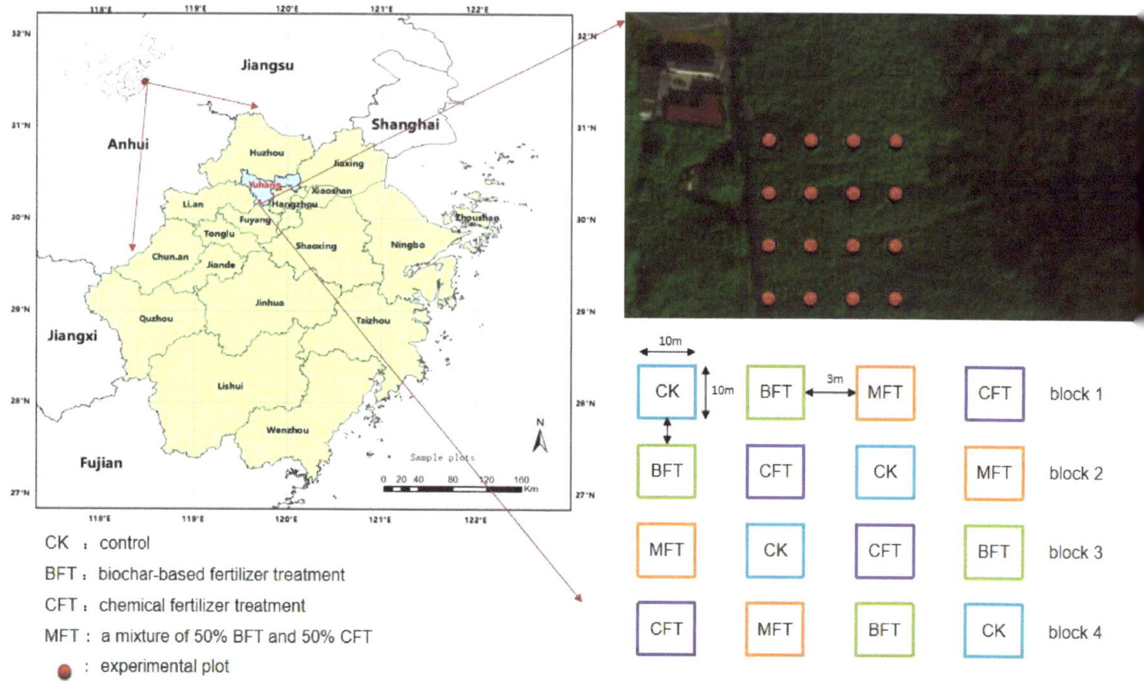

Figure 1. Location of the study area and the scheme of the experimental approach.

In April 2021, the experimental plots were treated with fertilizers, then tilled to 20 cm depth. After fertilization, a static PVC chamber was placed at the diagonal midpoint of each plot to collect soil GHG samples. The gas sampling was carried out on the first month's first, seventh, and fourteenth days. In the following months, the gas sampling were to be conducted on sunny days, and the GHG samples were collected once a month. For the purpose of collecting soil samples from each plot, the three-point sampling approach was used, and the sampling soil layer was the surface soil (0–20 cm).

2.3. Production of Biochar-Based Fertilizer

The biochar-based fertilizer used in this experiment was produced and sponsored by Qinfeng Zhongcheng Biomass New Materials Nanjing Co., Ltd., Nanjing, China. The raw material of biochar-based fertilizer was wheat straw biochar, which was made by anoxic pyrolysis at 500 °C for 3 h. According to the formula, the biochar and chemical fertilizer were mixed and crushed, stirred evenly, and dried. Finally, the biochar-based fertilizer was made after cooling and screening treatment. The C contents in the biochar-based fertilizer were 200 g kg^{-1}. The properties of biochar used to make biochar-based fertilizers are described in Table 1.

Table 1. The properties of biochar.

Properties	Surface Area (m^2 g^{-1})	pH	N (g kg^{-1})	K (g kg^{-1})	P (g kg^{-1})
biochar	11.4	9.07	5.6	8.6	0.64

2.4. Measuring Soil GHG Emissions

The soil N_2O, CH_4, and CO_2 fluxes from the plots for the experiment were evaluated by closed static chamber and gas chromatography analysis technology [20]. The stati chamber for collecting GHG consists of a base box (length 0.3 m, width 0.3 m, and depth 0.1 m) and a cover box (length 0.3 m, width 0.3 m, and depth 0.3 m) with a U-shaped slot (width 5 cm, depth 5 cm). The immovable chambers were uniformly constructed of polyvinyl chloride (PVC) panels (Figure 2).

Figure 2. The static chamber for GHG flux measurement.

On days without rain, samples of soil GHG were taken between 9:00 and 11:00 in the morning. Before collecting gas, weeds inside the base box were clipped along the roots, and then the groove of the base box was stuffed with 2 cm high distilled water to form a seal between the base and the lid. A fan was placed to mix the air in the box evenly, and then, the gas samples were obtained using a 100 mL syringe with a tee tube attached. The samples were taken at 0, 10, 20, and 30 min after the static box lid was closed. Finally, the collected gas was injected into a 100 mL gas collection bag for storage, and soil temperature and soil moisture at 5 cm depth were measured simultaneously.

The gas samples were returned to the lab for measurement of GHG flux using a gas chromatograph (GC-2014, Shimadzu Corporation, Kyoto, Japan). The measurement time and sample collection time should not exceed two days. We measured the gas samples collected in the experiment according to the standard curve made by the concentration value of GHG under standard references (N_2O: 5.0 × 10^{-6} mol/mol, CH_4: 20.4 × 10^{-6} mol/mol, CO_2: 302 × 10^{-6} mol/mol), which was provided by Shanghai Weichuang Standard Gas Analysis Technology Co., Ltd., Shanghai, China.

Formula (1) was applied to calculate soil CO_2, N_2O, and CH_4 emissions.

$$F = \rho \frac{V}{A} \frac{P}{P_0} \frac{T_0}{T} \frac{dC_t}{d_t} \quad (1)$$

where F is the emission rate of soil GHG (mg N_2O m^{-2} h^{-1}, mg CH_4 m^{-2} h^{-1} mg CO_2 m^{-2} h^{-1}), ρ is the density of GHG under ideal conditions (N_2O: 1.964 $\times$ 10^3 g m^{-3} CH_4: 7.163 $\times$ 10^2 g m^{-3}, CO_2: 1.98 $\times$ 10^3 g m^{-3}), A denotes the area of the chamber's bottom portion (m^2), V denotes the chamber's volume (m^3), and $(dC_t)/d_t$ is the slope of the concentration of the GHG in the sampling box per unit time (ppm h^{-1}). P_0 and T_0 represent the standard circumstances' absolute air pressure and temperature. Furthermore, P and T denote the chamber's atmospheric air pressure and air temperature when sampling, respectively.

Formula (2) was applied to calculate the annual cumulative GHG flux.

$$M = \sum F_i(t_{i+1} - t_i) \times 24 \times 10^{-5} \quad (2)$$

where M denotes the annual cumulative emission of GHG (Mg N_2O ha^{-1} yr^{-1} Mg CH_4 ha^{-1} yr^{-1}, Mg CO_2 ha^{-1} yr^{-1}), F denotes the flux of soil GHG (mg m^{-2} h^{-1}) i and t denote the sampling number and sampling time, respectively.

To estimate the combined effect of fertilization treatments on soil GHG emissions Formula (3) was used to determine the global warming potential (GWP):

$$\text{GWP} = M_{CO_2} + 298M_{N_2O} - 25M_{CH_4} \quad (3)$$

where GWP denotes the total global warming potential of GHG emission ($CO_{2\text{-eq}}$ Mg ha^{-1} yr^{-1}) M_{N_2O}, M_{CH_4}, and M_{CO_2} are the annual cumulative N_2O emission, CH_4 uptake, and CO_2 emission (Mg ha^{-1} yr^{-1}), respectively. The radioactive forcing potentials for CH_4 and N_2O expressed as CO_2 equivalent on a 100-year period are represented by coefficients 25 and 298, respectively [35].

2.5. Measuring the Physicochemical Properties of Soil

Soil physicochemical properties of the experimental plots were analyzed using standard methods [36]. The soil corer used a predetermined volume to determine the bulk density of soil. The soil pH value was analyzed by mixing the soil sample with water in a ratio of 1:2.5 using a pH meter (FE28, Mettler Toledo, Shanghai, China). The soil moisture content was determined by drying the fresh samples at 105 °C for 24 h. We used an elemental analyzer (Flash EA1112, Thermo Finnigan, Italy) to analyze soil organic C (SOC) concentrations. Soil total N was determined using an elemental analyzer (Flash EA1112, Thermo Finnigan, Italy). Soil available phosphorus concentration was measured by the approach offered by Bray and Kurtz [37]. Soil available potassium concentration was measured by a flame photometer (FP6410, Shanghai Co., Ltd., Shanghai, China) after soil was extracted with ammonium acetate solution.

When collecting GHG samples every month, soil samples from three places near the static chamber (0–20 cm depth) per treatment were randomly collected and mixed The soil physicochemical properties were measured after sieving (<2 mm) in the laboratory. The sieved soil samples were divided into two parts: one part was stored in the refrigerator within the following three days for analyzing microbial biomass C (MBC) water-soluble organic C (WSOC), water-soluble organic N (WSON), microbial biomass N (MBN), NO_3^--N, and NH_4^+-N concentrations, and either part for measuring soil pH available P and available K.

The methods outlined by Vance et al. [38] were used to examine the soil MBN and MBC concentrations using a Total Organic Carbon Analyzer (TOC-L CPN, Shimadzu Corporation, Kyoto, Japan). The soil WSOC and WSON concentrations were measured based on the approaches described in Singh et al. [39]. The concentrations of NH_4^+-N and NO_3^--N were analyzed using a double-beam spectrophotometer (UV-8000PC, Shanghai China) according to the method of Zhang et al. [12].

2.6. Statistical Analyses

All data used for the analysis were the average of four replicates. Microsoft Excel 2013 and SPSS 19.0 software were used for statistical analyses. One-way ANOVA with the least significant difference (LSD) was applied to determine the significance of the annual cumulative soil GHG emission and annual average values of soil physical and chemical properties under CK, BFT, CFT, and MFT treatments. In the pre-processing, the data should be tested for normality and homogeneity, and if necessary, it would be logarithmically transformed. Origin 2018 software was used to make graphs, and a stepwise regression analysis was utilized to analyze the connection between soil GHG emissions and soil properties, such as pH, temperature, MBN, MBC, WSON, WSOC, NO_3^--N, NH_4^+-N concentrations, and moisture content.

Structural equation modeling (SEM) was used to reveal the mechanisms involving GHG emissions under different fertilization measures. According to the results of stepwise regression analysis, we selected factors from soil carbon pool and nitrogen pool related to individual GHG flux as the inputs factors of the model.

The AMOS 21.0 software was used to explore the impact mechanism on soil GHG emissions using soil labile carbon pool and labile nitrogen pool with BFT and CFT. In the SEM, we used maximum likelihood (χ^2) goodness of fit test, goodness of fit index (GFI), normed fit index (NFI), and comparative fit index (CFI) to test goodness of fit. The standard basis of the best model used in this study is: (1) not significant χ^2 test statistics ($p > 0.05$), (2) GFI, NFI, and CFI > 0.90.

3. Results

3.1. Soil GHG Emissions

In the 12-month experiment, there were distinct seasonal changes in soil GHG emissions and soil temperature at 5 cm depth, which were basically consistent with the dynamics of air temperature. The minimum soil GHG emission appeared in winter, and the peak appeared in summer among all treatments (Figure 3).

The soil N_2O emission ranged from 15.40 to 55.06 μg m^{-2} h^{-1} in CK, from 12.73 to 41.73 μg m^{-2} h^{-1} in BFT, from 16.39 to 74.66 μg m^{-2} h^{-1} in CFT, and from 17.81 to 70.59 μg m^{-2} h^{-1} in MFT, respectively (Figure 3a). The annual cumulative soil N_2O emission under CK, BFT, CFT, and MFT were 2.58 ± 0.13, 2.16 ± 0.10, 3.38 ± 0.05, and 3.18 ± 0.07 kg ha^{-1} yr^{-1}, respectively (Figure 4a). Compared with the control, BFT reduced the annual cumulative emission of soil N_2O by 16.3% ($p < 0.01$), while CFT and MFT dramatically enhanced the emission by 31.0% and 23.3% ($p < 0.01$), respectively, on a yearly cumulative basis (Figure 4a).

Soil CH_4 uptake ranged from 33.91 to 100.76 μg m^{-2} h^{-1}, from 38.81 to 114.13 μg m^{-2} h^{-1}, from 26.75 to 93.33 μg m^{-2} h^{-1}, and from 35.87 to 128.20 μg m^{-2} h^{-1}, respectively, under the CK, BFT, CFT, and MFT. The annual cumulative soil CH_4 uptake were 5.50 ± 0.14, 5.82 ± 0.16, 5.46 ± 0.05 and 5.91 ± 0.11 kg ha^{-1} yr^{-1}, respectively, for the treatments CK, BFT, CFT, and MFT (Figure 4b). CFT had no significant effect compared with the CK, while BFT and MFT significantly increased the annual cumulative soil CH_4 uptake by 5.8% ($p < 0.01$) and 7.5% ($p < 0.01$), respectively (Figure 4b).

In the CK, BFT, CFT, and MFT, the soil CO_2 emission ranged from 63.25 to 461.58 mg m^{-2} h^{-1}, from 61.66 to 487.67 mg m^{-2} h^{-1}, from 71.21 to 583.02 mg m^{-2} h^{-1}, from 82.84 to 569.89 mg m^{-2} h^{-1}, respectively (Figure 3c). Under the treatments of CK, BFT, CFT, and MFT, the annual cumulative soil CO_2 emission were 20.96 ± 0.60, 22.93 ± 1.31, 23.68 ± 0.42, and 26.52 ± 1.04 Mg ha^{-1} yr^{-1}, respectively (Figure 4c). Compared with the CK, the BFT, CFT, and MFT significantly increased the cumulative CO_2 emission by 9.4% ($p < 0.05$), 13.0% ($p < 0.01$), and 26.5% ($p < 0.01$), respectively (Figure 4c). In summary, the GWP values under CK, BFT, CFT, and MFT were 21.59 ± 0.57, 22.82 ± 0.85, 24.55 ± 0.43, 27.76 ± 0.40 CO_2-eq Mg ha^{-1} yr^{-1}, respectively. In contrast, the GWP of BFT did not change significantly, while CFT and MFT increased it by 13.7% and 28.6% ($p < 0.05$), respectively (Figure 4d).

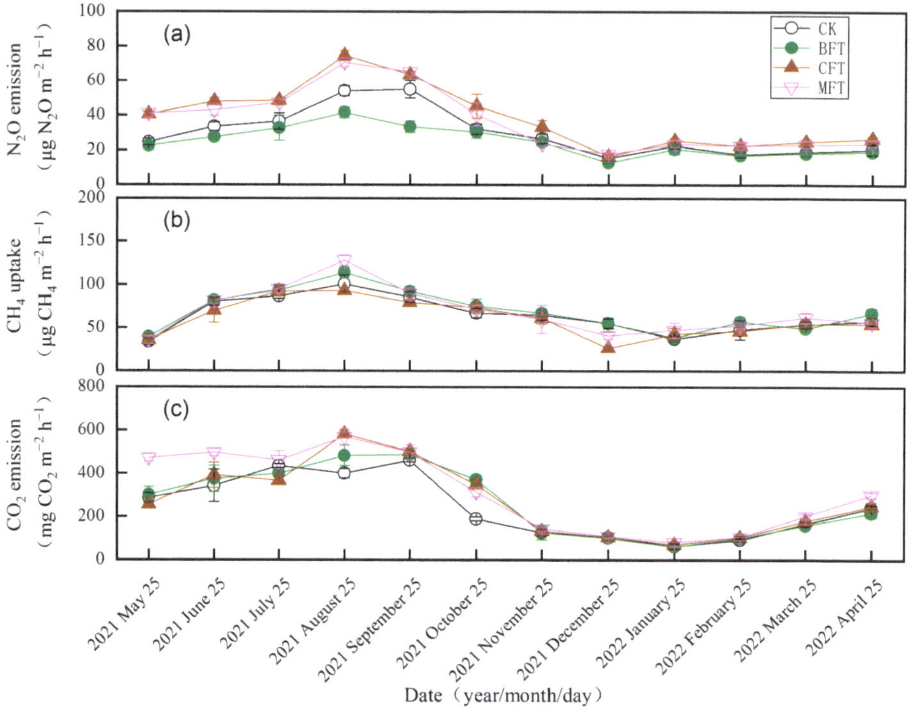

Figure 3. Effects of various fertilization strategies on soil (**a**) N_2O emission (**b**) CH_4 uptake (**c**) CO_2 emission in the *Pleioblastus amarus* (Keng) Keng f. plantation. The standard deviation is indicated by vertical bars.

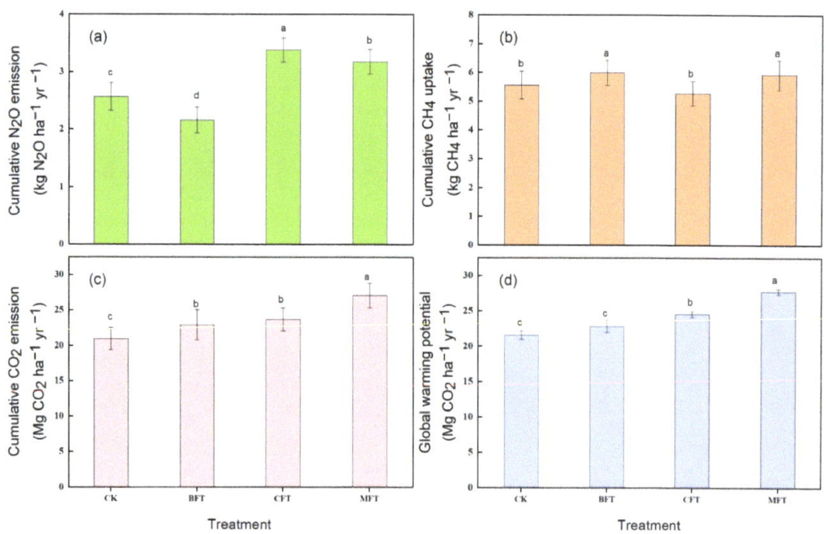

Figure 4. Effects of various fertilization treatments on annual cumulative soil: (**a**) N_2O emission (**b**) CH_4 uptake, (**c**) CO_2 emission, and (**d**) global warming potential in the *Pleioblastus amarus* plantation. The standard deviation is indicated by vertical bars. The letters above the bars denote significance in CK, CFT, CFT, and MFT).

3.2. Soil Environmental Factors

The temperature of the soil at 5 cm depth showed a distinct seasonal variation (Figure 5a). The soil temperature peaked in July–August and dropped to the lowest value during the period of January–February. Fertilizer applications had no discernible effects on the soil temperature and soil moisture at 5 cm depth (Figure 5). In contrast, when compared to the control, BFT treatment considerably enhanced the soil pH (Figure 5c).

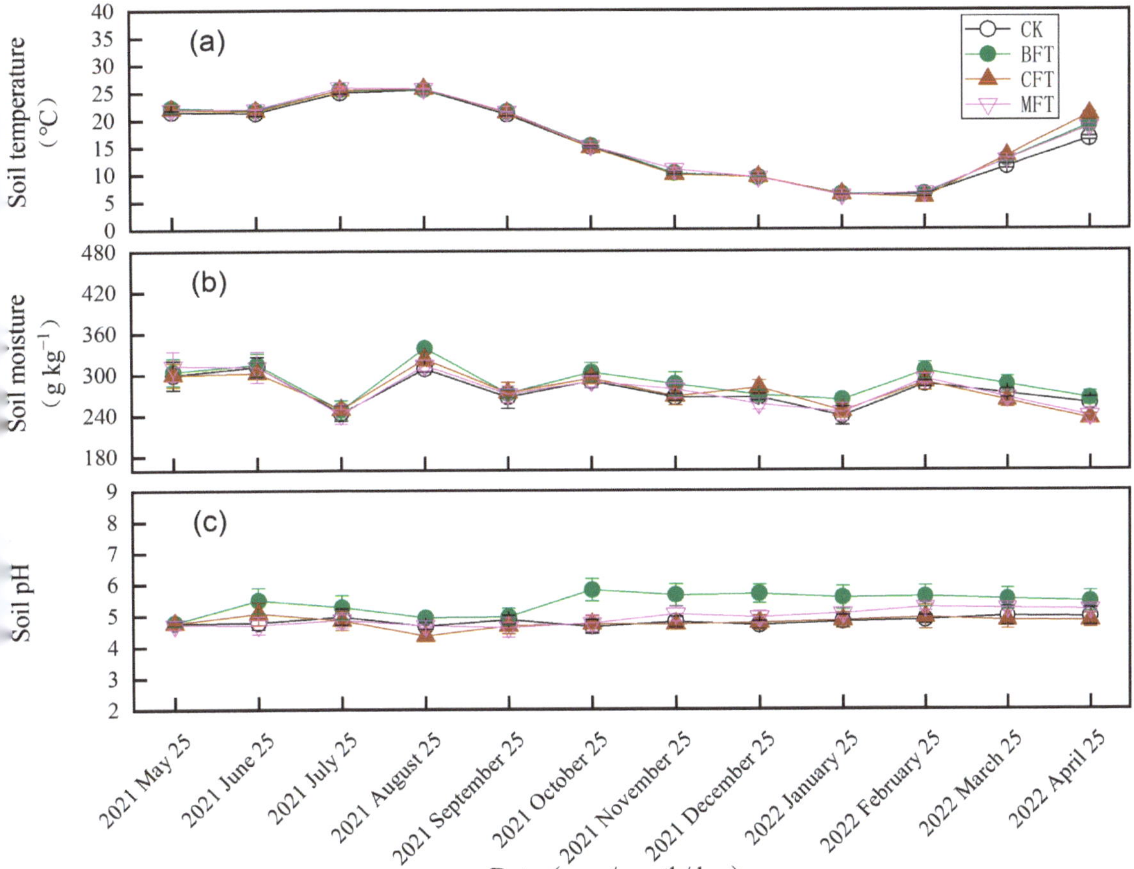

Figure 5. Effects of various fertilization strategies on soil (**a**) temperature, (**b**) moisture, (**c**) pH in the *Pleioblastus amarus* plantation. The standard deviation is indicated by vertical bars.

For soil carbon (C) pool, both soil MBC and WSOC concentrations exhibited a seasonal variation (Figure 6). Compared with CK, there were significant differences among different treatments. BFT, CFT, and MFT significantly increased the annual mean soil MBC concentration by 14.9% ($p < 0.01$), 6.4% ($p < 0.05$), and 9.9% ($p < 0.01$), respectively (Figure 6a). Similarly, BFT, CFT, and MFT increased the annual mean soil WSOC concentration by 10.7% ($p < 0.01$), 5.1% ($p < 0.05$), and 4.3% ($p < 0.01$), respectively (Figure 6b).

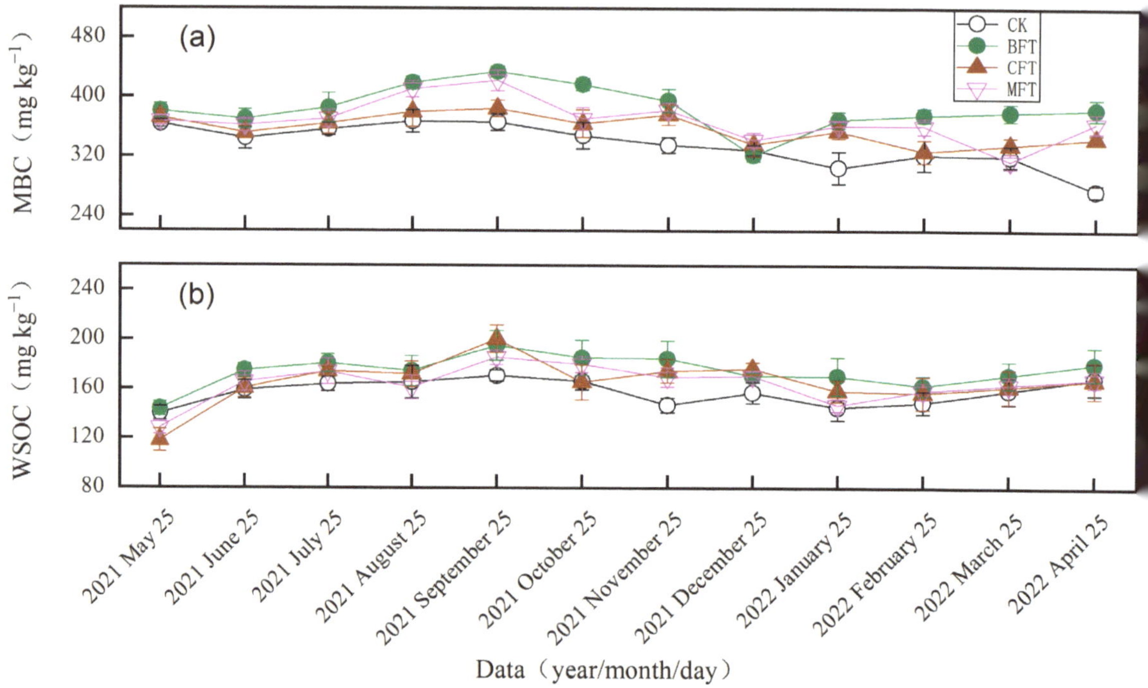

Figure 6. Effects of various fertilization strategies on soil (**a**) microbial biomass C (MBC) and soil (**b**) water-soluble organic C (WSOC) concentration in the *Pleioblastus amarus* plantation. The standard deviation is indicated by vertical bars.

As for the soil nitrogen (N) pool, the soil NO_3^--N and NH_4^+-N concentrations showed distinct seasonal variations. Compared with CK, BFT decreased the annual mean soil NO_3^--N concentration by 13.4%, while CFT and MFT increased the concentration by 20.7% and 13.7%, respectively (Figure 7a). BFT decreased the annual mean soil NH_4^+-N concentration by 9.4%, while CFT and MFT increased it by15.4% and 28.1%, respectively (Figure 7b). However, compared with CK, BFT reduced the annual mean WSON concentration by 6.4% while the WSON concentration increased by 15.0% and 3.9%, respectively, following CFT and MFT (Figure 7d).

3.3. Relationship between Soil Environmental Factors and GHG Emissions

There were significantly positive correlations between soil WSON concentration, MBN concentration, and soil N_2O emissions (Table 2). In contrast, regardless of the treatments soil N_2O emission was not significantly related to soil moisture and WSOC concentration. All treatments showed a positive correlation between soil CH_4 absorption and soil WSOC concentration (Table 3). In BFT, soil CH_4 uptake was positively linked with soil MBC, WSOC, and WSON concentrations. Regardless of the treatments, the soil CO_2 flux was significantly and positively correlated with soil temperature at 5 cm depth (Table 4). In addition, under CFT and MFT treatment, there was a positive correlation between soil CO_2 emission and soil moisture.

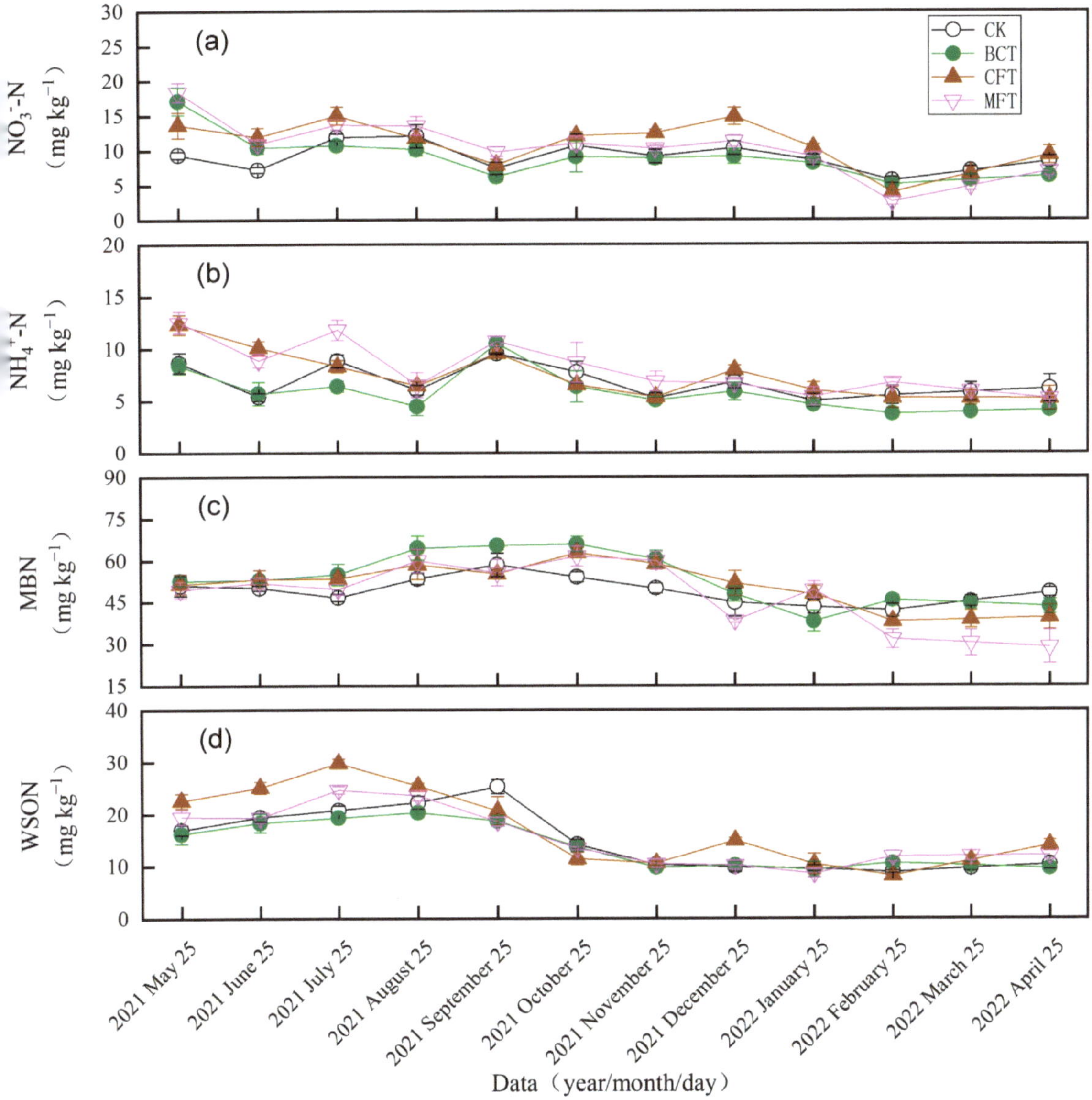

Figure 7. Effects of various fertilization strategies on (**a**) soil NO_3^--N, soil (**b**) NH_4^+- N, soil (**c**) microbial biomass N (MBN), and soil (**d**) water-soluble organic N (WSON) concentration in the *Pleioblastus amarus* plantation. The standard deviation is indicated by vertical bars.

Table 2. Model for stepwise regression analysis between N_2O flux ($\mu g\ m^{-2}\ h^{-1}$) and soil temperatur (T, °C), soil moisture (M, g kg^{-1}), microbial biomass C (MBC, mg kg^{-1}), water-soluble organic C (WSOC, mg kg^{-1}), microbial biomass N (MBN, mg kg^{-1}), water-soluble organic N (WSON, mg kg^{-1} NO_3^--N, and NH_4^+-N under CK, BFT, CFT, and MFF treatments. (The coefficient in the model i standardized.) R^2 denotes the percentage of variance revealed by Model.

GHG	Treatment	Model	df	R^2	*p*
	CK	Y = 0.827WSON	48	0.678	**
		Y = 0.614WSON + 0.342MBN	48	0.745	**
	BFT	Y = 0.777MBC	48	0.595	**
		Y = 0.547MBC + 0.486WSON	48	0.777	**
		Y = 0.467MBC + 0.401WSON +0.199MBN	48	0.793	**
		Y = 0.438MBC + 0.470WSON +0.252MBN + 0.164NH_4^+-N	48	0.807	**
	CFT	Y = 0.729T	48	0.521	**
N_2O		Y = 0.556T + 0.415MBC	48	0.659	**
		Y = 0.544T + 0.369MBC + 0.207M	48	0.694	**
		Y = 0.607T + 0.408MBC + 0.202M + 0.181NO_3^--N	48	0.714	**
		Y = 0.638T + 0.250MBC + 0.109M + 0.363NO_3^--N +0.379MBN	48	0.769	**
		Y = 0.649T + 0.248MBC + 0.393NO_3^--N +0.436MBN	48	0.764	**
	MFT	Y = 0.857WSON	48	0.729	**
		Y = 0.710WSON +0.331MBC	48	0.814	**
		Y = 0.681WSON +0.223MBC + 0.199MBN	48	0.836	**

** indicate significance at $p < 0.001$.

Table 3. Model for stepwise regression analysis between CH_4 flux ($\mu g\ m^{-2}\ h^{-1}$) and soil temperatur (T, °C), soil moisture (M, g kg^{-1}), microbial biomass C (MBC, mg kg^{-1}), water-soluble organic C (WSOC, mg kg^{-1}), microbial biomass N (MBN, mg kg^{-1}), water-soluble organic N (WSON, mg kg^{-1} NO_3^--N, and NH_4^+-N under CK, BFT, CFT, and MFF treatments. (The coefficient in the model i standardized.) R^2 denotes the percentage of variance revealed by Model.

GHG	Treatment	Model	df	R^2	*p*
	CK	Y = 0.596WSOC	48	0.341	**
		Y = 0.476WSOC + 0.375WSON	48	0.458	**
	BFT	Y = 0.722MBC	48	0.511	**
		Y = 0.464MBC + 0.417WSON	48	0.611	**
		Y = 0.327MBC + 0.481WSON + 0.344WSOC	48	0.716	**
		Y = 0.399MBC + 0.587WSON + 0.325WSOC − 0.278NH_4^+-N	48	0.766	**
		Y = 0.523MBC + 0.646WSON + 0.255WSOC − 0.391NH_4^+-N − 0.227M	48	0.796	**
CH_4	CFT	Y = 0.538T	48	0.274	**
		Y = 0.529T + 0.402WSOC	48	0.427	**
		Y = 0.658T + 0.386WSOC + 0.297NO_3^--N	48	0.490	**
		Y = 0.630T + 0.326WSOC + 0.103NO_3^--N +0.350MBN	48	0.557	**
	MFT	Y = 0.570WSON	48	0.310	**
		Y = 0.561WSON + 0.495WSOC	48	0.550	**
		Y = 0.836WSON + 0.487WSOC + 0.120NH_4^+-N	48	0.648	**

** indicate significance at $p < 0.001$.

The main soil factors affecting soil GHG emission were predicted and explained b SEM. Upon biochar-based fertilizer and chemical fertilizer treatments, soil carbon an nitrogen pool can drive soil GHG emission (Figure 8). The SEM results showed soi MBC concentration and WSON concentration were the main factors driving soil N_2O emission in biochar-based fertilizer treatment (Figure 8a). However, the concentration o soil MBC concentration and MBN concentration were important factors controlling N_2O emission in CFT treatment (Figure 8d). In addition, soil CH_4 uptake was driven by soi MBC concentration and WSON concentration in BFT (Figure 8b). However, the effect o soil MBN concentration on soil CH_4 flux is positively significant in CFT (Figure 8e). Th

SEM indicated soil MBC concentration drove soil CO_2 emission in biochar-based fertilizer treatment and chemical fertilizer treatment (Figure 8c,f).

Table 4. Model for stepwise regression analysis between CO_2 flux (mg m^{-2} h^{-1}) and soil temperature (T, °C), soil moisture (M, g kg^{-1}), microbial biomass C (MBC, mg kg^{-1}), water-soluble organic C (WSOC, mg kg^{-1}), microbial biomass N (MBN, mg kg^{-1}), water-soluble organic N (WSON, mg kg^{-1}), NO_3^--N, and NH_4^+-N under CK, BFT, CFT, and MFF treatments. (The coefficient in the model is standardized.) R^2 denotes the percentage of variance revealed by Model.

GHG	Treatment	Model	df	R^2	p
CO_2	CK	Y = 0.942T	48	0.885	**
		Y = 1.028T−0.200NO_3^--N	48	0.916	**
		Y = 0.982T−0.198NO_3^--N +0.116WSOC	48	0.927	**
	BFT	Y = 0.920T	48	0.844	**
		Y = 0.772T + 0.272MBC	48	0.895	**
		Y = 0.637T + 0.216MBC + 0.209WSON	48	0.907	**
		Y = 0.667T + 0.206MBC + 0.266WSON−0.160NO_3^--N	48	0.927	**
	CFT	Y = 0.900T	48	0.806	**
		Y = 0.868T + 0.212M	48	0.847	**
		Y = 0.862T + 0.224M + 0.177WSOC	48	0.877	**
		Y = 0.941T + 0.226M + 0.166WSOC + 0.182NO_3^--N	48	0.904	**
		Y = 0.934T + 0.149M + 0.122WSOC + 0.317NO_3^--N +0.230MBC	48	0.930	**
	MFT	Y = 0.966T	48	0.932	**
		Y = 0.934T + 0.135M	48	0.948	**
		Y = 0.984T + 0.136M + 0.085NO_3^--N	48	0.952	**
		Y = 0.951T + 0.127M + 0.103NO_3^--N +0.083NH_4^+-N	48	0.956	**

** indicate significance at $p < 0.001$.

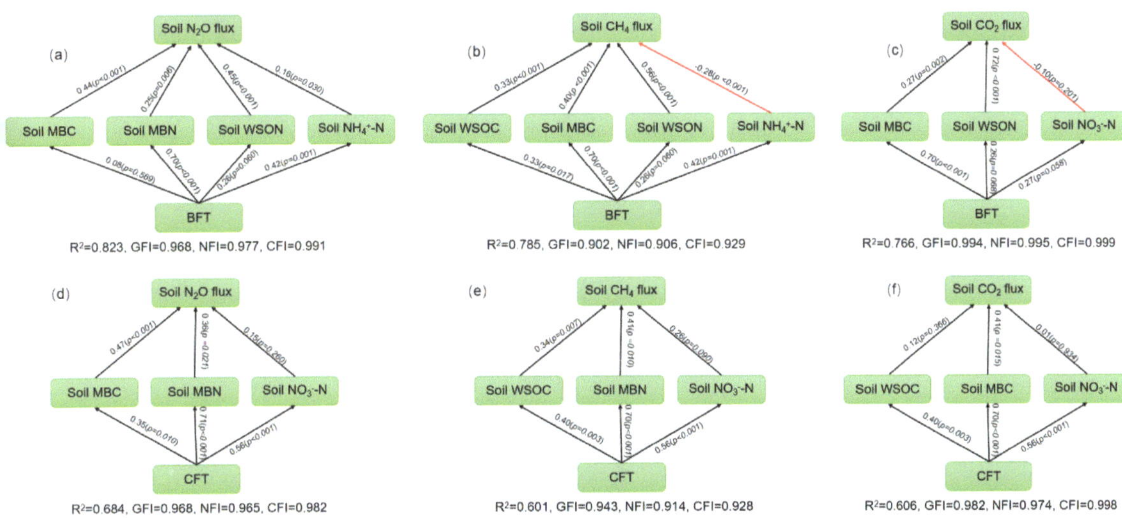

Figure 8. Structural equation modeling illustrating the impacts of soil MBC, WSOC, NO_3^--N, NH_4^+-N, WSON, and MBN concentration on (**a**,**d**) soil N_2O emission, (**b**,**e**) CH_4 uptake, and (**c**,**f**) CO_2 emission after applying biochar-based fertilizer treatment (BFT) and chemical fertilizer treatment (CFT). The numbers next to the arrows denote the correlation coefficient and statistical significance. R^2 denotes the percentage of variance revealed by SEM, GFI denotes the goodness of fit index, NFI denotes the normed fit index, and CFTI denotes the comparative fit index. Black and red lines denote positive and negative relationships, respectively.

4. Discussion

4.1. Effects of Fertilization on Soil N_2O Emission

Our findings demonstrated the application of biochar-based fertilizer dramatically decreased annual cumulative soil N_2O emission throughout the one-year trial period, which supported our first hypothesis (Figures 3a and 4a). In addition, this also supports the observation in Yang et al. [40], which reported biochar decreased N_2O flux by 35.90% in an experiment with farmland soil in the Sonnen Plain. Similarly, Song et al. [21] found biochar treatment considerably decreased soil N_2O emission via decreasing soil labile N concentration, and the reduction rate was positively correlated with the biochar application rate in a subtropical Moso bamboo forest. Thus, our analysis also provided additional and substantial evidence the application of biochar-based fertilizer can reduce soil N_2O emission from managed lands. The following explanations may account for the impact mechanism of reduction in N_2O emissions in the BFT-treated *Pleioblastus amarus* plantation soil.

Firstly, biochar contains a large amount of carbon and nutrients and has a rich pore structure, which has a good fixation effect on carbon and nitrogen. Applying biochar-based fertilizers promotes soil aeration and porosity, thereby increasing the oxygen concentration in the soil, which in turn inhibits soil denitrification processes and anaerobic microbial activity [41–44]. Moreover, our results demonstrated BFT treatment reduced soil NH_4^+-N and NO_3^--N concentrations when compared to CK (Figure 7). Previously, it was thought that one of the most important ways of reducing soil N_2O emissions was to restrict the supply of mineral nitrogen to denitrifying bacteria [45]. Our experimental results also support the above statement, as BFT treatment considerably lowered soil WSON concentration. Moreover, soil WSON concentration was positively correlated with N_2O emission in our investigation and study (Table 2). In addition, we found N_2O emissions were affected by soil WSON concentration in SEM (Figure 8a). Therefore, one key method to lower soil N_2O emissions may be the involvement of biochar-based fertilizers in restricting mineral N. The annual cumulative N_2O emission significantly increased after CFT and MFT compared to the control (Figure 4a), which was mainly due to the significant increase in soil NO_3^--N and NH_4^+-N concentrations. Since half of the nitrogen in MFT comes from CFT, it may be concluded this caused the higher soil N_2O emission in the MFT treatment. Different fertilizer treatments could alter soil N_2O emission through changing soil labile carbon and labile nitrogen pools, according to the results of stepwise regression models and SEM (Figure 8a,d, Table 2).

Secondly, the soil C pool is also an important factor affecting N_2O emissions. MBC and WSOC, as unstable parts of SOC carbon components, take part in a number of biochemical soil microbial activities that are correlated with soil N_2O emissions [46,47]. According to this study, BFT treatment increased MBC concentration and WSOC concentration while dramatically reducing soil N_2O emission (Figure 6). This indicated that using BFT did not reduce soil N_2O emission by changing the concentration of active organic carbon components. Furthermore, soil N_2O emission was not correlated with WSOC concentration. Therefore, we can rule out the possibility fertilization treatments increased soil N_2O emission by increasing WSOC concentration. At the same time, this result also confirmed our second hypothesis.

Finally, we observed the pH of BFT-treated soil was increased (Figure 5c), which could promote N_2O to N_2 conversion throughout soil denitrification, therefore lowering emissions of soil N_2O [41]. Additionally, we found there is no significant linear relationship between N_2O emission and soil moisture at 5 cm depth (Table 2). This supports the view of [48], who found regardless of fertilization, soil N_2O flux and soil moisture were not significantly associated in Moso bamboo forests in Hangzhou, China. Xu et al. [42] also revealed no correlation between soil N_2O emission and soil moisture with silicate fertilizer application in Phyllostachys Pubescens. Different studies had various conclusions about the association between soil moisture and soil N_2O emission. Vargas et al. [49] concluded N_2O emission increased linearly with soil moisture regardless of the application of sugarcane crop residues on the soil.

4.2. Effects of Fertilization on Soil CH_4 Uptake

The BFT rose the annual cumulative soil CH_4 uptake and the annual mean value (Figure 4b). Our findings were consistent with previous findings. For example, Lv et al. [50] found the application of biochar increased significantly ($p < 0.05$) CH_4 uptake in subtropical plantations through a two-year experimental study in Hangzhou, China. In addition, Fang et al. [51] showed one year after applying different amounts of biochar to Moso bamboo forests, the annual soil cumulative CH_4 uptake increased significantly. Adding biochar-based fertilizer to *Pleioblastus amarus* plantation soil for one year increased the annual cumulative soil CH_4 uptake. According to the experimental data and SEM analysis (Figure 8), the possible reasons are as follows.

First, because of its special biological structure, biochar can absorb and diffuse more CH_4 into the soil, thereby promoting the growth of CH_4-nutrient microorganisms. Under the conditions of good aeration, methanogen activity in the soil declines while the activity of methanotrophs increases, thereby promoting methane absorption in the soil [52]. Second, biochar treatment increases soil porosity, and soil water content decreases the anaerobic soil environment and increases CH_4 oxidation. Soil methanotrophs use soil oxygen to oxidize CH_4 to CO_2 in the soil [53]. Third, we found that compared with CK, CFT treatment had no significant effect, while BFT treatment increased the soil CH_4 uptake. In addition, the BFT, MBC, and WSOC concentrations in the soil were closely correlated with the soil CH_4 uptake (Table 3), and the BFT treatment increased the annual mean soil MBC and WSOC concentrations (Figure 6). Similarly, the SEM results showed the CH_4 uptake was affected by the soil MBC concentration in biochar-based fertilizer treatment (Figure 8b). This indicated BFT treatment promoted soil CH_4 uptake by increasing the concentrations of soil MBC and WSOC. This result was also confirmed by Liu et al. [54], who found soil MBC concentration increased when straw charcoal was applied.

However, some studies revealed the utilization of biochar either has no impact on soil CH_4 uptake or increases soil CH_4 emission. For example, in the growing choy sum and amaranth experiment, Jia et al. [55] discovered the use of maize straw biochar modifier had no discernible impact on CH_4 emission. The differences in these results may be caused by differences in study subjects, fertilization methods, land types, experimental designs, soil pH, and regional climates. Therefore, we need specific experimental studies to focus on the processes and internal mechanisms involving the impact of biochar on soil CH_4 uptake and translocation.

4.3. Effects of Fertilization on Soil CO_2 Emission

Some previous studies revealed soil CO_2 emissions decreased or had no effect following the input of biochar [17,56]. In contrast, this study demonstrated BFT, CFT, and MFT all substantially increased the annual cumulative CO_2 emissions compared with the control. Applying biochar-based fertilizers increased the annual cumulative emission (Figure 4c), which partially refuted our first hypothesis. Our results also complied with some other studies.

For example, Troy et al. [57] and Kalu et al. [58] established that incorporating biochar to soil increased CO_2 emission. Using biochar-based fertilizer in the *Pleioblastus amarus* plantation increased the annual cumulative soil CO_2 emissions, which may be caused by the following reasons.

Firstly, using fertilizers incorporated with biochar may raise the enzyme activity of the soil, which promotes the accumulation of active organic carbon and decomposition of soil organic carbon, which in turn affects soil CO_2 emissions [59]. Secondly, biochar as a soil amendment can increase soil pH [60]. Our experimental data found the application of biochar- based fertilizer increases soil moisture and soil pH (Figure 5). Appropriate water conditions and pH provide a suitable living environment for soil microorganisms, and the changes of microorganisms in biological activities result in increased CO_2 emission [56]. Thirdly, significantly favorable correlations were discovered between soil temperature, WSOC concentration, and soil CO_2 emission (Table 4). In addition, the SEM results showed there is a positive correlation between soil MBC concentration and soil CO_2 emission i biochar-based fertilizer and chemical fertilizer treatments (Figure 8c,f). Biochar, as a stabl organic carbon, was input into the soil of the *Pleioblastus amarus* plantation, which increase the contents of soil MBC, and thus, increased the soil CO_2 emission. Finally, the utilizatio of biochar-based fertilizers in the plantation soil greatly improved the soil environmen Moreover, the input of biochar reduced the loss of nutrients, and more nutrients wer retained in the soil. As a result, plants can absorb more nutrients from the soil to promot vegetation growth, vegetation roots' respiration is also enhanced, and more CO_2 is emitte

5. Conclusions

The addition of BFT greatly decreased the soil N_2O emission of the *Pleioblastus amaru* plantation, while applying CFT or MFT with nitrogen, phosphorus, and potassium nutrient significantly increased the N_2O emission. Similarly, BFT treatment significantly increase soil CH_4 uptake, but MFT treatment significantly decreased the uptake. Conversely, fe tilization constantly increases soil CO_2 emissions regardless of treatments. However, th soil CO_2 emission of biochar-based fertilizer treatment was lower than that of the chemica fertilizer treatment. In addition, this study reported soil temperature at 5 cm, MBC, WSOC and WSON concentrations were the important factors regulating soil GHG emissions. Th SEM results showed soil WSON concentration drives soil N_2O emission, and soil CH uptake and CO_2 emission are affected by soil MBC concentration.

The production process and technique of biochar-based fertilizers are mature, an the price is relatively low. Therefore, considering the economic cost and the yields c the *Pleioblastus amarus* plantation, our findings suggest utilizing biochar-based fertilizer instead of traditional chemical fertilizers can be promoted as an environment-friendly so management measure by reducing GHG emissions. Meanwhile, an essential part of th process of soil GHG emissions is played by the fungal and bacterial species in the so from the *Pleioblastus amarus* plantation. Therefore, in future studies, a longer trial period i needed to better understand how soil microbial communities engaged in C and N cyclin react to the administration of biochar-based fertilizers in *Pleioblastus amarus* plantation so

Author Contributions: Conceptualization, Y.S., L.X., G.Z. and Y.Z.; data curation, E.W.; form analysis, E.W. and N.Y.; investigation, E.W., N.Y., S.L., X.T., L.W. and G.W.; funding acquisition, Y. and L.X.; writing—original draft preparation, E.W. All authors have read and agreed to the publishe version of the manuscript.

Funding: This research was funded by the National Natural Science Foundation of China (Gra number: U1809208; 31870618; 32001315); Scientific Research Development Fund of Zhejiang A& University (Grant number: 2020FR008); the Key Research and Development Program of Zhejian Province (Grant number: 2021C02005; 2022C03039).

Data Availability Statement: Not applicable.

Acknowledgments: This study was supported by the Natural Resources Defense Council (NRDC We would like to thank the editor and anonymous reviewers for their contribution to the peer revie of our study.

Conflicts of Interest: The authors declare no conflict of interest.

References

1. Vrrousek, P.M. Beyond Global Warming: Ecology and Global Change. *Ecology* **1994**, *75*, 1861–1876.
2. Wardle, D.A.; Bardgett, R.D.; Callaway, R.M.; Van Der Putten, W.H. Terrestrial Ecosystem Responses to Species Gains and Losse *Science* **2011**, *332*, 1273–1277. [CrossRef] [PubMed]
3. Keenan, T.F.; Gray, J.; Friedl, M.A.; Toomey, M.; Bohrer, G.; Hollinger, D.Y.; Munger, J.W.; O'Keefe, J.; Schmid, H.P.; Wing, I. Net Carbon Uptake Has Increased through Warming-Induced Changes in Temperate Forest Phenology. *Nat. Clim. Chang.* **201** *4*, 598–604. [CrossRef]
4. Fu, W.J.; Fu, Z.J.; Ge, H.L.; Ji, B.Y.; Jiang, P.K.; Li, Y.F.; Wu, J.S.; Zhao, K.L. Spatial Variation of Biomass Carbon Density in Subtropical Region of Southeastern China. *Forests* **2015**, *6*, 1966–1981. [CrossRef]
5. Dou, Y.; Yu, X.J. Development and Comparison of Bamboo Industry in the World. *World Agric.* **2008**, *7*, 18–21.

6. Zhou, G.; Jiang, P.; Xu, Q. *Carbon Sequestration and Transformation in Bamboo Forest Ecosystems*, 1st ed.; Science Press: Beijing, China, 2010; pp. 2–3.
7. Wu, H. Analysis on Fertilization Test of *Pleioblastus amarus* in Different Growth Periods. *Green Sci. Technol.* **2021**, *23*, 164–166.
8. Shen, G.C.; Zhang, X.D.; Zhang, L.; Gao, S.H.; Zhang, R.; Zhu, W.S.; Tang, S.Q. Estimating the Carbon Stock and Carbon Sequestration of the *Pleioblastus amarus* Forest Ecosystem in Southern of Sichuan. *Sci. Silvae Sin.* **2013**, *49*, 78–84.
9. He, Y.P.; Fei, S.M.; Jiang, J.M.; Chen, X.M.; Yu, Y.; Zhu, W.S.; Tang, S.Q. The Spatial Distribution of Organic Carbon in Phyllostachyspubescens and *Pleioblastus amarus* in Changning City. *Sichuan For. Sci. Technol.* **2007**, *28*, 10–14.
10. Shen, G.C. Estimating the Carbon Sequestration of *Phyllostachys Edulis* and *Pleioblastus amarus* Plantations Based on Net Ecosystem Productivity. Master's Thesis, Chinese Academy of Forestry, Beijing, China, 2012.
11. Xiong, Y.M.; Xia, H.P.; Li, Z.A.; Cai, X.A.; Fu, S.L. Impacts of Litter and Understory Removal on Soil Properties in a Subtropical Acacia Mangium Plantation in China. *Plant Soil* **2008**, *304*, 179–188.
12. Zhang, M.; Fan, C.H.; Li, Q.L.; Li, B.; Zhu, Y.Y.; Xiong, Z.Q. A 2-Yr Field Assessment of the Effects of Chemical and Biological Nitrification Inhibitors on Nitrous Oxide Emissions and Nitrogen Use Efficiency in an Intensively Managed Vegetable Cropping System. *Agric. Ecosyst. Environ.* **2015**, *201*, 43–50. [CrossRef]
13. Li, Y.F.; Zhang, J.J.; Chang, S.X.; Jiang, P.K.; Zhou, G.M.; Fu, S.L.; Yan, E.R.; Wu, J.S.; Lin, L. Long-Term Intensive Management Effects on Soil Organic Carbon Pools and Chemical Composition in Moso Bamboo (*Phyllostachys Pubescens*) Forests in Subtropical China. *For. Ecol. Manag.* **2013**, *303*, 121–130. [CrossRef]
14. Tian, H.Q.; Yang, J.; Xu, R.T.; Lu, C.Q.; Canadell, J.G.; Davidson, E.A.; Jackson, R.B.; Arneth, A.; Chang, J.F.; Ciais, P.; et al. Global soil nitrous oxide emissions since the preindustrial era estimated by an ensemble of terrestrial biosphere models: Magnitude, attribution, and uncertainty. *Glob. Chang. Biol.* **2019**, *25*, 640–659. [CrossRef] [PubMed]
15. Jones, D.L.; Murphy, D.V.; Khalid, M.; Ahmad, W.; Edwards-Jones, G.; DeLuca, T.H. Short-Term Biochar-Induced Increase in Soil CO_2 Release Is Both Biotically and Abiotically Mediated. *Soil Biol. Biochem.* **2011**, *43*, 1723–1731. [CrossRef]
16. Wu, Y.; Xu, G.; Lv, C.Y.; Shao, H.B. Effects of Biochar Amendment on Soil Physical and Chemical Properties: Current Status and Knowledge Gaps. *Adv. Earth Sci.* **2014**, *29*, 68–79.
17. Wu, M.X.; Feng, Q.B.; Sun, X.; Wang, H.L.; Gielen, G.; Wu, W.X. Rice (*Oryza Sativa* L.) Plantation Affects the Stability of Biochar in Paddy Soil. *Sci. Rep.* **2015**, *5*, 10001. [CrossRef]
18. Sohi, S.P.; Krull, E.; Lopez-Capel, E.; Bol, R. A Review of Biochar and Its Use and Function in Soil. *Adv. Agron.* **2010**, *105*, 47–82.
19. Li, Y.F.; Hu, S.D.; Chen, J.H.; Müller, K.; Li, Y.C.; Fu, W.J.; Lin, Z.W.; Wang, H.L. Effects of Biochar Application in Forest Ecosystems on Soil Properties and Greenhouse Gas Emissions: A Review. *J. Soils Sediments* **2018**, *18*, 546–563. [CrossRef]
20. Ge, X.G.; Cao, Y.H.; Zhou, B.Z.; Wang, X.M.; Yang, Z.Y.; Li, M.H. Biochar Addition Increases Subsurface Soil Microbial Biomass but Has Limited Effects on Soil CO_2 Emissions in Subtropical Moso Bamboo Plantations. *Appl. Soil Ecol.* **2019**, *142*, 155–165. [CrossRef]
21. Xu, L.; Fang, H.Y.; Deng, X.; Ying, J.Y.; Lv, W.J.; Shi, Y.J.; Zhou, G.M.; Zhou, Y.F. Biochar Application Increased Ecosystem Carbon Sequestration Capacity in a Moso Bamboo Forest. *For. Ecol. Manag.* **2020**, *475*, 118447. [CrossRef]
22. Song, Y.Z.; Li, Y.F.; Cai, Y.J.; Fu, S.L.; Luo, Y.; Wang, H.L.; Liang, C.F.; Lin, Z.W.; Hu, S.D.; Li, Y.C.; et al. Biochar Decreases Soil N_2O Emissions in Moso Bamboo Plantations through Decreasing Labile N Concentrations, N-Cycling Enzyme Activities and Nitrification/Denitrification Rates. *Geoderma* **2019**, *348*, 135–145. [CrossRef]
23. Hawthorne, I.; Johnson, M.S.; Jassal, R.S.; Black, T.A.; Grant, N.J.; Smukler, S.M. Application of Biochar and Nitrogen Influences Fluxes of CO_2, CH_4 and N_2O in a Forest Soil. *J. Environ. Manag.* **2017**, *192*, 203–214. [CrossRef]
24. Lin, Z.B.; Liu, Q.; Liu, G.; Annette, L.C.; Bei, Q.C.; Liu, B.J.; Wang, X.J.; Ma, J.; Zhu, J.G.; Xie, Z.B. Effects of Different Biochars on *Pinus elliottii* Growth, N Use Efficiency, Soil N_2O and CH_4 Emissions and C Storage in a Subtropical Area of China. *Pedosphere* **2017**, *27*, 248–261. [CrossRef]
25. Liu, H.; Ding, Y.; Zhang, Q.; Liu, X.; Xu, J.; Li, Y.; Di, H. Heterotrophic Nitrification and Denitrification Are the Main Sources of Nitrous Oxide in Two Paddy Soils. *Plant Soil* **2019**, *445*, 39–53. [CrossRef]
26. Liu, Q.; Liu, B.J.; Zhang, Y.H.; Hu, T.; Lin, Z.; Liu, G.; Wang, X.; Ma, J.; Wang, H.; Jin, H. Biochar Application as a Tool to Decrease Soil Nitrogen Losses (NH_3 Volatilization, N_2O Emissions, and N Leaching) from Croplands: Options and Mitigation Strength in a Global Perspective. *Glob. Chang. Biol.* **2019**, *25*, 2077–2093. [CrossRef] [PubMed]
27. Zhang, S.B.; Li, Y.F.; Singh, B.P.; Wang, H.L.; Cai, X.Q.; Chen, J.H.; Qin, H.; Li, Y.C.; Chang, S.X. Contrasting Short-Term Responses of Soil Heterotrophic and Autotrophic Respiration to Biochar-Based and Chemical Fertilizers in a Subtropical Moso Bamboo Plantation. *Appl. Soil Ecol.* **2021**, *157*, 103758. [CrossRef]
28. Nguyen, T.T.N.; Xu, C.Y.; Tahmasbian, I.; Che, R.X.; Xu, Z.H.; Zhou, X.H.; Wallace, H.M.; Bai, S.H. Effects of Biochar on Soil Available Inorganic Nitrogen: A Review and Meta-Analysis. *Geoderma* **2017**, *288*, 79–96. [CrossRef]
29. Chen, W.; Meng, J.; Han, X.; Lan, Y.; Zhang, W. Past, Present, and Future of Biochar. *Biochar* **2019**, *1*, 75–87. [CrossRef]
30. Sim, D.H.H.; Tan, I.A.W.; Lim, L.L.P.; Hameed, B.H. Encapsulated Biochar-based Sustained Release Fertilizer for Precision Agriculture: A Review. *J. Clean. Prod.* **2021**, *303*, 127018. [CrossRef]
31. Yi, S.; Chang, N.Y.; Imhoff, P.T. Predicting Water Retention of Biochar-amended Soil from Independent Measurements of Biochar and Soil Properties. *Adv. Water Resour.* **2020**, *142*, 103638. [CrossRef]

32. Lin, Q.; Tan, X.F.; Almatrafi, E.; Yang, Y.; Wang, W.J.; Luo, H.Z.; Qin, F.Z.; Zhou, C.Y.; Zeng, G.M.; Zhang, C. Effects of Biochar-based Materials on the Bioavailability of Soil Organic Pollutants and Their Biological Impacts. *Sci. Total Environ.* **2022**, *826*, 153956. [CrossRef]
33. Knoblauch, C.; Maarifat, A.A.; Pfeiffer, E.M.; Haefele, S.M. Degradability of Black Carbon and Its Impact on Trace Gas Fluxes and Carbon Turnover in Paddy Soils. *Soil Biol. Biochem.* **2011**, *43*, 1768–1778. [CrossRef]
34. Xu, L.; Shi, Y.J.; Zhou, G.M.; Xu, X.J.; Liu, E.B.; Zhou, Y.F.; Zhang, F.; Li, C.; Fang, H.Y.; Chen, L. Structural Development and Carbon Dynamics of Moso Bamboo Forests in Zhejiang Province, China. *For. Ecol. Manag.* **2018**, *409*, 479–488. [CrossRef]
35. Wang, S.W.; Ma, S.; Shan, J.; Xia, Y.Q.; Lin, J.H.; Yan, X.Y. A 2-Year Study on the Effect of Biochar on Methane and Nitrous Oxide Emissions in an Intensive Rice–Wheat Cropping System. *Biochar* **2019**, *1*, 177–186. [CrossRef]
36. Lv, R.K. *Soil Agrochemical Analysis Methods*, 3rd ed.; China Agricultural Science and Technology Press: Beijing, China, 2000; pp. 1–638.
37. Bray, R.H.; Kurtz, L.T. Determination of Total, Organic and Available Forms of Phosphorus in Soils. *Soil Sci.* **1945**, *59*, 39–45. [CrossRef]
38. Vance, E.D.; Brookes, P.C.; Jenkinson, D.S. An Extraction Method for Measuring Soil Microbial Biomass C. *Soil Eiol. Biochem.* **1987**, *19*, 703–707. [CrossRef]
39. Singh, B.P.; Hatton, B.J.; Singh, B.; Cowie, A.L.; Kathuria, A. Influence of Biochars on Nitrous Oxide Emission and Nitrogen Leaching from Two Contrasting Soils. *J. Environ. Qual.* **2010**, *39*, 1224–1235. [CrossRef]
40. Yang, X.C.; Liu, D.P.; Fu, Q.; Li, T.X.; Hou, R.J.; Li, Q.L.; Li, M.; Meng, F.X. Characteristics of Greenhouse Gas Emissions from Farmland Soils Based on a Structural Equation Model: Regulation Mechanism of Biochar. *Environ. Res.* **2022**, *206*, 112303. [CrossRef]
41. Li, Y.C.; Li, Y.F.; Chang, S.X.; Yang, Y.; Fu, S.; Jiang, P.; Luo, Y.; Yang, M.; Chen, Z.; Hu, S. Biochar Reduces Soil Heterotrophic Respiration in a Subtropical Plantation through Increasing Soil Organic Carbon Recalcitrancy and Decreasing Carbon-Degrading Microbial Activity. *Soil Biol. Biochem.* **2018**, *122*, 173–185. [CrossRef]
42. Xu, L.; Deng, X.; Ying, J.Y.; Zhou, G.M.; Shi, Y.J. Silicate Fertilizer Application Reduces Soil Greenhouse Gas Emissions in a Moso Bamboo Forest. *Sci. Total Environ.* **2020**, *747*, 141380. [CrossRef]
43. Tang, Z.M.; Liu, X.R.; Li, G.C.; Liu, X.W. Mechanism of Biochar on Nitrification and Denitrification to N_2O Emissions Based on Isotope Characteristic Values. *Environ. Res.* **2022**, *212*, 113219. [CrossRef]
44. Zhong, L.; Li, G.Y.; Qing, J.W.; Li, J.L.; Xue, J.M.; Yan, B.B.; Chen, G.Y.; Kang, X.M.; Rui, Y.C. Biochar Can Reduce N_2O Production Potential from Rhizosphere of Fertilized Agricultural Soils by Suppressing Bacterial Denitrification. *Eur. J. Soil Biol.* **2022**, *109*, 103391. [CrossRef]
45. Zwieten, L.V.; Singh, B.P.; Kimber, S.W.L.; Murphy, D.V.; Macdonald, L.M.; Rust, J.; Morris, S. An Incubation Study Investigating the Mechanisms that Impact N_2O Flux from Soil Following Biochar Application. *Agric. Ecosyst. Environ.* **2014**, *191*, 53–62. [CrossRef]
46. Zhang, J.J.; Li, Y.F.; Chang, S.X.; Jiang, P.K.; Zhou, G.M.; Liu, J.; Wu, J.S.; Shen, Z.M. Understory Vegetation Management Affected Greenhouse Gas Emissions and Labile Organic Carbon Pools in an Intensively Managed Chinese Chestnut Plantation. *Plant Soil* **2014**, *376*, 363–375. [CrossRef]
47. Fracetto, F.J.C.; Fracetto, G.G.M.; Bertini, S.C.B.; Cerri, C.C.; Feigl, B.J.; Neto, M.S. Effect of Agricultural Management on N_2O Emissions in the Brazilian Sugarcane Yield. *Soil Biol. Biochem.* **2017**, *109*, 205–213. [CrossRef]
48. Zhou, J.S.; Qu, T.H.; Li, Y.F.; Van, Z.L.; Wang, H.; Chen, J.; Song, X.; Lin, Z.; Zhang, X.; Luo, Y.; et al. Biochar-based Fertilizer Decreased While Chemical Fertilizer Increased Soil N_2O Emissions in a Subtropical Moso Bamboo Plantation. *Catena* **2021**, *202*, 105257. [CrossRef]
49. Vargas, V.P.; Cantarella, H.; Martins, A.A.; Soares, J.R.; Carmo, J.B.; Andrade, C.A. Sugarcane Crop Residue Increases N_2O and CO_2 Emissions under High Soil Moisture Conditions. *Sugar Technol.* **2014**, *16*, 174–179. [CrossRef]
50. Lu, X.H.; Li, Y.F.; Wang, H.L.; Singh, B.P.; Hu, S.D.; Luo, Y.; Li, J.W.; Xiao, Y.H.; Cai, X.Q.; Li, Y.C. Responses of Soil Greenhouse Gas Emissions to Different Application Rates of Biochar in a Subtropical Chinese Chestnut Plantation. *Agric. For. Meteorol.* **2019**, *271*, 168–179. [CrossRef]
51. Fang, H.Y. Effects of Biochar Application on Carbon Sequestration and Greenhouse Gases Emission in Moso Bamboo Forests. Master's Thesis, Zhejiang A&F University, Hangzhou, China, 2019.
52. Sonoki, T.; Furukawa, T.; Jindo, K.; Suto, K.; Aoyama, M.; Sánchez-Monedero, M.Á. Influence of Biochar Addition on Methane Metabolism during Thermophilic Phase of Composting. *J. Basic Microbiol.* **2013**, *53*, 617–621. [CrossRef]
53. Powers, J.S.; Schlesinger, W.H. Relationships among Soil Carbon Distributions and Biophysical Factors at Nested Spatial Scales in Rain Forests of Northeastern Costa Rica. *Geoderma* **2002**, *109*, 165–190. [CrossRef]
54. Liu, Y.X.; Yang, M.; Wu, Y.M.; Wang, H.L.; Chen, Y.X.; Wu, W.X. Reducing CH_4 and CO_2 Emissions from Waterlogged Paddy Soil with Biochar. *J. Soils Sediments* **2011**, *11*, 930–939. [CrossRef]
55. Jia, J.X.; Li, B.; Chen, Z.Z.; Xie, Z.B.; Xiong, Z.Q. Effects of Biochar Application on Vegetable Production and Emissions of N_2O and CH_4. *Soil Sci. Plant Nutr.* **2012**, *58*, 503–509. [CrossRef]
56. Wang, J.Y.; Zhang, M.; Xiong, Z.Q.; Liu, P.L.; Pan, G.X. Effects of Biochar Addition on N_2O and CO_2 Emissions from Two Paddy Soils. *Biol. Fertil. Soils* **2011**, *47*, 887–896. [CrossRef]

7. Troy, S.M.; Lawlor, P.G.; O'Flynn, C.J.; Healy, M.G. Impact of Biochar Addition to Soil on Greenhouse Gas Emissions Following Pig Manure Application. *Soil Biol. Biochem.* **2013**, *60*, 173–181. [CrossRef]
8. Kalu, S.; Kulmala, L.; Zrim, J.; Peltokangas, K.; Tammeorg, P.; Rasa, K.; Kitzler, B.; Pihlatie, M.; Karhu, K. Potential of Biochar to Reduce Greenhouse Gas Emissions and Increase Nitrogen Use Efficiency in Boreal Arable Soils in the Long-Term. *Front. Environ. Sci.* **2022**, *10*, 1–16. [CrossRef]
9. Epron, D.; Ngao, J.; Granier, A. Interannual Variation of Soil Respiration in a Beech Forest Ecosystem over a Six-Year Study. *Ann. For. Sci.* **2004**, *61*, 499–505. [CrossRef]
0. Obia, A.; Cornelissen, G.; Mulder, J.; Dörsch, P. Effect of Soil PH Increase by Biochar on NO, N_2O and N_2 Production during Denitrification in Acid Soils. *PLoS ONE* **2015**, *10*, e0138781. [CrossRef]

forests

MDPI

Article

Effects of Nitrogen and Phosphorus Additions on Soil N_2O Emissions and CH_4 Uptake in a Phosphorus-Limited Subtropical Chinese Fir Plantation

Binjie Li [1,2], Guangsheng Chen [1,2,*], Xiaochen Lu [1,2] and Hongbo Jiao [3]

1 State Key Laboratory of Subtropical Silviculture, Zhejiang A&F University, Hangzhou 311300, China; libinjie19971128@163.com (B.L.); 13298290218@163.com (X.L.)
2 College of Environmental and Resource Sciences, Zhejiang A&F University, Hangzhou 311300, China
3 The Forestry Industry Development Administration, Xinyu 338000, China; xycyhgk@163.com
* Correspondence: auburncgs@126.com

Abstract: Increased nitrogen (N) inputs in subtropical forest ecosystems were widely reported. Extr N additions were reported to cause nutrient imbalance and phosphorus (P) limitation in many tropica and subtropical forests, and further result in changes in soil nitrous oxide (N_2O) and methane (CH fluxes. Here, we conducted experiments with N (high N addition: 15 g N/m^2, HN), P (low: 5 g P/m LP; high: 15 g P/m^2, HP) and their interactive (HNLP and HNHP) treatments to investigate how N and P additions affected CH_4 and N_2O exchanges in an N-rich Chinese fir plantation (*Cunninghami lanceolata*), and further explored the underlying mechanisms through the structural equation mode (SEM) analysis. The results indicated that N addition alone (HN) significantly ($p < 0.05$) increase the soil N_2O emissions by 30.15% and 80.47% over annual and 4-month periods, mainly owing t the elevated NH_4^+-N content. P addition alone (LP and HP) did not significantly affect the so N_2O emissions as compared with the control. The SEM analysis indicated that increased N_2O emissions under N addition were primarily explained by the increase in available N and contribute more to the stimulated NH_4^+-N contents. N and P interactive additions slightly (not significan stimulated the N_2O emissions as compared with that under the N addition alone treatment. High dose P addition significantly increased the soil CH_4 uptake by 15.80% and 16.23% under the HP an HNHP treatments, respectively, while N addition alone and low P addition (LP and HNLP) did no significantly affect CH_4 uptake as compared with the control. The increased water-soluble organi carbon and microbial biomass carbon explained the increased CH_4 uptake under high P additio The fertilization effects on N_2O emissions and CH_4 uptake mainly occurred in the first 4 months an diminished after that. Our results suggested that the direction, magnitude and timing of the N and addition effects on N_2O emissions and CH_4 uptake would depend on the soil nutrient status an plant–microbial competition for N and P in subtropical forests.

Keywords: nitrogen addition; phosphorus addition; N_2O emissions; CH_4 uptake; Chinese fir fore

check for updates

Citation: Li, B.; Chen, G.; Lu, X.; Jiao, H. Effects of Nitrogen and Phosphorus Additions on Soil N_2O Emissions and CH_4 Uptake in a Phosphorus-Limited Subtropical Chinese Fir Plantation. *Forests* **2022**, *13*, 772. https://doi.org/10.3390/f13050772

Academic Editor: Xingkai Xu

Received: 12 April 2022
Accepted: 15 May 2022
Published: 17 May 2022

Publisher's Note: MDPI stays neutral with regard to jurisdictional claims in published maps and institutional affiliations.

1. Introduction

Global warming has led to frequent extreme climates and natural disasters world wide [1]. Carbon dioxide (CO_2), methane (CH_4) and nitrous oxide (N_2O) are the thre most important greenhouse gases that contribute to global warming [2–5]. The methan concentration in the atmosphere increases at 5.7 ppb/year [6], while the N_2O concentratio increases at a rate of 0.8 ppb/year [7]. At present, methane contributes about 20% to globa warming [8], while N_2O contributes more than 5% to global warming [1]. It is important t identify and quantify the source, sink and factors controlling the levels of atmospheric CH and N_2O.

Most of the atmospheric N_2O is produced and consumed by soil processes [9,10]. So N availability is the main driver of N_2O production. Many previous studies indicated tha terrestrial soil N_2O emissions have been increasing since the industrial era and is primarily due to the elevated N deposition and fertilization [3,11–13]. The potential mechanisms regulating the responses of soil N_2O production and consumption to elevated N inputs were well summarized [10,14,15]. Phosphorus (P) availability is another important factor controlling N_2O emissions in forest ecosystems [12,16]. P is generally found to be the main limiting nutrient in the tropical and subtropical forest ecosystems due to the low P supply from highly weathered soil and relatively high N input, which stimulates more P uptake by plants and microbes [17]. Microbial processes, including nitrification or denitrification, were also suggested to be limited by P availability [16,18]. Many previous studies indicated that P addition can effectively reduce soil N_2O emissions stimulated by N addition [12,16,19–21]. These studies indicated that P addition may result in increases in plant N uptake or microbial N immobilization, and thus, reduce the soil N availability for N_2O production. However, there are a few reports showing that P addition only slightly reduces or even enhances N_2O emissions under an N addition condition. These studies argued that P addition can stimulate soil N cycling and alleviate P limitation on nitrifying and denitrifying bacteria, and thus, enhance the N_2O emissions [18,19,22–26]. Shen and Zhu [15] conducted a meta-analysis of the effects of both N and P additions on N_2O emissions and concluded that P addition significantly reduced soil N_2O emissions under elevated N input; however, their meta-analysis only collected 30 studies and a few of them indicated P addition increased N_2O emissions under N addition. More studies are needed to uncover the underlying mechanisms controlling N_2O emissions under both N and P additions.

At present, the location and strength of soil CH_4 sources and sinks are poorly constrained [27]. In well-drained soils, CH_4 oxidation normally dominates [28]. Soil environmental factors, the soil microbe community, and carbon and nutrient availability generally control soil CH_4 oxidation [29–31]. The ammonium-oxidizing and methanotrophic bacteria are responsible for CH_4 oxidation [27]. Nitrogen input was been found to inhibit soil CH_4 uptake in forests [32–37]. High concentrations of inorganic N, especially ammonium (NH_4^+), were reported to inhibit the methanotrophic activity and high contents of nitrate (NO_3^-) can inhibit methanogenesis [28,34,38,39]. In a meta-analysis, Liu and and Greaver [10] concluded that N addition reduced CH_4 uptake by 38% based on a meta-analysis of N addition experiments due to changes in the activity of methanotrophs and methanogens. However, N availability can also limit the metabolic activities of methanotrophic bacteria [40,41]; therefore, some studies also showed that N addition can stimulate CH_4 uptakes in N-limited forest soils [42,43].

In addition to N availability, P is a critical nutrient for methanotrophic bacteria and methanogenic archaea. Much less information is available on the effects of P availability on soil CH_4 fluxes, especially for forests [27,44]. Both stimulation and inhibition were found for P addition effects on forest CH_4 uptake [35,45–47]. For example, in a heavily fertilized subtropical forest, P addition alleviated the effects of added N on CH_4 uptake and stimulated root growth and soil methanotrophy activity [35,37,46], and thus, increased the CH_4 uptake. Furthermore, inhibitive effects on CH_4 uptake may be also due to the decreases in soil P availability caused by increases in soil acidity following N inputs [45,47].

Forest soil releases about 2.7–5.4 Tg N/year as N_2O, accounting for about 38% of terrestrial emissions [48], among which, about 14% is from subtropical forests [49]. In addition, upland subtropical forests are important CH_4 sinks [11]. Recent studies indicated that N_2O emissions and CH_4 fluxes from subtropical forests may be underestimated due to insufficient field measurements [50–52]. Direct evidence of soil N and P availability on soil N_2O and CH_4 emissions in the subtropical forests is even less [16,19]. Subtropical China has received very high atmospheric N deposition [53], and a recent meta-analysis study [15] indicated that this region is mostly N-rich and P-limited. It is necessary to understand how N and P additions affect N_2O and CH_4 fluxes under this continuous high-dose N input. Several studies were conducted for addressing the effects of N, P and their combined additions on forest soil N_2O and CH_4 fluxes in subtropical China [12,19,26,35,47,50,54–58].

Contrasting results were found for the effects, suggesting that our understanding of the mechanisms that lead to elevated N and P inputs affecting the direction, magnitude and timing of N_2O and CH_4 fluxes in N-rich and P-limited subtropical forests are not clearly understood. More experimental studies are needed to unravel these mechanisms.

In this study, we conducted a manipulative experiment to investigate N and P addition and their interactions on soil CH_4 and N_2O emissions in a subtropical plantation forest in southern China. Based on the results of previous studies, we hypothesized that (1) N addition inhibits CH_4 uptake due to increased soil N availability, (2) P addition reduces soil N_2O emissions due to stimulated activities of soil microbes from the relief of P limitation under ambient N condition, (3) P addition stimulates CH_4 uptake due to stimulated microbial activities and (4) P addition alleviates N_2O emissions under N addition.

2. Materials and Methods

2.1. Site Description

The study site was located at Wulitou forest station (119.67° E and 30.21° N), Hangzhou City, Zhejiang Province, China (Figure 1). The study site has a subtropical monsoon climate with an average annual temperature of 16.4 °C and rainfall of 1614 mm. The mean altitude is 175 m and the annual sunshine hours are 1847.3 h [59]. The study plots were selected from a planted pure Chinese fir (*Cunninghamia lanceolata*) forest (dominating at 90%), with a few Moso bamboo (*Phyllostachys Edulis)* trees scattered around. The aspect of the sample plot is southwest, and the slope is less than 15°. The mean stand density is 1311 stems·ha^{-1}, the average DBH (diameter at breast height) and tree height are 10.9 cm and 7.7 m, respectively. The content of soil organic carbon (SOC) in the 20 cm layer is 15.40 g·kg^{-1}, soil total P is 0.36 g·kg^{-1}, soil total N (TN) is 0.86 g·kg^{-1}, pH is 4.45 and soil bulk density is 1.32 g/cm^3. The leaf N:P ratio is 23.89. According to the N:P ratio [60], this forest can be determined as N-rich and P-limited since the N:P ratio is greater than 16 [17]. The soil NH_4^+-N and NO_3^--N concentrations are 12.62 and 3.71 mg.kg^{-1}, respectively. The soil belongs to the Ferralsols (Oxisols) soil type and the texture is composed of coarse sand (>250 μm; 67%), fine sand (53 μm~250 μm; 19%) and silty clay (<53 μm; 14%). The annual atmospheric N deposition was 3.16 g N/m^2/year in 2016 at the study site [53].

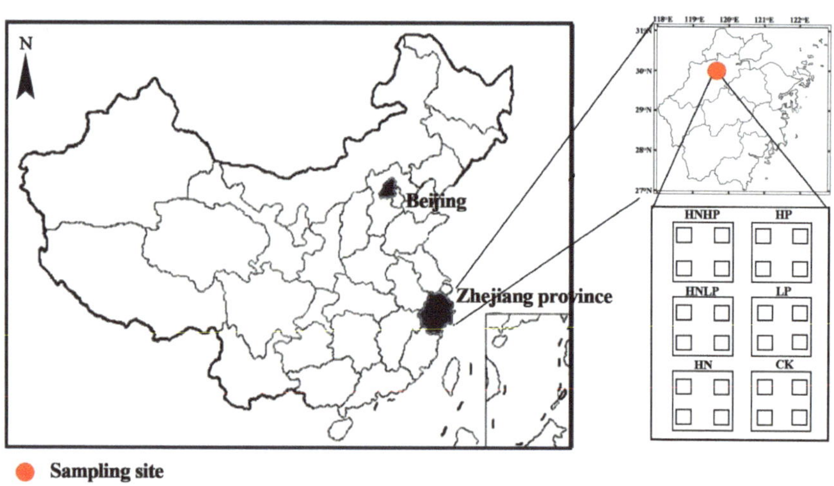

Figure 1. Location of the study area and experimental design (six treatments with four replicates).

2.2. Experimental Design

Six plots with a 20 m × 20 m area were established on 15 February 2019 in the selected 10-year-old Chinese fir forest (Figure 1). All plots were far away from the forest edge and were separated by a 5 m buffer zone to avoid the impacts of adjacent treatments. The static chamber method was used to collect CH_4 and N_2O gases. The static chamber is divided into two parts, a top chamber (30 cm for length, width and height) and a base (20 cm height; 10 cm width), both of which were made of PVC hard plastic board. Four static chambers were established at the four corners of each plot, which served as four replicates in each treatment. The static chamber base was deployed on 15 February 2019 and inserted 10 cm into the ground. The chamber design and techniques were described in many previous studies [12,35].

Six treatments were applied:

(1) Control (CK);
(2) Add 15 g N/m^2 (HN);
(3) Add 15 g N/m^2 and 5 g P/m^2 (HNLP);
(4) Add 5 g P/m^2 (LP);
(5) Add 15 g N/m^2 and 15 g P/m^2 (HNHP);
(6) Add 15 g P/m^2 (HP).

The fertilizer types were urea ($CO(NH_2)_2$) for N and sodium hydrogen phosphate (NaH_2PO_4) for P. We chose analytical grade P chemicals to prevent the addition of micronutrients, such as molybdenum, that may influence other processes not attributable to P. According to our preliminary test before the experimental design, the ambient soil available N, atmospheric N deposition and leaf N:P ratios were very high. According to the C:N:P stoichiometric balance, the soil in the study site could be determined as N-rich and P-limited, which was also indicated by Du et al. [17]. Therefore, only a high N addition level (15 g N/m^2) was applied. The low and high P treatment levels were determined according to the previous studies in subtropical China [12,26,55]. Each plot was divided into four 5 m $\times$ 5 m grids, and fertilizer was evenly applied in each grid. During the fertilizer application, each chamber base area received the respective amount of fertilizer. The fertilizer was applied on 20 March 2019. No rainfall occurred within 5 days after fertilization.

2.3. Soil Sampling and Chemical Analyses

Soil samples were taken at three-month intervals from 19 March 2019 (1 day before treatments) to 20 May 2020. Three soil cores at a 0–20 cm soil depth were collected around each chamber using a soil corer (2.5 cm inner diameters), and mixed as one soil sample; therefore, four soil samples were taken for each plot. Soil samples were sieved using a 2 mm sifter to remove litter, rocks and roots. A portion of the soil samples was stored at 4 °C, and the rest was air-dried for chemical analysis.

Soil pH was measured using solutions with a soil–water ratio of 1:2.5 using a pH meter (Mettler Toledo, Switzerland). SOC was measured using dichromate oxidation and titration with ferrous ammonium sulfate [61]. Soil microbial biomass C (MBC) and N (MBN) were measured using the chloroform fumigation–K_2SO_4 extraction method [62]. Water-soluble soil organic carbon (WSOC) was extracted with 0.5 M K_2SO_4 and analyzed using a total carbon analyzer (Shimadzu model TOC-500, Kyoto, Japan) [63]. The total N concentration was measured using semimicro-Kjeldahl digestion, and the total P concentration was measured spectrophotometrically after $HClO_4$-H_2SO_4 digestion [61]. The KCl extraction–indophenol blue colorimetric method was used to extract and determine the soil NH_4^+-N, and the KCl extraction–ultraviolet spectrophotometric method was used to extract and determine the NO_3^--N concentration.

2.4. Leaf and Litter Sampling and Chemical Analysis

The fresh leaves and floor litter were sampled twice (before and after treatments) during the study period. The leaf and litter samples were dried for 48 h at 80 °C to a constant mass and ground into powder. The leaf and litter C concentrations were determined using the potassium dichromate–sulfuric acid digestion method. The leaf and litter N concentrations were extracted using $HClO_4$-H_2SO_4 digestion and determined by the Kjeldah method. The leaf and litter P concentrations were determined by $HClO_4$-H_2SO_4 method.

2.5. Greenhouse Gases (GHGs) Data Collection and Measurements

The first GHGs sampling was conducted on 20 March 2019 (before treatments), an then gas sampling was conducted monthly at the end of each month. Due to the influence of COVID-19, the gas collections were stopped in January, February and March 202 During each sampling, the removable chamber top was covered on the fixed base to form closed environment. A running fan was placed at the bottom of each static chamber to sti the air to evenly mix the gas. Gas samples were collected between 9:00 and 11:00 am. Ga samples were collected four times at 5, 15, 25 and 35 min after chamber deployment usin a 100 mL plastic syringe; 20 mL gas samples were injected into sealed pre-evacuated glas vials (12 mL). The CH_4 and N_2O were simultaneously measured using a gas chromatograp (model 7890A, Agilent, Wilmington, NC, USA) equipped with a flame ionization detecto and an electron capture detector.

CH_4 and N_2O fluxes were estimated using their respective linear changes in conce trations over time. The gas flux calculation formula is [64]:

$$F = \rho \cdot \frac{V}{A} \cdot \frac{P}{Po} \cdot \frac{To}{T} \cdot \frac{dCt}{dt} \quad (1$$

where F is the net gas exchange flux (mg/(m^2·h)); ρ is the gas density under standar conditions (kg/m^3); V is the effective volume of the sampling chamber (m^3); A is the so area covered by the sampling chamber (m^2); T and To are the chamber temperature an standard state temperature (°C) during the observation, respectively; and P and Po are th atmospheric pressure and standard atmospheric pressure (kPa), respectively, during th observation. $\frac{dCt}{dt}$ is the slope of the gas mass concentration in the chamber with respect t time [65,66].

2.6. Measurement of Microclimatic Factors

During each sampling of GHGs, the soil temperature and moisture, as well as a mospheric temperature and humidity, at each chamber were measured simultaneousl A thermometer and hygrometer was used to measure the atmospheric temperature an humidity next to each static chamber. Soil moisture (SM) and temperature (ST) at a dept of 0–10 cm were monitored with a TRIME pico32 monitor (Ettlingen, Germany).

2.7. Data Analysis Method

We used Microsoft Excel for the preliminary data processing and R 4.0.2 (R Core Tear 2016, Auckland, New Zealand) for data processing and statistical analysis. Before th statistical analysis, all data were tested for normality (Kolmogorov–Smirnov's test) an homoscedasticity (Levene's test). ANOVA analysis [67] and Tukey's HSD test were used t test the treatment effects on soil properties and CH_4 and N_2O fluxes.

The structural equation model (SEM) was used to analyze different hypothetica approaches to explain the effects of treatments on CH_4 and N_2O fluxes. In the SEM analysi the maximum likelihood estimation method was used to fit the model. The adequacy c the model was determined using the χ^2 test and root-square-mean errors of approximatic (RMSEA). An insignificant χ^2 test ($p < 0.05$) and lower RMSEA ($p > 0.05$) showed that th model fits were appropriate. The basic steps of the SEM model analysis were as follow First, we established the model and preliminarily checked the fitting structure of the mod and the significance of the regression coefficient. Second, we adjusted the model repeated until the fitting standard was within the standard range (the smaller the RMSEA, the bette Finally, the results of the model were analyzed and explained. IBM SPSS Amos 23.0 (SPS Inc., New York, NY, USA) was used for all statistical analyses and SEM analysis. A p-valu < 0.05 was used to indicate statistically significant differences in this study.

3. Results

3.1. Changes in Soil Physical and Chemical Properties

The soil moisture showed large seasonal variability and the variation patterns were similar between different treatments, with generally lower moisture in the HNLP plots (Figure 2). The soil temperature showed the same monthly variations under the CK and treatments. Overall, the treatments had no significant ($p > 0.05$) impacts on soil moisture and temperature. Compared with the control, soil NH_4^+-N concentration significantly increased under the HN and HNLP treatments, while no significant changes occurred under the LP, HNHP and HP treatments (Table 1). The soil NO_3^--N significantly increased under the HNHP treatment, while there was no significant difference between other treatments. Compared with the control, the soil microbial biomass carbon (MBC) concentrations under all treatments except for the HN and HNLP significantly increased. The microbial biomass nitrogen (MBN) was significantly lower under all treatments except for the HN treatment, indicating that the P addition significantly stimulated N uptake by plant roots and reduced the N availability for microbial immobilization. After one year of fertilization treatments, leaf N concentrations did not significantly change for all treatments, with higher concentrations under HNLP and HNHP treatments. The leaf P concentrations significantly increased under the P addition treatments (LP, HP, HNLP and HNHP), while they slightly ($p > 0.05$) increased under the HN treatment. The HP and HNHP treatments had the largest increase in leaf P concentration, indicating a P limitation on plant growth at the study site.

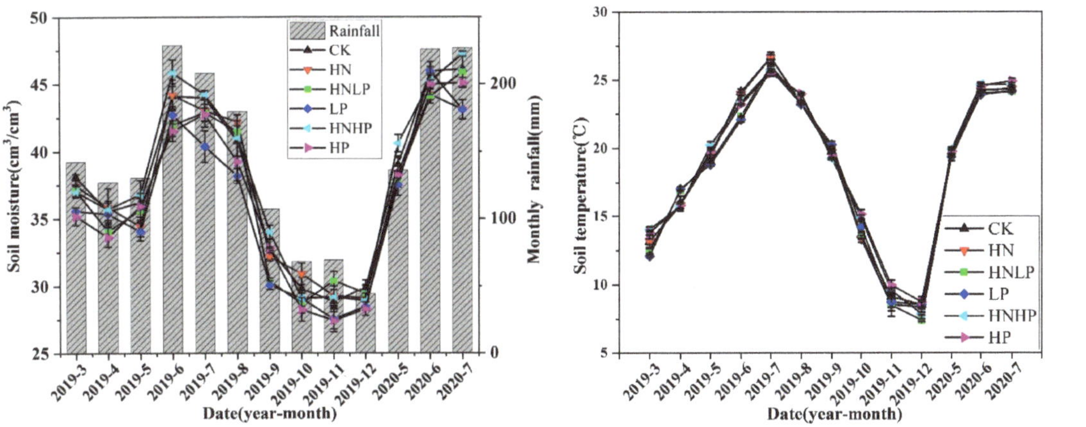

Figure 2. Monthly soil moisture (cm^3/cm^3) and rainfall (mm; **left**) and soil temperature (°C; **right**) under different treatments. Error bars denote the standard deviations (n = 4).

Table 1. The carbon and nutrient concentrations in the Chinese fir plantation under different treatments. Note: the numbers in the parenthesis are the standard deviations (n = 5). Different letters between treatments denote significant ($p < 0.05$) differences.

Variables	CK	LP	HP	HN	HNLP	HNHP
NH_4^+ (mg/kg)	12.62 (0.76) b	11.61 (0.86) b	12.80 (1.17) b	14.22 (1.97) a	14.87 (2.08) a	12.88 (1.36) b
NO_3^- (mg/kg)	3.71 (0.076) b	4.78 (0.42) ab	3.86 (0.18) b	4.46 (0.36) b	4.80 (0.33) ab	5.89 (0.30) a
MBN (mg/kg)	54.21 (9.47) a	22.23 (1.35) b	24.08 (1.17) b	54.53 (9.26) a	32.39 (4.36) b	28.55 (4.18) b
MBC (mg/kg)	137.55 (9.13) b	173.94 (18.94) a	206.76 (15.56) a	158.65 (11.44) ab	159.07 (7.74) ab	198.59 (12.70) a
WSOC (mg/kg)	46.93 (2.88) c	80.20 (3.52) ab	96.48 (5.85) a	62.69 (5.16) c	63.37 (2.02) bc	91.50 (3.85) a
Leaf N (g/kg)	7.55 (0.11) a	8.00 (0.27) a	7.05 (0.30) a	7.96 (0.81) a	8.75 (0.99) a	9.22 (0.86) a
Soil TN (g/kg)	0.86 (0.036) a	0.90 (0.040) a	0.91 (0.043) a	1.04 (0.095) a	1.00 (0.097) a	1.00 (0.082) a
SOC (%)	1.55 (0.031) ab	1.67 (0.10) ab	1.86 (0.12) a	1.47 (0.06) b	1.81 (0.11) ab	1.57 (0.048) ab
Soil TP (g/kg)	0.13 (0.014) b	0.14 (0.010) b	0.23 (0.024) a	0.13 (0.011) b	0.13 (0.008) b	0.15 (0.025) ab

3.2. *Monthly Dynamics of Soil CH_4 and N_2O Fluxes*

The soil N_2O emissions showed large variations between the months (Figure 3a). Compared with the control, soil N_2O emissions significantly increased within the first 4 months (from April to July 2019) under the HN, HNLP and HNHP treatments (Table 2), and the effects stabilized after that. The N_2O emissions and their monthly patterns were similar under the HN, HNLP and HNHP treatments, while they were similar under the CK, LP and HP treatments. The P addition alone (i.e., LP and HP) did not significantly affect the monthly N_2O dynamics as compared with the control.

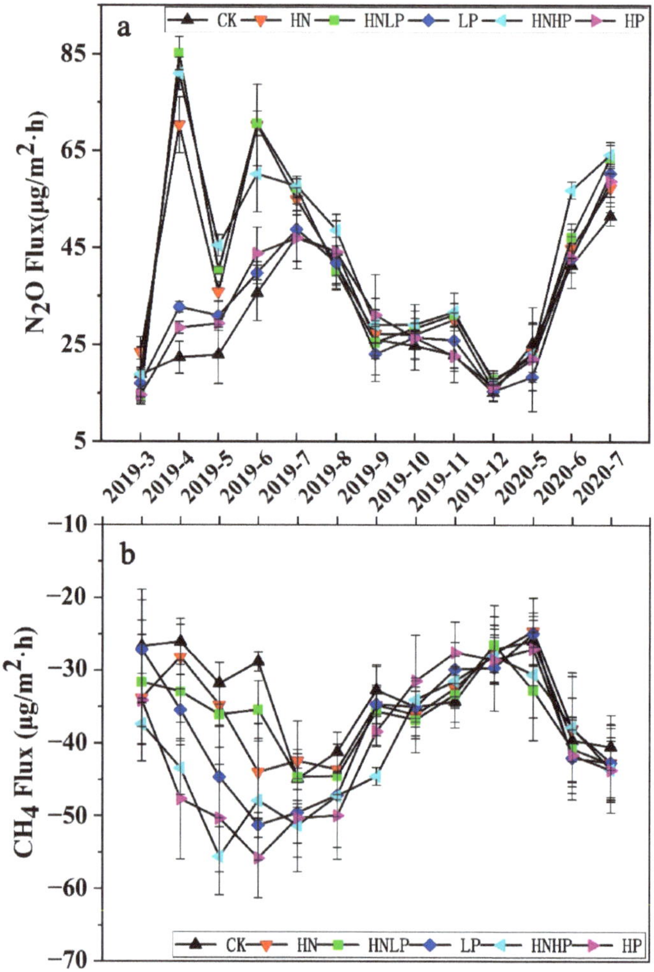

Figure 3. Monthly dynamics of soil N_2O ((**a**); $\mu g \cdot N_2O/m^2 \cdot h$) and CH_4 ((**b**); $\mu g \cdot CH_4/m^2 \cdot h$) fluxes under different treatments. Error bars represent standard deviations of mean CH_4 and N_2O fluxes (n = 4).

Table 2. Average N_2O emissions and CH_4 uptakes in a subtropical Chinese fir plantation over the first 4 months and the entire study period. Different letters denote significant ($p < 0.05$) differences between treatments.

Treatments	N_2O (μg/m²·h)	CH_4 (μg/m²·h)	N_2O (μg/m²·h)	CH_4 (μg/m²·h)
	4 Months		All Months	
CK	32.06 (11.69) b	−34.92 (7.74) a	30.63 (12.42) a	−35.06 (7.22) a
HN	57.86 (15.13) a	−35.37 (6.47) a	40.17 (17.98) b	−35.58 (7.12) a
HNLP	63.30 (16.73) a	−39.74 (7.03) ab	41.83 (21.23) b	−37.23 (8.07) ab
LP	38.08 (8.02) b	−43.42 (8.51) bc	32.57 (13.93) a	−38.28 (9.04) ab
HNHP	61.09 (13.52) a	−49.94 (6.52) c	43.346 (19.55) b	−40.75 (9.76) b
HP	37.15 (9.09) b	−51.03 (7.37) c	32.88 (13.39) a	−40.60 (10.83) b

The monthly soil CH_4 uptakes displayed even larger monthly variations than the N_2O emissions (Figure 3b). The monthly variation patterns under different treatments were similar. Compared with the control, the soil CH_4 uptakes increased within the first 4 months under all treatments and the effects were stabilized after that. Our results indicated that the N and P addition effects on the N_2O and CH_4 fluxes only lasted for a short period (about 4 months) at the study site.

3.3. Overall Effects of N and P Addition Treatments

Compared with the control, the annual mean N_2O emissions significantly ($p < 0.05$) increased by 31.15%, 36.57% and 41.51% under the HN, HNLP and HNHP treatments, respectively, while there were no significant changes under the LP and HP treatments (Table 2). This indicated that P addition alone did not significantly alter the soil N_2O emissions in this N-rich Chinese fir plantation (Table 3). N_2O emissions were higher under the HNLP and HNHP treatments; however, the differences were not significant. The effects of N and P additions on N_2O emissions were more obvious during the first 4 months after the treatments. Two-way ANOVA indicated that only N addition promoted N_2O emissions, and there was no significant interactive effect between the N and P additions (Table 3).

Table 3. Two-way ANOVA for mean N_2O emissions and CH_4 uptakes over the first 4 months and all months under the N and P treatments in the subtropical Chinese fir plantation. F: mean square ratio, P: p-value.

Treatments		N Effect		P Effect		N × P Interaction	
		F	P	F	P	F	P
N_2O	4 months	86.44	<0.05	1.62	0.20	0.04	0.96
	All months	25.92	<0.05	0.69	0.50	0.04	0.96
CH_4	4 months	0.87	0.35	33.30	<0.05	0.61	0.54
	All months	0.02	0.90	9.52	<0.05	0.22	0.80

The impacts of N and P additions on CH_4 uptake were different from the N_2O emissions. Compared with the control, the annual mean CH_4 uptake significantly ($p < 0.05$) increased by 16.23% and 15.80% under the HNHP and HP treatments, respectively, while there were no significant changes under the HN, LP and HNLP treatments. This indicated that the N addition alone and low-dose P addition did not significantly affect the CH_4 uptake in the Chinese fir forest. There was no significant difference between HNHP and HP treatments, further implying that the increased CH_4 uptake was mainly due to P addition, and the ANOVA analysis also demonstrated that the effects of N addition and N-P interaction on CH_4 absorption were not statistically significant (Table 3).

3.4. Pathway Determining N_2O Emissions and CH_4 Uptakes under N and P Additions

The SEM analysis was conducted to identify the causal effects of different contributin factors on CH_4 and N_2O fluxes after the N and P additions (Figure 4). The SEM resul indicated that about 64% of the soil N_2O variations under the N and P treatments wer explained by the soil nutrient and environmental factors, namely, the ST (soil temperature SM (soil moisture), NO_3^--N, NH_4^+-N, MBC, MBN and WSOC, which indicated that othe factors not included in this study contributed about 36% of the variations. The N additio directly increased NO_3^--N and indirectly increased NH_4^+-N, which further significantl enhanced N_2O emissions. The P addition significantly reduced MBN but increased th MBC, and both effects counteracted each other and resulted in non-significant impact on the N_2O emissions. The N addition only slightly interacted with the P addition t affect the NO_3^--N concentrations, resulting in non-significant interactive effects on th N_2O emissions.

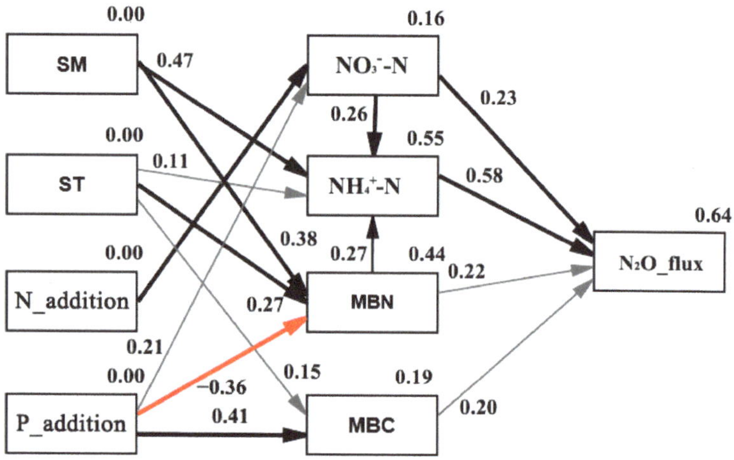

Figure 4. The structural equation model explaining the direct and indirect effects of N and additions and soil properties on soil N_2O fluxes across all treatments. SEM model parameter $\chi^2 = 18.11$, $df = 18$, $RMSEA = 0.008$ and $GFI = 0.90$. Black and wide arrows indicate significar positive relationships and red arrows indicate significant negative relationships ($p < 0.05$). The thi and light-colored arrows indicated insignificant effects. Numbers on arrows are standardized pat coefficients. The width of an arrow indicates the relative strength of the causal influence.

The SEM results indicated that about 70% of the soil CH_4 variations under the N and treatments were explained by the considered factors (Figure 5). The N addition significantl and positively affected NO_3^--N and further indirectly affected the NH_4^+-N concentratior but had non-significant impacts on the CH_4 uptake. It was interesting that the P additic had no significant direct impact on the soil NH_4^+-N. The P addition significantly increase the WSOC and further increased the MBC, resulting in significant increases in the CH uptake; therefore, the P addition indirectly increased the CH_4 uptake. The P addition als significantly reduced the MBN and further affected the NH_4^+-N concentrations, resultin in reduced CH_4 uptake. No significant N and P addition interactive effects on the CH uptake were found.

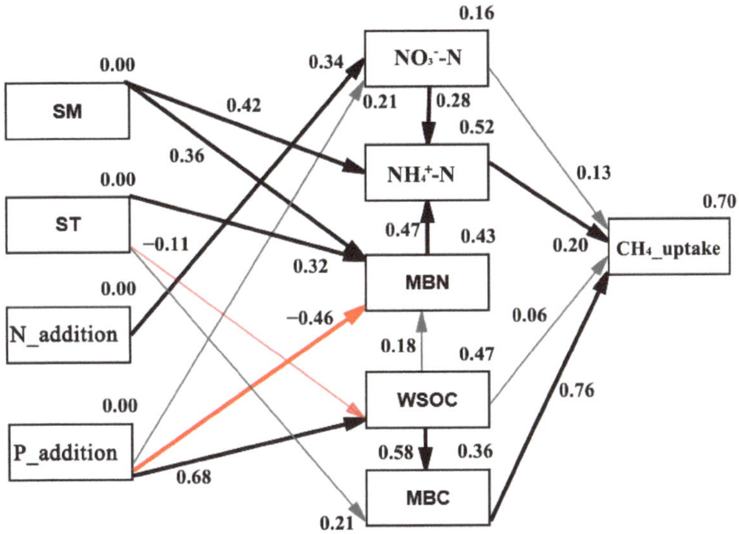

Figure 5. The structural equation model explaining the direct and indirect effects of N and P additions and soil properties on soil CH_4 uptakes across all treatments. The SEM model parameters: $\chi^2 = 15.27$, $df = 25$, $RMSEA = 0.000$ and $GFI = 0.93$. Black and wide arrows indicate significant positive relationships and red arrows indicate significant negative relationships ($p < 0.05$). The thin and light-colored arrows indicate insignificant effects. Numbers on arrows are standardized path coefficients. The width of an arrow indicates the relative strength of the causal influence.

4. Discussion

4.1. Effect of N Addition on N_2O and CH_4 Fluxes and the Controlling Mechanisms

We found that the N addition alone could significantly increase N_2O emissions, but the effects only lasted for about 4 months. The high N addition caused 31.15% and 80.47% increases in N_2O emissions over the annual and first 4-month periods. Under a similar N addition level, Li et al. [7] found that N addition increased N_2O emissions by 21.1%, which was significantly lower than our result. This may have been because our study site was N-rich and contained forests with high soil N availability, while their study site was N-limited. Under a similar N-rich soil condition and N addition level, Zhang et al. [56] and Wang et al. [19] found that N addition stimulated 58% and 44.5% increases in N_2O emission, respectively. In several meta-analyses, Liu and Greaver [10], Shen and Zhu [15] and Deng et al. [68] reported that N addition can globally cause about 216%, 101% and 91% increases in N_2O emissions, respectively. Our results were more close to the studies from Shen and Zhu [15] and Deng et al. [68]. N addition levels, fertilizer types and the length of the study period strongly affect the drawn conclusions [15]. Under the HN and HNLP treatments, N addition significantly increased the soil NH_4^+-N concentrations; while under HNHP, N addition significantly increased the soil NO_3^--N, suggesting that the increased N_2O emissions could result from either nitrification or denitrification processes. The SEM analysis further indicated that the increased NO_3^--N contributed less than the NH_4^+-N to the N_2O emissions. This was also shown in many previous studies [15,29,33,45]. We found that the N_2O emissions increased by 250% during the first month, and then the effects stabilized, indicating that the observation period significantly affected the responses of N_2O emissions to N addition. The N-rich condition and high N leaching due to acid soil and high rainfall may explain the short-term responses of increased N_2O emissions under N addition.

The first hypothesis stating that N addition inhibits CH_4 uptake was not supported. Our results indicated that N addition did not significantly influence the CH_4 uptake for both the first 4-month and annual periods. Many previous studies also showed that N addition had no significant effect on soil CH_4 uptake [57,69,70]. In a plantation forest in subtropical China, Zheng et al. [47] also found that N addition did not significantly affect CH_4 uptake. Studies found that increased NH_4^+ and NO_3^- under N addition can inhibit the bacterial oxidation activities, and thus, reduce the CH_4 uptake [10,33,36]. In our study, both the NH_4^+ and NO_3^- concentrations under N addition alone (Table 1) did not significantly differ from the control plot because the soil was N-rich. In addition, the WSOC and MBC were the precursors for CH_4 oxidation microbes [34,39,40], which our SEM analysis confirmed. Previous studies indicated that N addition can provide carbon compounds to the soil through the root-priming effect, which may change the WSOC content in the soil [39]. Li et al. [71] also found that N addition can significantly affect the WSOC in a bamboo forest. In addition, methanotrophs consume plant-derived carbon compounds through root exudation, resulting in an increased soil MBC [72]. Wang et al. [73] indicated that long-term N addition can significantly reduce the MBC and MBN. However, in our study, we found that N addition alone did not significantly affect both the WSOC and MBC (Table 1). In summary, our study did not find both inhibitive effects from the increased available N and stimulated effects from the increased WSOC and MBC under the N addition alone treatment.

4.2. Effect of P Addition on N_2O and CH_4 Fluxes and the Controlling Mechanisms

The second hypothesis stating that P addition alone could reduce N_2O emissions under ambient conditions was not supported in this study. Our study results were consistent with the study results of Zheng et al. [12] for three forests in tropical China. However, our study results were inconsistent with many previous studies. These studies either found that P addition increased [18,19,25,26,74] or decreased forest soil N_2O emissions [15,20,21,24,74,75]. The impact directions of P addition on soil N availability and MBN were the key explanations for the decreased or increased effects on N_2O emissions [27]. Our study found no significant changes in soil total inorganic N (NH_4^+ and NO_3^-) after the P addition alone treatment (Table 1). In addition, our SEM analysis also indicated that the MBN could slightly and negatively contribute to the N_2O emissions. We found that the P addition significantly reduced the MBN under P addition due to the stimulated growth of plant roots and reduced availability of N for microbial immobilization, which explained the slightly but not statistically higher N_2O emissions under the P addition alone.

The third hypothesis stating that P addition could enhance CH_4 uptake was partially supported in this study. We found that the high-dose P addition significantly stimulated CH_4 uptake, while the low-dose P addition only slightly (not significant) enhanced the CH_4 uptake over both the annual and first 4-month periods (Table 2). Our study was consistent with many previous studies [16,35,37,55,76], in which they argued that P addition stimulates root growth and soil methanogenic bacteria activity, and thus, increases CH_4 uptake. Our study found that the stimulated CH_4 uptake was mainly due to the stimulated MBC and WSOC concentrations under the high-dose P addition treatments (HP and HNHP; Table 1). The SEM analysis (Figure 5) also confirmed that P addition could directly increase the WSOC and indirectly increase the MBC. The increased MBC explained about 76% of the increased CH_4 uptake. In addition, we found that the reduced MBN due to the P addition could further result in reduced NH_4^+-N, lessening the inhibitive effects of NH_4^+-N on the CH_4 uptake. This was also reported in many previous studies [10,47]. The reduced MBN explained about 20% of the increased CH_4 uptake. In summary, the combination of increased MBC and decreased MBN could mainly explain the increases in CH_4 uptake under high-dose P addition treatments (HP and HNHP). The WSOC concentrations under the LP treatment and the MBC under the HNLP treatment were not significantly different from the control (Table 1), which explained the non-significant effects of the low-dose P treatments (LP and HNLP) on the CH_4 uptake.

4.3. Interactive Effects of N and P Addition on N_2O and CH_4 Fluxes

The fourth hypothesis stating that P addition could reduce N_2O emissions and increase CH_4 uptake under N addition was not supported in this study. In contrast, HNHP and HNLP slightly increased the N_2O emissions as compared with the HN treatment, although the difference was not significant. Under similar N and P addition levels, Zheng et al. [12] found that N and P addition significantly reduced N_2O emissions as compared with the N addition alone treatment in an old-growth forest, but no interaction effects were found in the younger mixed and planted forests. In the meta-analysis, Shen and Zhu [15] found that P addition can slightly (not significant) reduce N_2O emissions by 6% under N addition, while it significantly reduces N_2O emissions by 14% under ambient N conditions. The previous studies indicated that the effect direction and magnitude of N and P interactive additions on soil N_2O emissions would mainly be determined by the two contrary processes: (1) increased plant N uptake or microbial N immobilization, and thus, reduced soil N availability for N_2O production; and (2) stimulated soil N cycling and alleviated P limitation on nitrifying and denitrifying bacteria, and thus, enhanced N_2O emissions. Through the SEM analysis (Figure 4), our study found that N and P additions interactively affected the NH_4^+-N concentrations. N addition indirectly stimulated NH_4^+-N concentrations, while P addition indirectly reduced NH_4^+-N concentrations by decreasing the MBN concentration. The positive effects from the N addition counteracted the negative effects from the P addition, resulting in no significant interactive effects between P and N additions on N_2O emissions. This was also confirmed by the two-way ANOVA results (Table 3). However, due to a lack of enough long-term studies, our understanding of the universal mechanisms controlling the interactive effects between P and N additions is still unclear [12].

Our study also did not observe an interactive effect between the N and P additions on the CH_4 uptake. Our study was inconsistent with some previous studies. For example, Zhang et al. [35] suggested that nitrogen addition would inhibit CH_4 uptake, while P addition would promote CH_4 uptake. After combined nitrogen and phosphorus fertilization, P counteracted the inhibitory effect of N; therefore, their study concluded that N and P additions had no significant effect on CH_4 uptake. In another study, Gao et al. [26] indicated that CH_4 uptake in N and P addition plots was lower than that under N alone and P alone addition plots, implying an additive effect on inhibiting CH_4 uptake in a mixed forest in subtropical China. Martinson et al. [27] indicated that the N and P additions increased soil CH_4 uptake in tropical montane forests. Through the SEM analysis (Figure 5) and ANOVA (Table 3), we found that the N and P interactive effects on the CH_4 uptake were similar to those on the N_2O emissions. Namely, the N and P additions also interactively affected CH_4 uptake via their impacts on the NH_4^+-N concentrations. The negative effects from the N addition counteracted the positive effects from the P addition, resulting in no significant interactive effects between the P and N additions on the CH_4 uptake. The inhibiting effects of NH_4^+-N on CH_4 uptake was widely observed [26,35,36,41]. The change magnitude and direction in soil pH values and microbial community under N and P co-additions were suggested to be other main causes for the different responses of CH_4 uptakes [77–79]; unfortunately, our study did not include these two factors. Due to fewer previous studies, more and longer-term experimental data are needed to confirm our inferences on the controlling mechanisms of their interactive effects.

5. Conclusions

Under four hypotheses, our study conducted one and half years of manipulative experiments to investigate the effects of different levels of N and P additions on soil N_2O and CH_4 fluxes in an N-rich and P-limited subtropical Chinese fir plantation. Unlike many previous reports, we observed that the P addition slightly increased the N_2O emissions after the N addition. Although the stimulation effect was not significant due to the long-lasting effects of P addition, longer-term experiments are needed to further validate our conclusions. The ANOVA analysis indicated that the N and P additions did not have significant interactive effects on both N_2O and CH_4 fluxes. Based on the structural equation model, our study further revealed the controlling mechanisms from the perspectives of soi available N, WSOC and microbial activities. Our results were partially consistent with an partially contradicted previous reports. The previous studies in subtropical and tropica forests displayed very different directions and magnitudes for the N and P addition effect on both N_2O and CH_4 fluxes. Caution is needed when considering P fertilization as management choice to reduce N_2O emissions and increase CH_4 uptake in this region. Ou findings provide complementary evidence and mechanisms to the research of forest soi N_2O and CH_4 fluxes and have important implications for adaptive forest managemen strategies to mitigate global warming potential. In addition, different research sites wit varying soil properties and forest types should be taken into consideration in the future.

Author Contributions: Conceptualization, G.C.; methodology, B.L. and G.C.; formal analysis, B.L and X.L.; investigation, B.L., X.L. and H.J.; data curation, B.L., X.L. and H.J.; writing—original dra preparation, B.L. and G.C; writing—review and editing, G.C., X.L. and H.J.; visualization, B.L supervision, G.C.; project administration, G.C.; funding acquisition, G.C. All authors have read an agreed to the published version of the manuscript.

Funding: This research was funded by the Natural Science Foundation of Zhejiang Province (gran number LY20C030001), the Scientific Research Foundation of Zhejiang A&F University (grant numbe 2034020080) and the Overseas Expertise Introduction Project for Discipline Innovation (111 Projec grant number D18008).

Institutional Review Board Statement: Not applicable.

Informed Consent Statement: Not applicable.

Data Availability Statement: The data presented in this study are contained within the article.

Conflicts of Interest: The authors declare no conflict of interest.

References

1. IPCC. *Climate Change 2014: Mitigation of Climate Change. Contribution of Working Group III to the Fifth Assessment Report of th Intergovernmental Panel on Climate Change*; Cambridge University Press: Cambridge, UK; New York, NY, USA, 2014.
2. Jassal, R.S.; Black, T.A.; Trofymow, J.A.; Roy, R.; Nesic, Z. Soil CO_2 and N_2O flux dynamics in a nitrogen-fertilized Pacif Northwest Douglas-fir stand. *Geoderma* **2010**, *157*, 118–125. [CrossRef]
3. Tian, H.; Lu, C.; Ciais, P.; Michalak, A.M.; Canadell, J.G.; Saikawa, E.; Huntzinger, D.N.; Gurney, K.R.; Sitch, S.; Zhang, B. Th terrestrial biosphere as a net source of greenhouse gases to the atmosphere. *Nature* **2016**, *531*, 225–228. [CrossRef] [PubMed]
4. Tian, H.; Yang, J.; Xu, R.; Lu, C.; Canadell, J.G.; Davidson, E.A.; Jackson, R.B.; Arneth, A.; Chang, J.; Ciais, P. Global soil nitrou oxide emissions since the preindustrial era estimated by an ensemble of terrestrial biosphere models: Magnitude, attribution, an uncertainty. *Glob. Chang. Biol.* **2019**, *25*, 640–659. [CrossRef]
5. Tian, H.; Xu, R.; Canadell, J.G.; Thompson, R.L.; Winiwarter, W.; Suntharalingam, P.; Davidson, E.A.; Ciais, P.; Jackson, R. Janssens-Maenhout, G. A comprehensive quantification of global nitrous oxide sources and sinks. *Nature* **2020**, *586*, 248–25 [CrossRef] [PubMed]
6. Nisbet, E.G.; Dlugokencky, E.J.; Manning, M.R.; Lowry, D.; Fisher, R.E.; France, J.L.; Michel, S.E.; Miller, J.B.; White, J.W.C Vaughn, B. Rising atmospheric methane: 2007–2014 growth and isotopic shift. *Glob. Biogeochem. Cycles* **2016**, *30*, 1356–137 [CrossRef]
7. Li, R.D.; Zhang, K.; Su, D.; Lu, F.; Wan, W.X.; Wang, X.K.; Zheng, H. Effects of nitrogen application on soil greenhouse gas fluxe during the growing season of Eucalyptus plantations. *Acta Ecol. Sin.* **2015**, *35*, 5931–5939.
8. Zhang, P.D.; Li, X.R.; Yang, Y.L.; Zheng, Y.H.; Wang, L.S. Analysis of the benefits of greenhouse gas emission reduction in China large and medium-sized biogas projects. *Trans. Chin. Soc. Agric. Eng.* **2008**, *24*, 1–8.
9. Mosier, A.R. Soil processes and global change. *Biol. Fertil. Soils* **1998**, *27*, 221–229. [CrossRef]
10. Liu, L.; Greaver, T.L. A review of nitrogen enrichment effects on three biogenic GHGs: The CO_2 sink may be largely offset b stimulated N_2O and CH_4 emission. *Ecol. Lett.* **2009**, *12*, 1103–1117. [CrossRef]
11. Tian, H.; Chen, G.; Lu, C.; Xu, X.; Ren, W.; Zhang, B.; Banger, K.; Tao, B.; Pan, S.; Liu, M. Global methane and nitrous oxid emissions from terrestrial ecosystems due to multiple environmental changes. *Ecosyst. Health Sustain.* **2015**, *1*, 1–20. [CrossRef
12. Zheng, M.; Zhang, T.; Liu, L.; Zhu, W.; Zhang, W.; Mo, J. Effects of nitrogen and phosphorus additions on nitrous oxide emissio in a nitrogen-rich and two nitrogen-limited tropical forests. *Biogeosciences* **2016**, *13*, 3503–3517. [CrossRef]
13. Li, D.; Liu, Q.; Yin, H.; Luo, Y.; Hui, D.F. Differential responses and controls of soil CO_2 and N_2O fluxes to experimental warmin and nitrogen fertilization in a subalpine coniferous spruce (Picea asperata Mast.) plantation forest. *Forests* **2019**, *10*, 808. [CrossRe

4. Butterbach-Bahl, K.; Baggs, E.M.; Dannenmann, M.; Kiese, R.; Zechmeister-Boltenstern, S. Nitrous oxide emissions from soils: How well do we understand the processes and their controls? *Philos. Trans. R. Soc. B* **2013**, *368*, 20130122. [CrossRef] [PubMed]
5. Shen, Y.; Zhu, B. Effects of nitrogen and phosphorus enrichment on soil N_2O emission from natural ecosystems: A global meta-analysis. *Environ.Pollut.* **2022**, *301*, 118993. [CrossRef]
6. Mori, T.; Ohta, S.; Ishizuka, S.; Konda, R.; Wicaksono, A.; Heriyanto, J.; Hamotani, Y.; Gobara, Y.; Kawabata, C.; Kuwashima, K. Soil greenhouse gas fluxes and C stocks as affected by phosphorus addition in a newly established Acacia mangium plantation in Indonesia. *For. Ecol. Manag.* **2013**, *310*, 643–651. [CrossRef]
7. Du, E.Z.; Terrer, C.; Pellegrini, A.F.A.; Ahlstrom, A.; van Lissa, C.J.; Zhao, X.; Xia, N.; Wu, X.H.; Jackson, R.B. Global patterns of terrestrial nitrogen and phosphorus limitation. *Nat. Geosci.* **2020**, *13*, 221–226. [CrossRef]
8. Mori, T.; Ohta, S.; Ishizuka, S.; Konda, R.; Wicaksono, A.; Heriyanto, J.; Hardjono, A. Effects of phosphorus addition on N_2O and NO emissions from soils of an Acacia mangium plantation. *Soil Sci. Plant Nutr.* **2010**, *56*, 782–788. [CrossRef]
9. Wang, F.; Li, J.; Wang, X.; Zhang, W.; Zou, B.; Neher, D.A.; Li, Z. Nitrogen and phosphorus addition impact soil N_2O emission in a secondary tropical forest of south China. *Sci. Rep.* **2014**, *4*, 5615. [CrossRef]
0. Mori, T.; Ohta, S.; Ishizuka, S.; Konda, R.; Wicaksono, A.; Heriyanto, J. Phosphorus application reduces N_2O emissions from tropical leguminous plantation soil when phosphorus uptake is occurring. *Biol. Fertil. Soils* **2014**, *50*, 45–51. [CrossRef]
1. Zhang, W.; Zhu, X.; Luo, Y.; Rafique, R.; Chen, H.; Huang, J.; Mo, J. Responses of nitrous oxide emissions to nitrogen and phosphorus additions in two tropical plantations with N-fixing vs non-N-fixing tree species. *Biogeosciences* **2014**, *11*, 4941–4951. [CrossRef]
2. Minami, K.; Fukushi, S. Effecfs of phosphate and calcium carbonate application on emission of N_2O from soils under aerobic conditions. *Soil Sci. Plant Nutr.* **1983**, *29*, 517–524. [CrossRef]
3. Falkiner, R.A.; Khanna, P.K.; Raison, R.J. Effect of superphosphate addition on N mineralization in some Australian forest soils. *Soil Res.* **1993**, *31*, 285–296. [CrossRef]
4. Baral, B.R.; Kuyper, T.W.; Van-Groenigen, J.W. Liebig's law of the minimum applied to a greenhouse gas: Alleviation of P-limitation reduces soil N_2O emission. *Plant Soil* **2014**, *374*, 539–548. [CrossRef]
5. White, J.R.; Reddy, K.R. Influence of nitrate and phosphorus loading on denitrifying enzyme activity in everglades wetland soils. *Soil Sci. Soc. Am.* **1999**, *63*, 1945–1954. [CrossRef]
6. Gao, W.; Yang, H.; Li, S.; Kou, L. Responses of soil CO_2, CH_4 and N_2O fluxes to N, P, and acid additions in mixed forest in subtropical China. *J. Resour. Ecol.* **2017**, *8*, 154–164.
7. Martinson, G.O.; Müller, A.K.; Matson, A.L.; Corre, M.D.; Veldkamp, E. Nitrogen and phosphorus control soil methane uptake in tropical montane forests. *J. Geophys. Res. Biogeosci.* **2021**, *126*, e2020JG005970. [CrossRef]
8. Conrad, R. Microbial ecology of methanogens and methanotrophs. *Adv. Agron.* **2007**, *96*, 1–63.
9. Mosier, A.R.; Parton, W.J.; Phongpan, S. Long-term large N and immediate small N addition effects on trace gas fluxes in the Colorado shortgrass steppe. *Biol. Fertil. Soils* **1998**, *28*, 44–50.
0. Zhuang, Q.; Chen, M.; Xu, K.; Tang, J.; Saikawa, E.; Lu, Y.; Melillo, J.M.; Prinn, R.G.; McGuire, A.D. Response of global soil consumption of atmospheric methane to changes in atmospheric climate and nitrogen deposition. *Glob. Biogeochem. Cycles* **2013**, *27*, 650–663. [CrossRef]
1. Jang, I.; Lee, S.; Zoh, K.D.; Kang, H. Methane concentrations and methanotrophic community structure influence the response of soil methane oxidation to nitrogen content in a temperate forest. *Soil Biol. Biochem.* **2011**, *43*, 620–627. [CrossRef]
2. Maljanen, M.; Jokinen, H.; Saari, A.; Strömmer, R.; Martikainen, P.J. Methane and nitrous oxide fluxes, and carbon dioxide production in boreal forest soil fertilized with wood ash and nitrogen. *Soil Use Manag.* **2006**, *22*, 151–157. [CrossRef]
3. Hawthorne, I.; Johnson, M.S.; Jassal, R.S.; Black, T.A.; Grant, N.J.; Smukler, S.M. Application of biochar and nitrogen influences fluxes of CO_2, CH_4 and N_2O in a forest soil. *J. Environ. Manag.* **2017**, *192*, 203–214. [CrossRef] [PubMed]
4. Reay, D.S.; Nedwell, D.B. Methane oxidation in temperate soils: Effects of inorganic N. *Soil Biol. Biochem.* **2004**, *36*, 2059–2065.
5. Zhang, T.; Zhu, W.; Mo, J.; Liu, L.; Dong, S. Increased phosphorus availability mitigates the inhibition of nitrogen deposition on CH_4 uptake in an old-growth tropical forest, southern China. *Biogeosciences* **2011**, *8*, 2805–2813. [CrossRef]
6. Fender, A.C.; Pfeiffer, B.; Gansert, D.; Leuschner, C.; Daniel, R.; Jungkunst, H.F. The inhibiting effect of nitrate fertilisation on methane uptake of a temperate forest soil is influenced by labile carbon. *Biol. Fertil. Soils* **2012**, *48*, 621–631. [CrossRef]
7. Xu, X.; Inubushi, K. Effects of N sources and methane concentrations on methane uptake potential of a typical coniferous forest and its adjacent orchard soil. *Biol. Fertil. Soils* **2004**, *40*, 215–221. [CrossRef]
8. Wang, Z.-P.; Ineson, P. Methane oxidation in a temperate coniferous forest soil: Effects of inorganic N. *Soil Biol. Biochem.* **2003**, *35*, 427–433. [CrossRef]
9. Aronson, E.L.; Helliker, B.R. Methane flux in non-wetland soils in response to nitrogen addition: A meta-analysis. *Ecology* **2010**, *91*, 3242–3251. [CrossRef]
0. Bender, M.; Conrad, R. Effect of CH_4 concentrations and soil conditions on the induction of CH_4 oxidation activity. *Soil Biol. Biochem.* **1995**, *27*, 1517–1527. [CrossRef]
1. Bodelier, P.L.E.; Laanbroek, H.J. Nitrogen as a regulatory factor of methane oxidation in soils and sediments. *FEMS Microbiol. Ecol.* **2004**, *47*, 265–277. [CrossRef]
2. Steinkamp, R.; Butterbach-Bahl, K.; Papen, H. Methane oxidation by soils of an N limited and N fertilized spruce forest in the Black Forest, Germany. *Soil Biol. Biochem.* **2001**, *33*, 145–153. [CrossRef]

43. Börjesson, G.; Nohrstedt, H. Fast recovery of atmospheric methane consumption in a Swedish forest soil after single-sho N-fertilization. *For. Ecol. Manag.* **2000**, *134*, 83–88. [CrossRef]
44. Veraart, A.J.; Steenbergh, A.K.; Ho, A.; Kim, S.Y.; Bodelier, P.L.E. Beyond nitrogen: The importance of phosphorus for CH oxidation in soils and sediments. *Geoderma* **2015**, *259*, 337–346. [CrossRef]
45. Hall, S.J.; Matson, P.A. Nitrogen oxide emissions after nitrogen additions in tropical forests. *Nature* **1999**, *400*, 152–155. [CrossRef
46. Xu, M.; Chen, S.L.; Fang, H.J.; Yu, G.R.; Gao, W.L.; Wang, Y.S.; Dang, X.S.; Li, L.S. Low-level nitrogen addition promotes ne methane uptake in a boreal forest across the great Xing'an mountain region. *Sci. Silvae Sin.* **2014**, *60*, 973–981. [CrossRef]
47. Zheng, M.H.; Zhang, T.; Liu, L.; Zhang, W.; Lu, X.K.; Mo, J.M. Effects of nitrogen and phosphorus additions on soil methan uptake in disturbed forests. *J. Geophys. Res. Biogeosci.* **2016**, *121*, 3089–3100. [CrossRef]
48. IPCC. *Climate Change 2007: The Physical Science Basis. Contribution of Working Group I to the Fourth Assessment Report of th Intergovernmental Panel on Climate Change*; Cambridge University Press: Cambridge, UK, 2007.
49. IPCC. *Climate Change 2001: The Scientific Basis, Intergovernmental Panel on Climate Change, Working Group 3, Third Assessment Repor* Cambridge University Press: Cambridge, UK, 2001.
50. Xie, D.; Si, G.; Zhang, T.; Mulder, J.; Duan, L. Nitrogen deposition increases N_2O emission from an N-saturated subtropical fores in southwest China. *Environ. Pollut.* **2018**, *243*, 1818–1824. [CrossRef]
51. Zhu, J.; Mulder, J.; Wu, L.P.; Meng, X.X.; Wang, Y.H.; Dörsch, P. Spatial and temporal variability of N_2O emissions in a subtropica forest catchment in China. *Biogeosciences* **2013**, *10*, 1309–1321. [CrossRef]
52. Li, M.; Cui, Y.; Fu, Y.; Li, N.; Tang, X.; Liu, X.; Run, Y. Simulating the potential sequestration of three major greenhouse gases i China's natural ecosystems. *Forests* **2020**, *11*, 128. [CrossRef]
53. Yu, G.R.; Jia, Y.L.; He, N.P.; Zhu, J.X.; Chen, Z.; Wang, Q.F.; Piao, S.L.; Liu, X.J.; He, H.L.; Guo, X.B.; et al. Stabilization o atmospheric nitrogen deposition in China over the past decade. *Nat. Geosci.* **2019**, *12*, 424–429. [CrossRef]
54. Mo, J.; Zhang, W.; Zhu, W.; Gundersen, P.; Fang, Y.T.; Li, D.J.; Wang, H. Nitrogen addition reduces soil respiration in a matur tropical forest in southern China. *Glob. Chang. Biol.* **2008**, *14*, 403–412. [CrossRef]
55. Yu, L.; Wang, Y.; Zhang, X.; Dörsch, P.; Mulder, J. Phosphorus addition mitigates N_2O and CH_4 emissions in N-saturate subtropical forest, SW China. *Biogeosciences* **2017**, *14*, 3097–3109. [CrossRef]
56. Zhang, W.; Mo, J.; Yu, G.; Fang, Y.; Li, D.; Lu, X.; Wang, H. Emissions of nitrous oxide from three tropical forests in souther China in response to simulated nitrogen deposition. *Plant Soil* **2008**, *306*, 221–236. [CrossRef]
57. Zhang, W.; Mo, J.; Zhou, G.; Gundersen, P.; Fang, Y.; Lu, X.; Zhang, T.; Dong, S. Methane uptake responses to nitrogen depositio in three tropical forests in southern China. *J. Geophys. Res. Atmos.* **2008**, *113*, D11116. [CrossRef]
58. Zhang, K.; Zhu, Q.; Liu, J.; Wang, M.; Zhou, X.; Li, M.; Peng, C. Spatial and temporal variations of N_2O emissions from globa forest and grassland ecosystems. *Agric. For. Meteorol.* **2019**, *266*, 129–139. [CrossRef]
59. Zhang, Q.X.; Chen, J.; Wang, X.T.; Wang, N.; Bai, S.B. Short-term response of Chinese fir plantation soil greenhouse gas emission to Phyllostachys edulis invasion and logging. *J. Zhejiang A F Univ.* **2021**, *38*, 703–711.
60. McGroddy, M.E.; Daufresne, T.; Hedin, L.O. Scaling of C:N:P stoichiometry in forests worldwide: Implications of terrestria redfield-type ratios. *Ecology* **2004**, *85*, 2390–2401. [CrossRef]
61. Bao, S.D. *Soil Agrochemical Analysis*; China Agricultural Publishing House: Beijing, China, 2010.
62. Vance, E.D.; Brookes, P.C.; Jenkinson, D.S. An extraction method for measuring soil microbial biomass C. *Soil Biol. Biochem.* **198** *19*, 703–707. [CrossRef]
63. Vargas, R.; Baldocchi, D.D.; Bahn, M.; Hanson, P.J.; Hosman, K.P.; Kulmala, L.; Pumpanen, J.; Yang, B. On the multi-tempora correlation between photosynthesis and soil CO_2 efflux: Reconciling lags and observations. *New Phytol.* **2011**, *191*, 1006–101 [CrossRef]
64. Wang, Y.; Liu, G.; Wang, Y.; Sun, Y.; Xue, M. Simultaneous measurement of CO_2, CH_4 and N_2O emission from terrestria ecosystem with one improved gas chromatography. *Tech. Equip. Environ. Pollut. Control* **2003**, *4*, 84–90.
65. Fang, S.; Lin, D.; Tian, Y.; Hong, S. Thinning intensity affects soil-atmosphere fluxes of greenhouse gases and soil nitroge mineralization in a lowland poplar plantation. *Forests* **2016**, *7*, 141. [CrossRef]
66. Liu, L.; Liu, Y.; Wen, X.; Wang, Y. Soil CH_4 emission flux from artificial coniferous forest in red soil hilly area of Qianyanzho *Chin. J. Plant Ecol.* **2008**, *32*, 431–439.
67. Hansen, M.C.; Potapov, P.V.; Moore, R.; Hancher, M.; Turubanova, S.A.; Tyukavina, A.; Thau, D.; Stehman, S.V.; Goetz, S. Loveland, T.R.; et al. High-resolution global maps of 21st-century forest cover change. *Science* **2013**, *342*, 850–853. [CrossRe [PubMed]
68. Deng, L.; Huang, C.B.; Kim, D.G.; Shangguan, Z.P.; Wang, K.B.; Song, X.Z.; Peng, C.H. Soil GHG fluxes are altered by N depositio New data indicate lower N stimulation of the N_2O flux and greater stimulation of the calculated C pools. *Glob. Chang. Biol.* **202** *26*, 2613–2629. [CrossRef] [PubMed]
69. Ambus, P.; Robertson, G.P. The effect of increased N deposition on nitrous oxide, methane and carbon dioxide fluxes fro unmanaged forest and grassland communities in Michigan. *Biogeochemistry* **2006**, *79*, 315–337. [CrossRef]
70. Mori, A.; Hojito, M.; Kondo, H.; Matsunami, H.; Scholefield, D. Effects of plant species on CH_4 and N_2O fluxes from a volcani grassland soil in Nasu, Japan. *Soil Sci. Plant Nutr.* **2005**, *51*, 19–27. [CrossRef]
71. Li, Y.; Jiang, P.K.; Liu, J.; Wang, X.; Wu, J.; Ye, G.; Zhou, G. Effect of fertilization on water-soluble organic C, N, and emission c greenhouse gases in the soil of Phyllostachys edulis stands. *Sci. Silvae Sin.* **2010**, *12*, 165–170.

'2. Kuzyakov, Y.; Friedel, J.K.; Stahr, K. Review of mechanisms and quantification of priming effects. *Soil Biol. Biochem.* **2000**, *32*, 1485–1498. [CrossRef]
'3. Wang, C.; Lu, X.; Mori, T.; Mao, Q.; Zhou, K.; Zhou, G.; Nie, Y.; Mo, J. Responses of soil microbial community to continuous experimental nitrogen additions for 13 years in a nitrogen-rich tropical forest. *Soil Biol. Biochem.* **2018**, *121*, 103–112. [CrossRef]
4. Mori, T.; Yokoyama, D.; Kitayama, K. Contrasting effects of exogenous phosphorus application on N_2O emissions from two tropical forest soils with contrasting phosphorus availability. *SpringerPlus* **2016**, *5*, 1237. [CrossRef]
'5. Schwenke, G.D.; Herridge, D.F.; Scheer, C.; Rowlings, D.W.; Haigh, B.M.; McMullen, K.G. Greenhouse gas (N_2O and CH_4) fluxes under nitrogen-fertilised dryland wheat and barley on subtropical Vertosols: Risk, rainfall and alternatives. *Soil Res.* **2016**, *54*, 634–650. [CrossRef]
'6. Liu, L.; Gundersen, P.; Zhang, T.; Mo, J. Effects of phosphorus addition on soil microbial biomass and community composition in three forest types in tropical China. *Soil Biol. Biochem.* **2012**, *44*, 31–38. [CrossRef]
7. Wang, J.H.; Liu, J.S.; Yu, J.B.; Wang, J.D. Effects of nitrogen and phosphorus fertilizers on soil microbial biomass carbon and nitrogen in black soil corn farm ecosystem. *J. Soil Water Conserv.* **2004**, *18*, 35–38.
8. Yue, P.; Li, K.; Gong, Y.; Hu, Y.; Mohammat, A.; Christie, P.; Liu, X. A five-year study of the impact of nitrogen addition on methane uptake in alpine grassland. *Sci. Rep.* **2016**, *6*, 32064. [CrossRef] [PubMed]
9. Zhang, L.H.; Yuan, F.H.; Bai, J.H.; Duan, H.T.; Gu, X.Y.; Hou, L.Y.; Huang, Y.; Yang, M.G.; He, J.S.; Zhang, Z.H.; et al. Phosphorus alleviation of nitrogen-suppressed methane sink in global grasslands. *Ecol. Lett.* **2020**, *23*, 821–830. [CrossRef] [PubMed]

MDPI AG
Grosspeteranlage 5
4052 Basel
Switzerland
Tel.: +41 61 683 77 34

Forests Editorial Office
E-mail: forests@mdpi.com
www.mdpi.com/journal/forests

www.ingramcontent.com/pod-product-compliance
Lightning Source LLC
Chambersburg PA
CBHW041351210926
51850CB00013B/3043
9783725830589